MW00345826

HOME BASES

MEMORIES & STORIES OF US MILITARY BASES AROUND LONDON

BY

SEAN C. KELLY

Cover design by Jeremy Hopes
Published through Bayberry Books, Brooklyn, New York, USA

Cover photo – On 2 September 1942 some 300 US military personnel marched through London past St Paul's Cathedral on their way to the Guildhall to a lunch hosted by the Mayor of London in an overt political display of Allied togetherness. Photo © Daily Herald/Getty Images

DEDICATION

Dedicated to:

Jacqueline

Cameron, Erin & Caitlin

Bob & Gail

Duff & Jesse

THANK YOU FOR BEING IN THE COALITION OF THE WILLING AND PATIENT EVEN WHEN
COMMON SENSE & SANITY SUGGESTED OTHERWISE!

&

FOR THOSE WHO SERVED 'OVER THERE'

Bayberry
books

HOME BASES

©2014

THIS BOOK IS COPYRIGHT

SEAN C. KELLY

www.ushomebases.com

ISBN 978-0-9892133-3-2

HOME BASES
The story of the US military bases and installations and facilities around London from WWII to the present day as told by some of those who were there.

Blenheim Crescent

Bushey Hall

Bushy Park

Carpenders Park

Columbia Club

Denham

Douglas House

Eastcote

Daws Hill, High Wycombe

Hendon

Medmenham

South Ruislip

US Navy Headquarters, London

West Drayton

West Ruislip

Uxbridge

CONTENTS

Paradoxes

A Dash of History

The Bases

The Stories

A plaque for the 7500th Air Base Group which had USAF South Ruislip in Middlesex as its headquarters.

We won't do much talking until we've done more fighting. After we've gone, we hope you'll be glad we came.

Ira C. Eaker, Lt. Gen. USAAF
High Wycombe, 1942

PARADOXES

Sean C. Kelly

USAF South Ruislip
Eastcote
USAF High Wycombe / RAF Daws Hill
1960 – 1978
1983 – 1986

Bushey Hall, Bushy Park, the Columbia Club, Denham, Daws Hill/High Wycombe Air Station, Blenheim Crescent, Douglas House, Eastcote, Medmenham, South Ruislip, West Ruislip, RAF Uxbridge and the US Naval Headquarters.

These were among some of the US military bases and associated support facilities that were part of my life growing up – even sometimes if just in passing mention.

But they were almost all paradoxes.

Paradoxes because they were all US military bases or facilities and, as such, 'Little Americas' with US military personnel and their families, American sports, food, money and customs. And they were all located either in or around London, England!

Paradoxes because while they were US Air Force or US Navy installations, they weren't readily identifiable as such. Hardly any had that key 'ingredient' you would normally associate with a USAF facility – a flightline. And not one had a frigate or battleship parked nearby.

However those bases and facilities, in their own small ways, each played important roles in the story of the US Military in Europe. In some cases they were key components in WWI or WWII. Others were either central or subordinated stories in post-war Europe and the Cold War.

Now these bases are gone or going – either in memory or in reality. Global priorities and changes have resulted in a post-Cold War redeployment and resulted in a 'swords to ploughshares' process whereby military installations have been – and still are being – sold and put to civilian use mostly as residential estates or business/mixed-use developments.

It's clear, however, that history was forged at many of these bases. Daws Hill in High Wycombe, along with Bushey Hall near Watford, Bushy Park in Teddington, South Ruislip in Middlesex and the US Navy Headquarters in London were major US military headquarters. It can be argued that it was from these installations that the US military was able to establish its World War and Cold War credentials.

Daws Hill, rising above the furniture-making town, was home of the 8th Air Force Bomber Command. Bushey Hall was home to the 8th Fighter Command in WWII. The US Navy Headquarters was located on Grosvenor Square just across the street from the US Embassy. Bushy Park, near Teddington to the southwest of London was, for a time, home to then General Eisenhower and his main Supreme Headquarters Allied Expeditionary Force.

They had military pedigree.

But no more. Largely with the exception of one small base – the 1.36 acre Blenheim

Crescent in West Ruislip, Middlesex – and one or two small office 'outposts' around London, as well as American representatives and exchange officers at NATO Maritime Command in Northwood and at the MOD in London, the bulk of the once formidable US Military presence in and around London has ebbed away.

Many of the old bases and their respective buildings have – or are – being razed to the ground. For those who are quick the exceptions remain at Bushey Hall and, for the time being until the bases are fully redeveloped, the former base at Daws Hill (aka High Wycombe Air Station/USAF High Wycombe and latterly RAF Daws Hill) and at RAF Uxbridge, itself once a major RAF base but, as of 2013, a place succumbing to the bulldozers and bricklayers.

Both of those bases have been purchased for housing redevelopment.

Bushey Hall in Watford still provides the best opportunity for the public to see still the old base buildings harking back, not only to the 1950s/1960s but to WWII. But even there it is a race against time, nature and limited budgets to preserve those remaining structures.

So now the bases have gone or are going. But they are not completely forgotten – at least in the minds of those who served, lived and visited these places over something like seven decades.

Why bother writing about them?

The answer lies in both a personal journey and in tracking the so-called 'swords to ploughshares' fates of these and other US military bases.

Personal in that I was born at one of the bases; in that I attended elementary school at another and in that I went to Junior High and High School and worked at a third one.

Personal in that I went to church with my family at yet another base where my brothers had gone to elementary school years before. Personal in that my mother taught at Bushy Park, Bushey Hall, Eastcote and High Wycombe. And personal in that we lived close to the Douglas House – the US Military enlisted club – and went, on occasion, to the Columbia Club – which was for officers and senior civilians.

They were all scenes of my youth – of our youth. They were places where history unfolded in its own little footnotes – many more surely lost than found these days.

You've heard the saying; *if these walls could talk*. They can't… but the people who were there can. And that's what has fascinated me most of all – the history as told by those who were there living it.

I don't *do* history, I wouldn't presume to do so. So do not expect the definitive book covering each little detail about every little installation. There will be some smaller or wartime installations and offices that I've missed or are not included. *Home Bases* isn't so much an accurately detailed historical record but more of a soupçon of 'mc-history' and then a series of personal experience *snapshots* by – and of – people who were there and some of the places, procedures and events that now largely exist in memory.

This is not a book for historians, military fanatics or detail merchants. Sure some history and stories of military operations will be served up – but it's largely the history and recollections according to those who were there and told from their own perspectives.

In that historical sense this book only scratches the proverbial surface. There are, literally, thousands more great stories and details out there. And there are some places – mostly dusty warehouses containing forgotten filing cabinets in the United States – where there is more information… and answers. Sadly, that search – for me, at least – won't happen in this lifetime.

There will, I know, be plenty of other stories. Indeed, if there are, I would commend those who can tell them to either email the book website *www.ushomebases.com* or to the already popular website www.ruislip.co.uk – where there is a considerable wealth of additional information. I'm hoping that those who read this book and have their own memories will get in touch and contribute.

Not only is the book limited in its historical scope and detail, but please expect it to go 'off the reservation' from time to time to follow some person or place in detail that may not be necessarily London area-centric. Usually this is, I would argue, where the story has relevance; but sometimes simply because it became of interest to me.

The book project really started with an interview many years ago with WWII veteran Dwin Craig – who, sadly, has since passed on. It continued, randomly, as I found time and also came across people – US and British military personnel, spouses, students, teachers, as well as American and British civilians. They had worked, lived or been in-volved with one or more of the bases and, for the most part, largely enjoyed their tours of duty; not just in WWII but in the 1950s, 60s, 70s, 80s, 90s and right through to 2008.

In one case Ollis Prentice had already passed on but had the foresight to record his interesting story. In another case it was a connection made after reading a letter to the editor of a British village church paper. In yet another it was after a chance discussion with a neighbor who knew a "Yank" who had stayed after WWII. Way, way too slowly for the last few years I have been gathering those stories, memories and details.

For the purposes of some semblance of order I was counseled by Ed Côté, the son of a WWII contributor Paul Côté and a leading historian and amazing collector of military memorabilia in his own right, to do the military thing and put the interviews into some parade ground-type order. And so I have tried to establish some loose formation via the date each contributor arrived at a London area base for 'duty'.

As you read, please remember that each of the main contributors has come at this with their recollections from a different era, place and point of view.

I have journeyed as far as California for an interview with Mr. Côté Sr. and I have traversed 'virtually' throughout the world for others – catching people where they lived or even during their own travels. My closest journey was to my family home where I sat down with my mother while she retold me three 'family lore' stories that I had recalled hearing only in scant detail at various times in my youth.

In three other cases – particularly Henry Farwell, Archie Tatum and Ed Brennan – it was purely by chance that I heard about each of them and then, amazingly, discovered that I needed only pop down the road from my home some 800 meters in three different directions to conduct interviews!

Henry had started his career in the UK in WWII, had flown on D-Day, had fallen in love and then come back again on another tour of duty and to stay. In his case, he didn't qualify technically as a candidate for the book until he arrived at USAF South Ruislip in 1967. However, on hearing his story, which he told me before he passed in September 2013, I just couldn't concentrate on the last few years of his career – I had to begin at the beginning!

Likewise, I spent – and continue to spend – happy times with Archie Tatum and Ed Brennan talking about sports, education, music and math – but essentially talking about lives lived.

The book finishes with, for me, the equally important recollections of several people involved with the closing down and demolition of some of the bases. Dave Dittmer, a

Captain in the US Navy, was the last commander of Naval Activities United Kingdom (COMNAVACTUK) and had the difficult job of closing down much of the remaining US military presence around London. I was honored to be invited to his retirement ceremony from the US Navy held on dockside at Portsmouth in the summer of 2012. And the main, chaptered memories conclude with British worker Danny Shrubb who knew nothing about the US bases until he became part of a demolition team who levelled the base at Eastcote – a team who kindly let me into their midst to photograph them at work.

I hope that their stories make interesting reading. I know they provide snapshots – and the *minutia* of life – of events, places and pastimes. They also tell the story, in part, of the swords-to-ploughshares transition of these locations and properties wrought by changing times and international relations.

The contributors not only come from different eras but from diverse backgrounds and jobs. They comprise both military and civilian personnel – both American and British people – who were there to do a job. In some cases willingly and in some others not. Some had happy memories, some not so happy, though admittedly most contributors by their very nature of contributing, have been enthusiastic about their tour(s) of duty in the UK.

Some saw and experienced sad and even terrible things. Some, like actor Larry Hagman, who passed away in November 2012, came home with a lover or a spouse. Some remained in the UK permanently or have come back to visit regularly. Some have never come back. And, in some cases, a few wish they could still make one last visit.

So this, if you will, is my own love-letter to those places and times.

Several people changed their minds in the course of interviews and decided not to proceed. I was not sure if it was down to lack of conviction about whether their story might be of interest or the feeling that 'weight' of what they recalled was too much to write about for personal reasons. Some struggled to reach back through the fog of fading memories but, when they did so, it was with amazing recollection for the details they served up with considerable grace in the face of my many questions. The challenges of looking back – 70 plus years in some cases – can be considerable!

And I haven't helped. This has been a labour of love, but a love interrupted by the day-to-day challenges of work, family . . . and life. I know, with great regret, more of my contributors will have passed on having hoped to see this in print. For that I have real regrets.

There is, as mentioned, only a scratching of the surface – certainly that is the case of the Douglas House, the enlisted men's club which had several locations in London (and where I suspect the term 'party-'till you drop' might have been invented). It is a location already celebrated in an eponymous book whose author kindly shared his own recollections for this book. However there are stories out there of near-mythical status when it comes to this location that I have failed to chase down or fully confirm involving celebrities and scandals of their day.

Likewise, the Columbia Club (the USAF Officers Club) and the US Navy Headquarters – certainly in the early days. Indeed – there's probably plenty missed. But someone else will have to tell those tales.

You can never recapture those times but you can recall them. I know that there are plenty of others out there with amazing stories, but the ones that follow are from the people who were, to put it bluntly, the coalition of the Found and the Willing! For that I thank them – it has been a privilege and honor to be able to interview them.

One thing that this book did was to encourage me to visit or re-visit those old locations – many of which qualify as scenes from my youth. What can be seen now is less and less and soon will only be in pictures and the occasion old newsreel or, in some cases, a bit of YouTube video imagery here and there and, of course, the fading memories of those who were there.

Along that trail I was most fortunate to be able to:

- Stand amidst the smoke as 'British', 'American' and 'German' forces re-enacted a battle for a WWII weekend at Bushey Hall, near Watford, in the summer of 2012. I also walked into the old base gym and poked my nose around some of the Quonset huts and deteriorating buildings that had been used by the USAF VIII Fighter Command in WWII and later acted as classrooms for London Central High School, one of the schools set up for US military dependent children.

- Stroll the parklands at Bushy Park and see the memorials to the US military. They are largely all that remain of the base where, among others, contributor Bill Cooper's father was Base Commander, and where another contributor, Bob Beeghley, stood guard.

- Walk the empty and eerie halls and stand atop the antennae-laden roof of the former US Navy Headquarters building adjacent to Grosvenor Square where contributor Maurice "Mo Gibbs" served not once but twice and where Dave Dittmer became the 'tail ender' whose job was to close the Headquarters down and arrange for the building's sale.

- Walk up and down the sloping hallway and into empty rooms of Eastcote – once my old elementary school when it was a USAF base and, more recently, a base for US Navy /US Marines offices and barracks. I was able to make several nostalgic journeys down memory lane and back to my pre-teen years recalling where my mom had been a teacher, where I had played four-square, had running races and where I had fallen in love for the first time – her name was Lisa.

- Poke around the old base buildings where I was, from time to time, interrupted by heart-stopping encounters with foxes. Neither of us took to the other. And more than once I experienced the up-close and personal flapping of wings of disturbed pigeons across my face. They had moved into the buildings once the humans had moved out. Later, at Eastcote, I was able to visit some of the buildings and rooms again – and become – literally the last person to stand in them before bulldozers from Syd Bishop & Sons ripped into and demolished the structures and turned them into hardcore and scrap.

- Watch as contractors demolished the Chapel of Faith at RAF West Ruislip where my own family had been fortunate enough (in later years when the Chapel had been open to the local British civilian community) to attend church, celebrate baptisms and then be a part of the Final Service and closing ceremony on 4 June 2006.

- Wander through the vacated Daws Hill, High Wycombe, base – once home to my high school and, long before, a major WWII command base for USAAF's 8th Air Force Bomber Command. Between 2007, when the US Navy handed the base back to the British, and early 2013, I was able to watch as the weeds –

some now small trees – sprouted from the gutters of the old base theatre and started pushing up through pavement cracks. The visits proved to be a living version of the best-selling book *The World Without Us*. On one occasion I stood in the old Non-Commissioned Officers (NCO) club where I had last attended officially a base deactivation in the bedecked building. Four years on I was standing atop a floor covered in fallen, mushy ceiling tiles themselves atop a soaked and decaying carpet. A few buildings away there was annoying bleep of emergency power failure fire alarms control boxes that were heard by no one other than the security foot patrols who, one suspects, had long ago roundly learned to ignore them.

- Have encounters with the wildlife of Daws Hill – the equally startled animals including occasional rats and muntjack deer – a number of whom had quickly taken up residence around the urban demi-Eden provided by the abandoned school and base buildings.

- Faced 'danger' in the form of unexpected trips and falls into collapsed and collapsing underground pipes. I'll particularly remember the 'pitfalls' I found and bumps and grazes I took at Eastcote, Uxbridge and High Wycombe.

- Been privileged to be granted access (in 2003) to spend five hours wandering through and around the once top-secret underground bunker at Daws Hill from which the USAF VIII Air Force Bomber Command was controlled during WWII and which had also had a Cold War role. I went subterranean with the noted Airfield Society historian John Hadfield and, after being locked in behind the bunker door by the RAF Station Commander of the time (believe it or not for safety – so nobody else could get in by accident and then become trapped in there when we left as it was only being checked on a irregular basis), we spent a happy, anorak-type day getting a fascinating look inside the once-secret bunker where so much history had been made.

- Watch as demolition company Cuddy ripped apart the bunker used by the USAF at RAF Uxbridge in early 2013.

In juxtaposition, I have also watched homes being built on the former Eastcote base where the developer, George Wimpey was, at time of writing, finishing up the 385 units on its Sandringham Estate, as the place is now generally known. Likewise at West Ruislip where the developer CALA and a housing association have created around 415 apartments and family homes (including one six-bedroom £1,000,000 (circa $1.55 million) home in the large family plot area known as The Chase which now sits atop the former sports facilities, barbeque area and baseball field). As of publication care home developer McCarthy & Stone is still to create an 80-bed care home for the elderly at what has generally been rebranded as Ickenham Park.

Sometime during late 2014 the base at Daws Hill should start going the same way with circa 500-1,000 homes and some other facilities expected to be built by major residential construction company Taylor Wimpey.

I never completely concurred with the drawdown of US military, not only around London but around the UK, in that I think the cuts may have gone too far. Of course, I'm biased perhaps, but maybe for good reason. The cost of obtaining and rebuilding a secure base will be far more, should one ever be required in the future rather than the option of keeping one 'mothballed' and ready to go. These days the average site assembly and planning for a major urban development site in the UK can take around 15 years if not longer. It can, of course, be argued that another US base will never be needed.

Housing remains, I suspect, an issue for the residual US military personnel stationed to London (they are still here in small numbers in liaison, support and exchange programs) and so, in fact, the cost of keeping either all or part of West Ruislip, which was ideally positioned immediately next to the end of the line for London Underground's Central Line, or even High Wycombe which combined both space and both base housing and operational buildings, might have been seen as cheap at the price.

But it was not to be. Of course 'battle lines' have moved – towards the Mediterranean and the Middle East and as events in 2013 have highlighted, Africa. And new strategic thinking means more rotational deployments from the Continental United States (CONUS). Of course, on the flip side, it's hoped that these type of facilities are never needed again.

However there is a legacy and, of course, it's not just in the bases and buildings. It's provided by people who worked there and, directly, in the children of Anglo-American relationships. And sometimes in the most surprising places: A stained glass window or a gravestone in a church here, a social club there or even by an old, sometimes rusty 'Yank Tank' driving past on a London suburb street.

The links are still there in surprising ways. Even the Mayor of the London Borough of Hillingdon wears a tiny part of that legacy to this day: On his full ceremonial mayoral chain is a pendant known as the US Navy Chief's Badge which was presented by the London Chief Petty Officers Association in May 2007. It was made by UT2 (SCW) Trickel M.D.

One of those many other legacies is the Outdoor Learning Centre at Abbotsfield School in Middlesex which was built with the generous help of the US Navy 'Seabees' (the nick-name for the Construction Battalion) who supported the local Karers4Kidz charity set up by local British carer Katrina Bijowski, who has sadly passed on herself.

And the list goes on.

In short, I have done my best to follow these bases in their respective swords-to-ploughshares transfers and have watched as new homes are built in their place – probably for people who may know little of what went before.

This book – hopefully – redresses a little bit of that.

A DASH OF HISTORY

Let's get the 'history bit' out of the way first:

Ignoring the 1776 War of Independence... 'and all that', the 'modern' military relationship has its roots in WWI when there was a Anglo-American exchange of training fields and military hardware and, of course, personnel towards the end of the War.

It helped immeasurably the Anglo-American relationship that the British intelligence agents decoded what has become known as the Zimmerman Telegram on 16 January 1917. The coded telegram contained a proposal German Foreign Minister Arthur Zimmerman to the German ambassador Johann von Bernstorff in Mexico with a proposal for the Mexicans to declare war on the US should the Americans declare war on Germany. Not only would the Germans apparently fund Mexico's war costs but they would also give them Texas, New Mexico and Arizona once the Americans had been defeated. Naturally the public revelation of the proposal outraged the Americans.

The British didn't want to admit they had 'snagged' the telegraph because they had been spying on the American cable link so, to cover their tracks, they went about getting a legitimate 'ciphertext' copy of it from a transmission station in Mexico before presenting a translated copy to the Secretary of the US Embassy on 19 February. He, in turn, passed it to the US Ambassador the next day.

Three days later the British Foreign Secretary met and formally showed the US Ambassador the original message and its translation. Word quickly reached US President Woodrow Wilson after that and then the story became public. On 29 March Zimmerman admitted that the telegram was genuine. The actual telegram is part of Britain's Government Communications Headquarters (GCHQ) historical collection.

The Anglo-American alliance had already been paved, some historians claim, by the sinking of the British liner *Lusitania* by a German torpedo in WWI – the liner went down in just 18 minutes on 7 May 1915. The sinking of that ship, with around 100 Americans out of 1,200 passengers on board, is claimed by some to have helped bring America into WWI or at least made them more sympathetic to the British cause.

Indeed, future British leader Winston Churchill (who was then First Lord of the Admiralty of the time) had earlier written to a fellow minister: "It is most important to attract neutral shipping to our shores in the hope of especially embroiling the US with Germany. For our part we want the traffic – the more the better – and if some of it gets into trouble, better still."

If anything, Churchill's expressed desire probably didn't include a flagged ship of his own nation, let alone a cruise liner. It's further worth noting that the *Lusitania* was carrying some 170 tonnes of American made munitions (including 4.2 million rounds of Remington .303 rifle bullets) to England in its cargo – some would say at variance with the neutrality laws of the time. The US conveniently overlooked the munitions issue and sided with Britain against Germany on the sinking – a stance taken by Woodrow Wilson which caused his Secretary of State William Jennings Bryan to resign and be replaced by Robert Lansing. The latter had become, as a result of the incident, convinced that America and Britain would side together.

So while the Zimmerman Telegram was given as the main reason, the foundations for the Anglo-American military alliance probably can be found in some 93 meters of water in the Atlantic in which the *Lusitania* rests on its side.

War was declared on 6 April 1917. The American military presence in Britain largely came in the shape of the US Naval Forces Operating in European Waters. They were under the command of Rear Admiral William Sowden Sims whom was ordered to London in March 1917 and arrived on 10 April. Sims spent four days on intelligence gathering in which time he met the King at St. Paul's Cathedral.

"It gives me great pleasure to meet you on an occasion like this," the King (whom would meet Sims again soon after at Windsor Castle) said at the Anglo-American memorial service. "I am glad to meet an American admiral on such a mission as yours. And I wish you all success."

Sims then wrote a four-page despatch and came to the conclusion that coal-fired destroyers were the way ahead (oil was in short supply and any larger oil-fired ships could not be supported logistically). Several days later he travelled to Brighton where US Ambassador Walter Hines Page happened to be staying to ask the Ambassador to provide additional written firepower with a separate cable to Washington DC. Page immediately sent a cable to the Secretary of State and the President in which he wrote: "I cannot refrain from most strongly recommending the immediate sending over of every destroyer and all other craft that can be of anti-submarine use." Page continued: "Thirty or more destroyers and other similar craft sent by us immediately would very likely be decisive. There is no time to be lost."

With the outbreak of war the pro-British Sims headed up the office of Commander of the US Naval Forces in the European Theatre first to be found within the US Embassy (Ambassador Page had offered to turn out the Embassy staff "into the street" if the Navy needed the space) for just Sims and an aide and then to be found six converted five-storey houses (of 25 rooms each) in Grosvenor Gardens, London as Sims' office would eventually grow to a staff of more than 1,200. *(See picture under Other Bases section).*

The Americans who arrived were quick to get down to…being Americans.. In particular the US Navy and US Army held a American Baseball League match in 1917 at Highbury, North London, and the first-pitch went to Admiral Sims. The Army won.

In response to his on-the-scene reports and requests, a squadron of destroyers was despatched from Boston on April 24 and shortly thereafter they arrived in Queenstown, Northern Ireland (on 4 May) to be greeted by a large crowd of American-flag waving residents. The squadron was headed up by the *USS Wadsorth* under Cmdr. Joseph K. Taussig, but it would fall under the Command of a British Admiral when it went into action just three days later. Taussig had received orders (only opened once he had sailed out to a point off the American coastline, to "assist naval operations" and "cooperate fully with the British Navy" and "even the French Navy" if such a mission were required.

A second squadron arrived on 17 May 1917 and by the middle of summer there were 34 destroyers based out of Queenstown, near Cork in Ireland. The initial basing location was not without some issues; there were reports of fights (resulting in at least one death) between US Navy personnel and representatives of Sein Feinn, the Irish Republican party wanting to end British rule in Northern Ireland. US Navy personnel were banned from Cork but that didn't stop the women from Cork catching the train to Queenstown – which they did in droves.

One of the key tactical changes impressed upon the British by Sims was the operation of convoys and the use of submarine avoidance tactics such as zig-zagging in formation. His tactical suggestion would result in an outcome-changing reduction of shipping losses from 875,000 to 101,168 tonnes a month from April 1917 to May 1918.

By June 1918 the American squadrons would relocate to Plymouth (there were 36 US Navy ships based there alone) and the US Navy would have many more ships and submarines (seven based out of Ireland), mine-making bases in Scotland and a seaplane base at Killinghome in England. During the War US Naval Aviation would grow to 24,500 officers and men across Europe during WWI with 500 planes. And the Navy would help ship upwards of two million American military personnel to France to take on the Germans.

By June 1917 General John Pershing, the US Commander of the Allied Expeditionary Force (AEF), arrived in Britain to be feted, receive numerous delegations and even be greeted by the King George V. Within the week General Pershing moved most swiftly on to France with the 1st Division AEF arriving in June 1917 with around 14,000 personnel and the 2nd Division on 31 October 1917 to create a US force of just over 87,000 personnel.

Most of the American soldiers in France by 1918 as part of the AEF had arrived there directly rather than stopping in England. Those numbers would escalate to around one million American military personnel by May 1918, but with relatively few (in their thousands or tens of thousands) stopping off in the UK. Those who were assigned here or came through the UK had the American YMCA in Aldwych on the Strand next to St. Mary Le Strand church. It comprised a series of large wooden hut buildings. Today the site is occupied by Bush House – the one-time home of the BBC World Service from 1941 to July 2012) and now home of businesses including law firm Dundas & Wilson.

It was no small facility. The Eagle Hut, as it was known, had been opened in September 1917 by Dr. Walter Page, the US Ambassador. The YMCA's own publicity reiterated the rumor that after the opening ceremony the key to its front doors of its main entrance (at the foot of Kingsway) had been thrown into the nearby Thames River to signify its emphasis on open-door service.

The Eagle Hut at Aldwych in London in World War I.

A stone marker which can be found on the Aldwych site today is the only recognition of the old YMCA facility which played such an important part for morale in World War I.

And serve it did – hosting something like 6,000 visitors a day and serving normally 3,000 meals a day (the number mushroomed to 7,662 meals on 4 July 1918 – US Independence Day). It was that same day that the US Army vs US Navy baseball game was played at Stamford Bridge (a football ground in London) and King George VI & Queen Mary paid a visit to the center.

Some 800 volunteers (including a group of four founding American businessmen known as 'The Eagle Hutters') bolstered the nine core staff and supported the activities. The majority were women.

Visitors were able to enjoy use of a range of facilities including a lounge, concert hall and billiards room, shop and information centre, showers and latrines, canteen, library and four dormitories with at least 300 beds in one. They could even attend concerts or take French classes there.

And those who did make it to England may have had their morale boosted by the likes of London-based camera-maker Sands Hunter & Co. who took out advertisements welcoming the "Boys We See in the Strand" and calling for readers to donate spare camera equipment to help fund the war effort as part of a Cameras for War Loan.

Under a drawing of a US Navy sailor the ad read: "As we see these clean-cut fellows we sometimes think of the fact that human nature is much the same the world over. This American sailor-boy has at home, probably a little sister just like yours, or like your little daughter. Think for a moment of that English child you love. Is she to grow up in a better and finer England, free from the horrors of war and from all the brutalities which are being imposed upon free countries by the Germans? These American boys are fighting for her!"

World War I ended with the Armistice on 11 November 1918 with the Armistice and peace was (supposedly) cemented with the Treaty of Versailles in 1919.

However fighting sadly did become the order of the day in March 9, 1919 around the Eagle Hut. Three American soldiers (two US Navy and one Army) were arrested by two British police for playing a card game next to the buildings after the Americans ignored their commands and instead boasted they had won the war for them. After their arrest a crowd gathered,more police were called and clubs were used. One American serviceman was injured as he tried to peace-make and the police were said to have then retreated to their nearby station with the arrested men.

Later the crowd (estimated at 1,000 – 2,000 and including American and Canadian service personnel), mistakenly believing the peacemaker had died, rallied and stoned the nearby Bow Street Police Station. The police responded with baton charges and later with mounted police as a three-hour long street-battle, forever known as the Battle of Bow Street, broke out. The incident ultimately resulted in 20 police and servicemen getting injured and the arrest of 11 Americans soldiers and sailors. US military personnel were subsequently ordered to disperse from the Eagle Hut and US Navy Admiral Sim ordered all US Navy personnel back to their ships afterwards.

The Eagle Hut shuttered its doors on 25 August 1919. Today a stone tablet marks in Aldwych marks the site of the Eagle Hut and its role in underscoring largely the Anglo-American friendship.

However that WWI military linkage was just a comparative first kiss compared to the full-on relationship that would blossom in WWII and the Cold War years following.

A revitalized Germany had declared war with the invasion of Poland on 1 September 1939 while Britain and France declared war on Germany two days later. In that same week the US President Franklin D. Roosevelt, under great pressure from the Isolationists within the US who wanted to stay out of the war, declared neutrality and banned arms sales to belligerent powers.

The result was that between September 1939 and well into October 1940 the US-UK relationship was somewhat strained by the detaining of US freighters by British authorities around the UK waters and also in Gibraltar as part of the contraband control. American ships detained included the freighter *SS Saccarappa* (on the day Britain declared war on Germany – its cargo of phosphates and other goods was unloaded and five days later it was allowed to leave), followed by the freighter *Black Osprey* the next day.

Among many others detained was the *Tulsa* (October 23-9 November), the *Sundance* (11 October – 25 October) and the *West Hobomac* (18-25 October). Under the Neutrality Act the ship's cargo was often seized.

One ship not given problems was the *SS Washington*, a 24,189 ton, 1,083 capacity United States Lines passenger liner that sailed to England to evacuate Americans out of there and Europe. One of its later trips back to the US on 11 June 1940 saw a capacity-stretching 1,787 passengers (700 of them children) face terror when a German U-Boat Commander gave them ten minutes to abandon ship before he realized that they had intercepted the wrong ship.

It was on 4 November 1939 that US President Roosevelt declared the waters around Britain a combat zone. By December there were diplomatic exchanges between the US and UK over an incident involving British warship *HMS Orion* which had lobbed a bow shot over the German freighter *Arauca*. The shot ended up landing inside US territorial waters off Florida.

The incident prompted a rebuke to the British by the US Secretary of State which was followed a week later by another protest to Britain's Foreign Office over the censoring and removal of US Mail from detained ships. A further protest followed in January 1940. The British response could be categorized as diplomatically dismissive.

That aside, the historical bedrock for both the modern-day so-called 'Special Relationship' between the US and the UK and for the bases covered in this book was largely underwritten by the US Congress in legislation formally known as 'An Act Further to Promote the Defense of the United States'. It was better known as the 'Lend Lease Act' and it was passed in March 1941 (some nine months before the Japanese bombed Pearl Harbor and the US formally entered WWII).

As a result, Britain received approximately $31.4 billion (out of a total 'pot' of $50 billion) in supplies in return for considerations such as the free lease of US bases. A second massive pot of money would follow in the immediate post-WWII years – one that would only get paid off in December 2007! It's worth noting that WWI debts of the UK to the US of about $835 million were never – and apparently will never be – paid off.

From January to March in 1941 US and British military personnel met in Washington DC to exchange information and protocols for military missions to be established in each other's countries under what was known as the ABC-1 Conversations. The Special Army Observer, US Army Air Corps Maj. General James E. Chaney, arrived on 18 May 1941 and established his HQ known as Special Observer Group the following day. His primary duty was to monitor Britain's use of Lend Lease equipment, but after the attack on Pearl Harbour and America's entry into the conflict the office was renamed the United States Army Forces in the British Isles (USAFBI).

On 24 January 19 1942 marked the beginning of United States Army Air Forces (USAAF) in Europe with the creation of the VIII Bomber Command, as it was originally known, which established its UK & European headquarters at Daws Hill in High Wycombe, on 23 February 1942. By June the first combat aircrews (97th Bombardment Group) arrived at RAF Polebrook in Northamptonshire.

Their arrival would be the vanguard for 250,000 American personnel (out of nearly two million who passed through the UK during WWII) serving in the US Air Force here. Of those in the Eighth Air Force alone some, 26,000 personnel were killed, 47,483 were injured and 28,000 were made POWs while some 6,292 USAF aircraft would be lost by VE (Victory in Europe) Day.

Belfast in Northern Ireland witnessed the first arrival of US Army personnel in Europe at the end of January 1942. led by Private First Class Milburn Henke of the 133rd Infantry Regiment, 34th Infantry Division. The USAAF's VIII Interceptor Command was formed in January 1942 and activated on 1 February. Eventually it became VIII Fighter Command in May and was headquartered first in the Michigan and then South Carolina before establishing at Daws Hill, High Wycombe, on 12 May 1942 and then relocating to Bushey Hall near Watford on 27 July 1942. At the end of the WWII it would return to High Wycombe (from 17 July – 25 October 1945) before being inactivated on 20 March 1946.

Subsequently both VIII Bomber and Fighter Commands were subsumed by the Eighth Air Force (8th Air Force), which was established in Savannah, Georgia, in early 1942 and became the formal organizational name in 1944.

Grosvenor Square, already home to the US Ambassador to the Court of St. James (as the job of US Ambassador to the UK is formally titled) since 1785, was to also become a hub of US military command activity during WWII. The US Navy Headquarters in London was established at 20 Grosvenor Square by Admiral Harold F. Stark in March 1942.

The building, which also fronted onto North Audley Street, was used by General Dwight D. Eisenhower, the Supreme Allied Commander, as one of his many headquarters. The whole area would become referred to in tongue-in-cheek fashion as "Eisenhowerplatz". In 1949, four years after the end of WWII, the US Navy set up its European Headquarters there and would remain in the building under various command names until the US Navy HQ and command was disestablished on 1 September 2007.

The Visiting Forces Act of 1942 set the legal framework for US Military personnel and bases in England. One of its primary elements was that visiting US military forces could not be subjected to court trial by any court in the United Kingdom, but could be courts-martialled by their own military.

The modern political foundation for the US Air Force bases in the UK had been laid by US Ambassador John G. Winant (who served as Ambassador in London from 1941 – 1946 and who had served in the US Army Air Force's 8th Aero Squadron in France during WWI). But it was given new impetus by General Henry 'Hap' Arnold, then Chief of the USAAF and the person who would oversee the creation of the 8th Air Force.

A key moment in the so-called 'Special Relationship' came on 1 June 1942. It was the day after the British RAF launched a devastating 1,000 bomber raid on the German city Cologne. Gen. Arnold had been on a tour of the UK along with John H. Towers, head of the US Navy's Bureau of Aeronautics, and both had been guests of British Prime Minister Winston Churchill when the news of the raid was announced. Gen. Arnold reportedly sent a game-changing message to President Roosevelt: "England is the place to win the war. Get planes and troops over here as soon as possible."

From that stemmed what came to be called the "Combined Bomber Offensive" (CBO) and that would lead to the establishment of around 70 main US Army Air Force bases in the UK.

Symbolically, the USAAF's first flight operation took place on 4 July 1942 when six crews used British "Bostons" which had been given US markings to attack targets in Holland.

It was not a day for celebration – only three crews made it back.

It would be a different story for the 8th Air Force's first B-17 bomber mission. The dozen plane bombing attack against railway marshalling yards in Rouen, France, on 17 August 1942 saw all aircrews make it back.

But they weren't the first Americans in aerial combat. From the early days of WWII, other Americans had already become involved with the UK military. Their involvement was unofficial, of course, as Americans could lose their citizenship by joining a foreign nation's armed forces. However a few Americans had risked it and joined the Royal Air Force.

One the best known of them, William 'Billy' Fiske, became the first American to join the RAF after having returned to England in 1939.

Fiske had been born in New York, had come to England first to study at Cambridge in 1928 – the same year he became the youngest winner of an Olympic gold medal in the five-man bobsled at St. Moritz, Switzerland. He would win a second gold medal at the 1932 Olympics in the four-man event and also go on to co-develop the first ski lift in Aspen, Colorado. He was working for New York investment banking company Dillon, Reed & Co. (today part of UBS) in their London offices and would marry Rose Bingham (better known as the Countess of Warwick) on 8 September 1938 following her divorce from the 6th Earl of Warwick that same year.

As the war clouds formed over Europe Fiske had been summoned back to New York, but he managed a quick return to London with a colleague and claimed to be a Canadian when he joined the RAF's Volunteer Reserve (because of the US restrictions). He was made a Pilot Officer on 12 April 1940 and would be assigned to the 601st Squadron (aka "The Millionaires Squadron" due to the aristocratic backgrounds of many of the pilots and because the squadron had been formed at the famous gentleman's club White's in London) at Tangmere in West Sussex as a Hawker Hurricane pilot. He had been initially considered an "untried American adventurer" in the official log book of the 601st.

Fiske would go on to become just one of 11 (there remains some debate on the numbers depending on how they qualified) Americans to fly in the famed Battle of Britain. Sadly he crash landed back at Tangmere on 16 August 1940 after a fight in which eight German Junkers Ju 87 aircraft were shot down. He had been badly burnt after his plane's fuel tank was struck by a bullet. Fiske was taken to the Royal Sussex Hospital in Chichester but would die of shock on 18 August and be buried in the St. Mary and St. Blaise churchyard at Boxgrove Priory Church in Sussex two days later in a coffin draped in both a British and American flag. A stained glass window in Fiske's honour was donated by the 601st Squadron Old Comrade's Association and dedicated at the Church in 2008.

The other Battle of Britain Americans who flew and are listed on the Battle of Britain Honour Roll to be found on the eponymous monument by sculptor Paul Day next to the Thames in London included:

- Minnesota-born Arthur G. Donahue came through the Canadian route and flew Spitfires for the RAF's 64 Squadron. However he was wounded after having to bail out from a Spitfire on 12 August. He would go on to receive a DFC after joining 64 Squadron (comprising of a large portion of American volunteers) in September 1940. He received orders to Singapore but was again wounded and sent back to England as Flight Commander. On 11 September 1942 he ditched his aircraft in the channel and is presumed drowned. His body was never recovered.

- Petty Officer John Haviland was a dual national (his father was a US Navy officer) and grew up in England. Having enlisted in 1939 he ended up flying in Hawker Hurricanes in 151 Squadron but suffered a mid-air collision on 24 September 1940 and thereafter became an instructor and involved in support of Bomber Command. Haviland, who received a DFC in February 1945 is the only one of the Battle of Britain Americans thought to have survived the war (he died in 2002 after a long career as a professor of Mechanical & Aerospace Engineering at the University of Virginia).

- Petty Officer Phillip Leckrone from Salem, Illinois, joined the RAF and was part of RAF's 616 Squadron on September 1940. Six weeks later he was transferred to 71 Squadron RAF to serve with the Eagles. He was killed on 5 January 1941 after failing to bail out from a mid-air collision during formation flying practice despite repeated radio calls from fellow formation pilot American Vernon Keough (*see entry below*). He is buried in Kirton-in-Lindsey, Lincolnshire.

- Experienced aviator Andrew Mamedoff had already been in the aviation business on both coasts of the US before he joined up with two other Battle of Britain American pilots, Eugene Tobin & Vernon Keough, to first try and help Finland in the war. Arriving too late the trio first joined the French Air Force before evacuating from France on the last ship out of St. Jean-de-Luz. Back in England they immediately joined the RAF. Mamedoff underwent a Spitfire conversion course before joining 609 Squadron in August 1940. He would go on with his two buddies to be the first to join 71 Eagle Squadron and then become Flight Commander of 133 Eagle Squadron at RAF Duxford. Mamedoff was killed when he crashed his Hurricane fighter near Ramsey, the Isle of Man, during bad weather on a transit flight to Northern Ireland on 8 October 1941. His body was laid to rest in the American section of Brookwood Military Cemetery in Surrey.

- Petty Officer Eugene Tobin had been a messenger boy working for MGM in Hollywood so that he could earn money to learn to fly. As with Mamedoff and Keough he also went to 609 Squadron on 8 August 1940 and was co-credited with shooting down a Dornier and a Messerschmitt 110 during the following six weeks. And like his friends he would move across to the 71 Eagle Squadron. The 24-year-old Tobin died on 7 September 1941 as the Squadron made its first flight over France. His remains are buried at Boulogne Eastern Cemetery in France.

- Like his two friends New Yorker Vernon Keough had a frustrating start to his volunteering efforts even though he had been in a flying show in the US in which he also parachuted from aircraft. Not only did he face the late arrival and then travails of an evacuation from France, but Keough's stature (he was just 4'10" tall) made him the smallest RAF pilot on record at that time and earned him the inevitable nickname 'Shorty'. His ground crew regularly fitted extra seat cushions in his Spitfire. Keough is credited with helping shoot down a German Dornier aircraft while serving with the RAF's 609 Squadron in August 1940 before going on to become one of the first pilots to help form the 71st Eagle Squadron. On 15 February 1941 he failed to return home after a convoy protection flight over the North Sea near Bridlington, East Yorkshire. His body was never found.

There were two other Americans also in the Battle of Britain, however, who came to the RAF through the Canadian route including:

- Flying Officer De Peyster Brown, who flew with a Canadian Squadron based at RAF Northolt in Middlesex at which he survived a post-battle crash during the Battle of Britain and later went on to join the one of the three American Eagle Squadrons (*see further on*).

- Flight Lt. Carl R. Davis, an American born in South Africa and who lived in London, became a pilot in 1936 and was called up to fly a Hurricane 601 Squadron in 1939. After shooting down and damaging around a dozen aircraft and receiving a Distinguished Flying Cross on 30 September 1940, he was shot down himself just a week later on 6 September and died after crashing into the garden of Canterbury Cottage in Matfield near Tunbridge Wells. He was the brother-in-law of 601 Squadron commander Sir Archibald Hope. Davis was buried at St. Mary's Church in Storrington, West Sussex.

Fiske's death had followed that of the first American-born man to die in WWII aerial combat, James William 'Elias' Davies. Davies had been born in the US but had grown up in Wales and was officially recorded as British by the Commonwealth War Graves Commission. He had joined the RAF in 1936 – long before the outbreak of war and during the Battle of Britain had been 'mentioned in despatches' for gallantry. On June 27, 1940 the day he was apparently due to have been presented with Britain's Distinguished Flying Cross (DFC) for actions resulting in the shooting down of six German aircraft and shared 'kills' in two others but sadly, on his return from an escort flight as part of a reconnaissance mission over a French port, his was one of three Hurricane aircraft attacked by three German Messerschmitts while over the English Channel. His body was never recovered. His award was made posthumously on 28 June and his name is recorded on the Air Forces Memorial at Runnymede along with 20,455 others from the British Empire who were lost in action but have no grave.

There was further direct participation in the Battle of Britain in the shape of an American aircraft. The Grumman Martlet replaced the Sea Gladiator of No 804 Squadron, Fleet Air Arm, late in October 1940. As the dates of the Battle of Britain are recognized as 10 July – 31 October 1940, the Grumman Martlet qualifies as a Battle of Britain aircraft. The Squadron flew in No 13 Group which was responsible for the air defence of the North East of England, Scotland and Northern Ireland.

And Americans were involved with the RAF in other guises. Joe McCarthy, who would join the RAF via the Canadian route, was one of the best known examples. The Long Island, New York-born Coney Island lifeguard who worked on the beach in order to save money to fly, was persuaded by a friend, Don Curtain, to join him and head to Canada where they both joined the Royal Canadian Air Force. RCAF training resulted in them being send to England (via ship to Liverpool) where they ended up in Bomber Command flying first Hampden bombers and then Lancasters on raids to Germany including Berlin and also to Italy – receiving Britain's Distinguished Flying Cross (DFC) for his efforts. McCarthy would eventually end up in the famed 617 Squadron, better known as the Dam Buster's at the personal request of legendary British Squadron Commander Guy Gibson. There would participate on the raid on the dams – his crew were the only flight crew out of five assigned to reach the Sorpe dam to bomb it. While the Mohne and Eder dams were successfully breached the Sorpe only received minor damage – though the Germans had to lower the dam waters to reduce stress on the all immediately after. McCarthy would go on to receive the Distinguished Service Order (DSO), survive the war, become a Canadian citizen and, eventually lecture at the USAF Maxwell Air College.

Don Curtain, who was also awarded a DFC, was killed in February 1943, on a bombing raid to Nuremberg, Germany.

It had been another American, Col. Charles Sweeny, who effectively bankrolled around 50 of his fellow countrymen to come to Europe initially to fight in Finland against the Soviets. Priorities changed, however, and it was his British-based nephew, also named Charles Sweeny, who set to work persuading the British that they needed American help. The expatriate Sweeny first created a 'Home Guard' unit comprising Americans living in London. Then he took the idea of American involvement further – knocking on the door of Britain's Air Ministry to suggest American flying squadrons.

In July 1940 the Air Ministry agreed and the famed Eagle Squadrons of the RAF were formed. There would be three squadrons (71, 121 and 133) comprising American volunteer pilots with the first forming on 19 September 1940. They would start by flying in training aircraft and then Brewster Buffalo fighters before eventually getting Hurricanes in November, becoming operational in February 1941 and starting to receive Spitfires to defend the UK in August 1941.

The work initiated by Sweeney was carried on by the Clayton Knight Committee that had also been formed in 1939. Headed up by the eponymous American WWI veteran, the Committee recruited some 7,000 Americans of whom around 250 became members of the Eagle Squadrons (others went to Canadian units).

The Eagle Squadrons eventually were subsumed by the USAAF in September 1942 to become the 4th Fighter Group. Of the 250 Eagle Squadron members, 107 were killed and 34 were captured either with the RAF or, subsequently, the USAAF.

Col. James Goodson was one of the (if not *the*) first to join the Eagle Squadrons. Goodson, who was aboard the *SS Athenia* when it was torpedoed on the first day of the WWII, was rescued by the Norwegians and then went on to joint the RAF serving under the famous Douglas Bader and his 45 Squadron before eventually moving to 133 Eagle Squadron at Debden flying Spitfires and Hurricanes.

Goodson would go on to fly more than 1,000 missions and shoot down 32 enemy aircraft and would, on the way, gain the Distinguished Service Cross, the Silver Star, the Distinguished Flying Cross with 8 Oak Leaf Clusters, the Air Medal and a Purple Heart, among other awards. He led the first P-51 flight to Berlin before becoming a 'guest' of the Third Reich in Stalag Luft III on June 20 1944 just two weeks after D-Day.

"That (the sinking) literally threw me in at the deep end," he told the author in the early 1990s. "The sensible thing to do was join the RAF."

Goodson was rescued and from Scandinavia he went back to Canada to begin fighter training before joining 133 Eagle Squadron.

"We were fortunate in Fighter Command and in the RAF," he said. "We only flew in daylight so as soon as it was dark we were free. We could be in Berlin in the afternoon and in the Crackers Club (near Piccadilly in London) by evening. London was a great place. There were Free French, the South Africans, the Canadians and the Americans. There was a great spirit of cooperation and camaraderie. We had an active social life."

So great in fact that on one occasion he and a few friends traveling on a familiarization flight aboard a B-29 were fogged out of their landing airport. But they were so desperate to get to London and meet their dates that they flew to another field and landed.

"We had only one thing on our mind – our dates in London," he said. "We jumped out and asked a controller where the next transport to London was. There was a truck just leaving and we all piled on board."

There was only one problem; when the pilots met at the Cracker Club the next morning.

"We'd forgotten to ask the name of the airfield," he recalled. "Our commander wasn't happy – it took three days to locate the plane!"

Without the red-tape restrictions of today, the girls came first – even in combat.

"We had a common frequency May-Day channel so that if we were in trouble on the way back we could contact the Air-Sea Rescue people," Goodson said. "But when you actually tuned into it, you would usually hear messages like: 'Would you call Daphne on Mayfair 6363 and tell her I'll be half an hour late for our date.'"

When Goodson and his fellow American fliers in the Eagle Squadron formally transferred to the USAAF they all had to report to the US Embassy in September 1942 to re-pledge formally their allegiance.

"We couldn't stop flying, though," Goodson, who sometimes flew up to five missions a day in the early days of the war, recalled. "They couldn't afford for us to be out of the war for too long at that point so two of us at a time would go down to the US Embassy in London and swear allegiance and then go right back to duty."

The transfer of the Eagle Squadron led to one other problem involving the aircraft markings, as Goodson recalled.

"For awhile, some of the aircraft had (USAAF) stars and some had the RAF Roundels," he said. "Painting a star was fairly difficult but I had a Jewish crew chief who had a Star of David he wore and he used that as a model. So in a way the first US Fighter over Europe actually flew under the Star of David!"

Goodson had one memorable encounter in London's own 'officer country' when he and other fliers were spotted in Grosvenor Square.

"We were very proud of our RAF Wings and when they asked us to change over (to USAAF insignia) we asked permission to keep our RAF Wings," he said. "Ultimately we wore our RAF Insignia on the left side and our US wings on the right side. Some dyed-in-the-wool general saw us in Grosvenor Square wearing both insignia and didn't like it. He asked just what we thought we were doing. "Major who gave you permission to wear foreign wings and foreign decorations?" Goodson recalls him demanding. "The President of the US and King George of England," I replied. He recovered mighty fast because the next thing he said was: "Mighty purdeee"."

Then Major (and later Colonel) James Goodson his P51 with 30 'kills' of German aircraft. He would actually get two more kills with the USAAF's 15th Air Force.

Meanwhile on the ground in London some 60 to 75 expatriate Americans were also doing their bit by forming the First American Motorized Squadron of the Home Guard in July of 1940 – just a couple of months after a call went out to the British populace to defend the country. The 1st AMS became the first company of foreign troops to be in the British Army.

The group of largely middle-aged 'expat' Americans were headed by General Wade H. Hayes, who had been working in London as an American banker. The squadron based themselves out of The Queen's Westminsters Territory Army HQ at Buckingham Gate and became nicknamed, according to Life Magazine, as "The Gangsters" thanks to their high-powered cars and American accents. In September 1940 they became the first foreigners to be incorporated in the British Army, though initially there were concerns form both British authorities and American Ambassador Joseph Kennedy about the status of their neutrality.

By 1942, however, they were formally inspected by Prime Minister Winston Churchill and his wife on Horse Guards Parade in Whitehall. It helped perhaps that Hayes had served in the 107th Infantry in World War I under General Pershing during 1917 Mexican Border campaign and gone on to serve as the escort to the Prince of Wales on a tour of the US in the early 1920s.

"We appreciate the generosity, hospitality and friendship of the British people and we believe we have a stake in this country hardly less than that of the people themselves," General Hayes, who had also been an editor of the New York Tribune, was quoted as saying in the Universal News video of the time.

Prime Minister Winston Churchill inspects the First American Motororized Squadron of the Home Guard in London in January 1941. To the left is US General Wade H. Hayes, who commanded the 1st AMS. Picture courtesy The Granger Collection/Topfoto.

And, of course, it was early on in WWII that the US Marine Corps would become well represented as the primary security force for the US Embassy in London which had operated in some form (as a mission or embassy) since 1785.

Significant involvement of US Marines in the UK commenced in the summer of 1941 at the request of Admiral Robert Ghormley, Assistant Chief of Naval Operations and the US Embassy's unexpectedly designated security officer, Major John McQueen, to involve them. It was Major McQueen who, along with Major Arthur T. Mason, who effectively became the first USMC representatives to be sent to the UK. They had arrived in 1940 reportedly during a German air-raid and had gone on to observe the British Royal Marines in training in Scotland. Mason would go on to carry out liaison work for British Admiral Lord Louis Mountbatten (the Supreme Commander for South East Asia), while McQueen had been given the embassy security officer job by US Ambassador John G. Winant after the Marine had discussed his security concerns with the diplomat.

Lt. Colonel Harry W. Edwards, in writing *A Different War: Marines in Europe & Africa* as part of the *Marines in World War II Commemorative Series* for the Marine Corps Historical Division (published by the US National Parks Service), records that Marines were stationed singly and in detachments at bases and embassies and on ships (including the single-carrier (*Ranger,* CV-4) and the four-strong battleship and destroyer squadron US Atlantic Fleet in 1941. Indeed a squadron of US Marines would be inspected by Britain's King George VI on 7 June 1942 aboard the *Washington.*

By November 1941 nearly 3,793 Marines were aboard 68 US Navy ships with around 24 of those ships forming the Atlantic Fleet.

The first US Marine Detachment – indeed the first US Military unit to be assigned to the UK – was the 12th Provisional Marine Company who were to become the vanguard US Marines bound to provide protection for the American Embassy then located at 1 Grosvenor Square. However their journey was far from straightforward. On the way over their Dutch transport ship was sunk by a German submarine on 26 June 1941. It meant that the 11-man unit, under the command of Colonel Walter Jordan, would be actually awaiting ashore in Iceland to welcome the 1st Provisional Marine Brigade which was supposed to be the first combat force sent into the Atlantic Region.

Eventually Jordan and his men would arrive and activate their Embassy mission on 15 July 1941 and link with 48 US Marines who had left the United States after them but managed to arrive without incident.

There would be 59 US Marines (four officers and 55 enlisted men) who would form the original Marine Detachment, American Embassy, as it was titled. In December 1941 the Embassy team had was joined by 64 more Marines from the 12th Provisional Marine Company). Marines would not only guard the building but also run fire watch on the embassy roof and the roof of 20 Grosvenor Square, the Embassy Annex in which they resided and ride Harley Davidson motorbikes on courier service and escort duties. From 17 March 1942 with the arrival of Admiral Stark, Commander Naval Forces Europe, the USMC's role would be refocused to prioritize the nearby US Navy Headquarters building and supporting flag officers, as well as providing ceremonial squads.

McQueen, who would be accorded the title of Assistant Naval Attache, would eventually depart the UK for the US on a somewhat dangerous mission to return to Washington DC via the Azores with a top-secret radar device and plans for it that the British were share with the Americans.

Lt. Col. Edwards, who himself would command the Marine Detachment, American Embassy in London from 1944-46, wrote of Anglo-American rifle matches in Portsmouth as early as 1942 and of a programme established by USMC Maj. General Ralph Mitchell to send Marines as observers to London, other parts of the UK and on board British warships on courses (learning about everything from radar control to demolition to abseiling) to learn from the British even prior to the outbreak of WWII. A number of US Marines would also go on the seven-week long British 'Commando' course. Several Marine pilots and ground control personnel ended up the RAF's 256 Night Fighter Squadron while collectively learning night fighter tactics with the RAF both in control, pilot and radio positions.

Such interaction actually resulted in new or changed policy back stateside. This was particularly the case US Marine aviators in interception and night fighter operations as well as on the determination not to go ahead with a USMC glider programme. US Marines also made alterations to landing craft tactics.

Several Marines met Prime Minister Churchill in the course of their observation and joint training duties. One, Colonel Franklin A. Hart, acted as a Special Naval Observer and joined the joining the Commander Naval Activities Europe (COMNAVEU) before being made a liaison to Combined Operations Commander Vice Admiral Lord Louis Mountbatten, along with US Marine Lt. Colonel Harold D. Campbell, where they participated in operations planning for the ultimately disastrous (3,648 British personnel were killed, wounded, capture or listed as missing) British raid on Dieppe in July 1942.

Others worked for the US Office of Strategic Services (OSS), the forerunner to the CIA. One of the latter, Capt. Peter "Pierre" Julien Ortiz, would receive the Navy Cross on two separate occasions for his WWII work – one of which was presented by Admiral Stark in London – for his behind-the-lines work between 12-20 May 1944 during which he helped rescue four downed RAF flight crew and got them to Spain.

Ortiz would later become a prisoner of war (POW). He would not only become the most decorated OSS officer (among his honours he received the Order of the British Empire), but would then go on to become an actor and work in several Hollywood films including the famous Western *Rio Grande*. His connection with Hollywood didn't end there. Two other Hollywood films – *Operation Secret* and *13 Rue Madeleine* – had storylines based in various degrees of accuracy on his wartime exploits.

During WWII, around US Marines also arrived in Londonderry, Northern Ireland (in the summer of 1942) to provide a security force (they earned the moniker "The Irish Marines") for the Naval Operating Base and for the US Naval Radio Station there. They were involved with the RAF in the retrieval of an Allied bomber and crew from Ireland with the unofficial tacit approval of the Irish Government.

Other US Marines were dispatched to Rosneath in Scotland (from October 1942 to January 1943) where a USMC training camp was established to train personnel for Operation Torch (the Allied landings in North Africa).

One US Marine, John H. Magruder III, was assigned as liaison officer to the staff of the commander of the 21st Army Group, British General Bernard Montgomery. Another, Colonel William T. Clement, spent time working on the plans for D-Day in the Intelligence Division of COMNAVEU. Two others participating directly in the D-Day codenamed Overlord planning were USMC Colonels James E. Kerr and Robert O. Bare. Colonel Kerr worked on plans and then joined in the D-Day landings with the 11th Amphibious Force. Colonel Bare was assigned to the Office of the Chief of Staff for the Supreme Allied Commander in June 1943. He received a Bronze Star for his actions while with the British Assault Force J during the D-Day Invasion a year later. Bare and his colleagues had helped select training beaches on the South Coast of England to prepare for the invasion including the infamous Slapton Sands training exercise on 28 April 1944.

The US disbanded the Marine Barracks in Londonderry in August 1944, though the 80-man detachment stayed on until November 1945 to protect the US Navy radio station, according to Lt. Harry Edwards.

The US Marines were formally given an Embassy security role in 1946 after having taking on full security of the US Navy HQ in 1945 and, the following year, were designated Marine Detachment US Naval Forces Europe.

Over the years the Marines have been quartered at various distances from the US Embassy. There were barracks at 20 Grosvenor Square, then at Elvaston Place in the 1950s and also in Kennington and later on Allitsen Road in Swiss Cottage (near West End Lane). Their quarters became known as Marine Barracks London (MBL). Today the Marine Security Guards for the Embassy are quartered at Marine House located in Westminster. It is close to the US Embassy and reportedly has 21 accommodation units with a ground floor bar (complete with ship's bell), adjacent pool table room and a TV room and a small internal courtyard. The Embassy Marines are likely to relocate again when the US Embassy moves to Nine Elms near Battersea in London from 2018.

London – and the US Navy Building at 7th North Audley, also became the HQ for the US Marine Force, Europe with 40 to 180 Marines operating from there an across Europe.

Among the MBL cadre over the years was a USMC legend James Everett Livingstone who came to command the Marines there as a Medal of Honor recipient (1968) for his actions on his first tour of duty in Vietnam. Livingstone, then a Lt. Col (he would rise to become a General) was the 'top-hat' for all the US Marines in England – including those at St Mawgan in Cornwall (guarding nuclear weapons) and likewise at the US Navy submarine base at Holy Loch in Scotland.

The separate Marine Force Security Company Detachment was based at Eastcote in Middlesex (see under Bases section). US Marines and other US military services personnel are believed to form part of the contingents at several other bases around the UK including the long-standing National Security Administration (NSA) site at Menwith Hill in Yorkshire. There was some recent suggestion following a July 2013 question in Parliament that joint services US military personnel may also be stationed at Britain's Government Communications HQ (GCHQ) at Bude in Cornwall among other locations.

The American Eagle Club, originally at 28 Charing Cross Road in London, was formally opened in December 1940 to serve the 5,000 or so Americans already thought to be in England as the US became involved in WWII. GIs were also able to talk to the folks and a general audience back home via the BBC's weekly American Eagle Club Program which started in November 1940.

The Club was founded by Robert H. Hutchinson and opened by General Wade H. Hayes in a celebrity-laden event. It would serve something like 300 lunches a day and Her Majesty would even pictured serving tea to an airman there in February 1941. The club was a place of respite run by the American Red Cross which later opened the Rainbow Corner/ Rainbow Club in Piccadilly.

Charter members included Capt. Jack Beckman of the 358th AAF Fighter Squadron (based at Royston in Hertfordshire) who flew 99 missions over Germany before being killed by German citizens after parachuting out over Bavaria on 1 March 1945.

The main US military influx into the UK really commenced with the 4,500 men from the 34th Division who arrived as part of Operational Bolero, the codename for the US troop build-up in Britain, in January 1942. The numbers of US personnel remained on an upward trajectory for the next 32 months until 1.5 million personnel were in the UK (just under half had arrived in the six months from January to June 1944).

There would be highs and lows. The former would include a parade by 300 US military personnel who, in March 1942, marched through London past St. Paul's Cathedral to Guildhall to a lunch hosted by the Mayor of London in an overt political display of togetherness (*see book cover image*).

The latter would include the devastating Schweinfurt-Regensburg raid on 17 August 1943 (on the anniversary of the USAAF's first raid on Europe from England) in which 60 of the 376 B-17s were lost along with 564 crew (of the 2900 who had participated in the mission to attack ball bearing and other manufacturing plants) being killed or captured and 21 wounded. And a return raid to Schweinfurt on October 1943 would see 77 B-17s fail to return with the loss of 590 air crew who were killed in action and a further 65 taken as POWs. Other daylight raids, for which the USAAF was tasked, resulted in horrible losses for the USAAF but those two missions were particularly bad.

Another low that was largely forgotten about until the 1980s was Exercise Tiger, the practice D-Day landings taking place at Slapton Sands in Devon on 28 April 1944. The exercise would see 638 American GIs (197 US Navy and 441 US Army) lose their lives as two German E-boats got amongst the formation as they preparing to land. The deaths were a result of enemy action and, also, friendly fire in the confusion and just plain bad luck as many of the soldiers, in preparation for landing , had heavy packs and lifebelts in the wrong position which caused them to flip upside down and drown once they went into the water as their landing craft caught fire or sank.

American service personnel quickly found themselves at home in England and certainly around London. There was a line from the famous British film *The Dambusters* that summed up a wartime London increasingly filled with American service personnel. Two Royal Air Force crew are talking about going up to London to see a theatre show. They think they can get tickets to a show but one cautions: "You can't get into the stalls because of the Americans."

British actor David Niven, in his acclaimed memoirs *The Moon's a Balloon*, also addressed the American invasion of Britain in WWII: "By the Spring of 1944, it was obvious that the Second Front would soon be opened on the Continent. Training increased in tempo and there were so many American troops in Britain that only barrage balloons kept the island from sinking beneath the waves under their weight."

The peak of the US military WWII 'relationship' came in 1944 when the United States Army Air Force (as it was) had almost 160 bases with many, many other encampments for US military personnel. All in it is reckoned that nearly two million US military personnel, including 350,000 with the 8th Air Force, either served in – or passed through – the UK during the WWII years.

Among them were celebrities of their day and icons of today. James Stewart, the Academy-award winning actor *(The Philadelphia Story)*, who flew 20 combat missions as pilot of a B-24 Liberator with the 445th BG out of Tibbenham, Norfolk, before becoming Group Operations Director of the 453rd BG out of Old Buckenham airfield nearby. Stewart would win a Distinguished Flying Cross with an Oak Leaf Cluster for his actions and would leave the UK as a full Colonel. It is said that when he lead the 455th, not a single aircraft or crewman was lost.

Another of celebrity ilk was Douglas Fairbanks Junior who also saw real-life action. Fairbanks served a stint at the US Naval Headquarters in Grosvenor Square during the second half of 1942, with part of that time spent working in the office of British Rear Admiral Lord Louis Mountbatten and his Combined Operations Command Headquarters. It necessitated commando training in the Inveraray, Scotland, before he joined the British Royal Marine's cross-channel amphibious raiders based in the Isle of Wight. Fairbanks was awarded a Distinguished Service Cross by King George VI.

Fairbanks and the twice Oscar-nominated and Tony award winning actor Robert Montgomery, who had enlisted for American Field Service while in London and drove ambulances in France in 1940, had both applied for commissions in the US Naval Reserve. When America entered WWII Montgomery served in the US Navy Headquarters Map Room in London and is also said to have served as an attache aboard several British destroyers involved in hunting German submarines. Montgomery would return from the WWII with a Bronze Star and, in his first post-WWII film, go on to star in *They Were Expendable* with John Wayne.

They were not, by far, the only ones.

First to lead the way in entertaining the troops was Al Jolson, the best-paid Hollywood star of the 1930s and star of *The Jazz Singer*. Jolson had lobbied to help entertain the troops and headed to England in 1942 on the first hundreds of performances for the troops.

Others who would go on to enjoy celebrity included Andy Rooney, who would eventually become the *CBS 60 Minutes* presenter. Rooney had been assigned to a US Army artillery regiment before becoming a reporter with US military paper *Stars & Stripes* in London (he would be a frequent visitor to the offices of leading British paper *The Times* in which *Stars & Stripes* located its own offices – it also had offices in Soho).

Rooney and Walter Cronkite (who had arrived in London in 1942 and would go on to become the legendary anchor of CBS *Evening News*), were among a select band of eight journalists (sometimes known as 'warcos' – shorthand for war correspondents) and five newsreel cameramen who were given permission to fly on a single combat mission over Germany (on February 23, 1943). They were dubbed 'The Writing 69th' by the 8th Air Force Public Relations team (and 'The Flying Typewriters' or the 'Legion of the Doomed' by others). One of them, Homer Bigart, would later fly on the very last combat mission of WWII over Kumagaya, Japan.

In the end only six journalists (including Cronkite and Rooney) flew on the mission to Germany. However the flight was weather-diverted from their main target at Bremen to the German submarine pens at Wilhelmshaven and the *New York Times'* Robert Post was killed when his B-17 blew up in mid-air following a German fighter attack.

Actor Clark Gable also came to England with the Army Air Corps in February 1943 at the behest of General Hap Arnold to make a propaganda film about aerial gunners. Gable joined the 351st Bomb Group stationed at RAF Polebrook to make a film *Combat America* and ended up flying five combat missions from May 4 to September 1943 to get footage for the film.

Bob Hope, who briefly appeared in that film, also undertook his first USO tour in 1943 in England along with actress Frances Langford – one that included numerous hospitals and bases with an ensemble known as the Hope Gypsies. While in London he stayed at Claridge's – there is a famous story of him arriving, checking into a suite and finding no soap in the bathroom. When he called the front desk he received apologies but was told not even the King of England could obtain soap in his own suite.

Clark Gable was reportedly in the audience for one of the USO shows at Polebrook but was apparently so popular with the 351st personnel that the story goes when Hope called him up to stage and he didn't go Gable's fellow GIs wouldn't give away his location in the watching crowd, though the two men met and talked afterwards and then at the Embassy Club in London.

Also accompanying Hope on his tour for awhile was one John Steinbeck who was then reporting for the *New York Herald* syndicate but who had already written *The Grapes of Wrath* and *Of Mice and Men* and who would himself ended up working for the US Office of Strategic Services (the forerunner to the CIA).

Steinbeck reiterated his assessment of Hope in his own 1958 book *Once There Was A War:* "When the time for recognition of service to the nation in wartime comes to be considered, Bob Hope should be high on the list. This man drives himself and is driven. It is impossible to see how he can do so much, can cover so much ground, can work so hard, and can be so effective. He works month after month at a pace that would kill most people."

During that tour Hope, who had actually been born in London suburb of Eltham, also made a brief appearance in a special British War Office film starring and part-written by Burgess Meredith (who would become The Penguin in the hit TV series *Batman*) entitled *Welcome to Britain*. The one-hour training film for American GIs was to tell them about British customs and life. In the film US Army Chief of Operations (European Theatre) Lt. Gen. Jacob Devers says: "We Americans know how to meet death but we are not always so good at meeting life." Hope is seen using and explaining British coinage.

What isn't seen in the film apparently is that both Hope and Meredith were evicted from the grounds when the film crew tried to film at Westminster Abbey.

Bill White, Jamie MacDonald, Joe Evans, Craig Thompson, Jean Nichol, Robert Post seen larking around at the Savoy Hotel where many American correspondents kept court in the American Bar during WWII in London. Nichol was the popular public relations officer for the Savoy. Post, a reporter for the *New York Times,* was killed while on a ride-along bombing mission to the submarine pens at Wilhelmshaven, Germany.

"Bob Hope and I got thrown out of the Westminster Abbey grounds," Meredith wrote to his future wife, actress Paulette Goddard, in a letter now held in The Jerome Lawrence and Robert E. Lee Theatre Research Institute of The Ohio State University Libraries. "My God, it was funny. We were making a scene for a governement[sic] film about good-will and the Dean walks up and tossed us out on our altars. Get Bob to tell you about it. I laughed for two days. Poor Bob, he just arrived back from Africa where he was bombed nine times and on his one day of rest he offers to do this picture – or this scene – and after five hours work the Dean walks right in front of the camera and said "stop". I thought the people were going to mob the Abbey. All I can say is we have NEVER been thrown out of better places than that. Ah, me…"

In another letter to Goddard, cited in an article by Liza Schalleart in the November 1943 edition of *Photoplay* magazine, Meredith wrote of Hope: "The most wonderful thing about England right now is Bob Hope… He is tireless and funny and full of responsibility, too, although he carries it lightly and gaily. There isn't a hospital ward that he hasn't dropped into and given a show; there isn't a small unit anywhere that isn't either talking about his jokes or anticipating them."

Hope, who met Prime Minister Winston Churchill at 10 Downing Street on that tour (and even pilfered some Prime Ministerial paper during the visit) would return to England in 1944 on a tour that included the base at Stansted in Essex (today one of London's major airports) and numerous other bases. And he came again after VE Day in 1945 when he arrived via the *Queen Mary* to play a series of gigs including one for 10,000 GIs at the Royal Albert Hall on 4 July 1945.

It was also in 1944 that Major Glen Miller arrived in England and went to perform hits like *In the Mood* and *Tuxedo Junction* at something like 800 concerts for the troops during that summer and also record at the famous Abbey Road Studios before his untimely and still-mysterious death after his plane was lost flying from Twinbrook to France on December 15, 1944.

"Next to a letter from home, the Major Glenn Miller Army Air Force Band was the greatest morale booster we had in the European Theatre of Operations in World War Two." 8th Air Force commander General James H. Doolittle would later write.

It was the day after Victory in Europe Day, 7 May 1945, that the 8th Air Force undertook its last combat mission – a propaganda leaflet drop flown by a dozen B-17s.

Throughout WWII in particular, there were issues, of course – not just cultural but ones of crime, damage to property, impacted and uprooted communities (particularly on the South Coast of England). They essentially boiled down to one well-worn summation: Yanks were "Over paid, Over sexed and Over Here". Or as British comedian Tommy Trinder used to joke "Overfed, Overpaid, Oversexed and Over Here!" In retort the Americans would respond that the British were "Underpaid, undersexed and under Eisenhower!" On a slightly more derogatory tone – the leading insult was, of course, "England will fight to the last American."

Sex wasn't the only cross-cultural issue of concern for both British and American authorities during WWII. As part of the war effort there were issues of race to do with up around Black American GIs – who were often referred to as "Tan Yanks", stationed largely in London and at dock cities including Liverpool, Cardiff and Hull. The British made a feeble attempt to exclude black GIs completely. But it was General Eisenhower who, while based in Washington DC in April 1942, countermanded a cable from US Army Chief of Staff Gen. Marshall to the US War Department that no black personnel be sent to the UK. Various reports – list figures between 15, 000 and 150,000 of the million plus GIs stationed in Britain as being black. The figure is actually thought to be around 100,000.

The 1987 book, *When Jim Crow Met John Bull* by Graham A. Smith and published by I.B. Tauris, explores the issues in depth and also highlights racial tensions between white and black soldiers (there had been at least one large riot between black and white GIs station at a Quartermaster unit (569) in the Lancashire town of Bamber Bridge on 23 – 24 June 1943 in which one soldier was killed and four injured following a siege and confrontation between black quartermaster personnel and white military police.) Some UK locations were particularly supportive of black GIs and refused local demands to segregate or bar them. But there were incidents including one in Newbury, Berkshire, in which a British civilian publican was killed in a gun battle between black and white GIs. There were other serious incidents in Cornwall and Devon.

Additionally, there was also an apparent near-fevered desire on the part of some British women to pursue black GIs in relationships.

The later aspect might be said to be underscored by two events:

- a riot at a US Army Barracks in Bristol as black GI's were being repatriated home at the end of the war. Several hundred local women were cited as besieging the barracks, singing and demanding: "We want our coloured sweethearts".
- the "Brown Baby" problem. A number of black/mixed-race babies born as a result of inter-racial conjugations prompted, not only a post-war backlash against the children and their mothers in many British communities but caused the leading minority rights advocate of the day, Harold Moody, to write to Britain's Ministry of Health in late 1945 and then again in March 1946 demanding that each baby to be treated as a "war casualty".

Britain's Health Minister of the day was minded to suggest that the babies should be able to grow up in the UK. However, there seemed to be a different point of view at Britain's Home Office (the equivalent of the US State Department) with one official writing in reply: "Provided it is clear that the mother does not want the child and there is a reasonably satisfactory home in the US the child will have a far better chance if sent at an early age to the US than if it brought up in this country."

Perhaps the core issue that matters often boiled down to – that of sex – been summed up in a song popular with GIs at the time: *Roll Me Over in the Clover*. The song had its etymology in the Victorian era but was quickly and happily adapted by the GIs to the even more popular *Roll Me Over Yankee Soldier*! And, of course, the hit song *Over There* written by George M. Cohan in 1917 with its classic refrain "The Yanks are Coming" was reprised by soldiers in WWII and probably didn't help quell the wariness of British fighting men wondering about their women back home.

It was no surprise that German propaganda sought to exploit fears by British soldiers that sex-starved American GIs of whatever colour were living it up with their women.

Nor, perhaps, is it any wonder given the situation, that venereal disease rates for GIs in the UK reportedly shot up by 300% in the first year after arrival, thanks to encounters with what were euphemistically described by some as the "good time girls".

While around two million US 'GI's' passed through Britain during the war years perhaps the only surprise is that only an estimated 20,000 babies were born as a result of Anglo-American wartime encounters with about one-in-20 being black. One might be tempted to suggest that, based on the figures available, Canadian forces appear to have had slightly greater 'success rate' than the Americans with an estimated 22,000 babies fathered. As a general demographic backdrop it's been estimated that of the 5.3 million babies born in Britain between 1939 and 1945 about around 30% were illegitimate.

David Reynolds, the historian and author of *Rich Relations* (*The American Occupation of Britain, 1942-1945*), best summed matters up succinctly by citing a 1945 Home Office report on the issue:

"To girls brought up on the cinema, who copied the dress, hair styles and manners of Hollywood stars, the sudden influx of Americans, speaking like the films, who actually lived in the magic country and who had plenty of money, at once went into the girls' heads," the Home Office report said. "The American attitude toward women, their proneness to spoil a girl, to build up, to exaggerate, talk big and to act with generosity and flamboyance, helped to make them the most attractive boyfriends."

Of course not all encounters were fleeting – various figures cite around 70,000 Anglo-American WWII marriages, with a further 150,000 to 200,000 GIs getting married across the European theatre of war. In 1946 the State Department estimated the number at 60,000 for the UK.

A 1945 War Brides Act (Public Law 271) enabled the new spouses, children and adopted children of United States military personnel to enter the U.S. at the end of the War. (The actor Cary Grant brought particular Hollywood attention to the matter when he starred in the 1949 comedy film *I Was A Male Warbride*. There had actually been cases of male warbrides.

Starting on January 27, 1946 with 585 'emergency cases' aboard the *SS Argentina* the War Brides (and, in some cases, their babies) headed to the US from Southampton and other UK ports aboard special ships in a logistics mission known variously, in typical military parlance, as 'Operation GI Bride', 'Operation Mother-in-Law' or 'The Diaper Run'. The Queen Mary would carry a further 2,300 Brides just a week later. A special brides' embarkation camp was even constructed in Hampshire. Program oversight was provided by the US Navy Headquarters on Grosvenor Square along with the US Army.

Sadly not all the trips back were as happy as the comic-sounding code-names would indicate. By early February 1946, just as the repatriation was gearing up, some 100 GIs had already requested cancellations of their wives' travel plans – pending divorces.

And that wasn't the worst of it. At the outset, and also later on, the military repatriated brides and babies on cruise liners with *The Queen Mary* and *The Santa Paula* among them. However, the US Military also converted Liberty cargo ships for the job. The latter vessels were clearly ill-equipped to deal with the babies (who had to be at about six months old before they could travel) with several reports that six babies died on one voyage and that nine died in one week in May 1946. The cause of death was often put down to a diarrhea epidemic.

There were other tragic transatlantic love stories. In one case recounted within the GI bride community, the British bride and baby arrived on US shores only to learn that her GI husband had died in a motorcycle accident at home as she travelled to the US to be with him.

It could be said that love continued hand-in-hand with war for the US Forces in the UK during the Cold War years, too. In fact, at one point in the 1960s, it was estimated that nearly 200 Americans a month from across all US military bases in the UK were requesting permission from their Commanding officers to marry British partners. To this day US military personnel stationed in the UK are still regularly getting married to British and other non US citizens. To do so, they must fill out form AFI36-2609 if they're stationed in England and Wales. A local registrar near the remaining major USAF bases at RAF Mildenhall and RAF Lakenheath in Suffolk told the author that the number of marriages went 'way up' every time American forces stationed at those bases in East Anglia deployed to potential combat.

The WWII US military presence and wartime efforts and exploits by the British and US military personnel are well documented elsewhere. The establishment of bases, air-fields, supply chains, training and missions during WWII and then during the Cold War created what the leading British investigative journalist and writer Duncan Campbell went on to describe as the '*Unsinkable Aircraft Carrier*' – the eponymous name he used for his seminal 1980s book about the US Military in the UK.

The term is said to have had its roots in British Prime Minister Winston Churchill's description of the island of Malta but, given the amount of US hardware and personnel that arrived during and after WWII and right up through the 1980s, at least, was aptly applied closer to home.

In essence it was Campbell's contention – and, indeed, the ascribed theory of many others since – that the US Military had, in metaphoric if not physical terms, paved the UK with runways and military bases.

Even as America's wartime presence was ending in the UK and troops numbers rap-idly dwindled, there was what historians refer to as a "pell-mell demobilization" of US troops across the global theatre post-war that saw the US military shrink in the two years starting in 1945 from 12 million to 1.5 million personnel.

However, the Yalta Conference of February 1945 between the leaders of Britain, the US and Russia to determine the future of Post-War Europe was already resulting in alarm bells ringing among the Americans and British about the Soviet intentions – even though President Truman kept faith in the Soviet promises.

In was in March 1945 that the astute W. Averell Harriman (the wartime special envoy to Europe who helped implement the Lend Lease program and who would become the US Ambassador in the USSR as the Soviet Union was formerly known) cabled President Truman: "…the Soviet program is the establishment of totalitarianism ending personal liberty and democracy as we know it."

The follow-on Potsdam Conference, which started on 26 July 1945 (just days before the Atomic bomb strikes on Japanese cities of Hiroshima and Nagasaki (6 and 9 August respectively), saw complaints that the Soviets were interfering with Bulgaria, Hungary and Romania.

It was in the summer of 1946 that plans were laid at the most senior levels between USAAF's General Carl A. "Tooey" Spaatz, the WWII commander of the 15th and 8th Air Forces, and Lord Tedder, Britain's Chief of Air Staff, to leave the way open for a return by US Forces to pre-designated UK bases including Bassingbourne, Lakenheath, Marham, Mildenhall and Scampton – in particular for a returning USAF Bomber Force of B-29 Superfortress bombers.

Later that year they first flew to the UK landing at Burtonwood, a former USAF and US Army base halfway between Manchester and Liverpool (and whose old main runway provided the groundworks for part of the present-day M62 freeway), before deploying on to what was then West Germany. More followed on the ubiquitous training deploy-ment and by 1947 – so-called '30 day deployments' became a regular feature of Strategic Air Command (SAC) operations which were formally launched on 21 March 1946 (along with the Tactical Air Command (TAC) and Air Defense Command (ADC)).

SAC was responsible for the strategic bomber aircraft and support aircraft (refueling rankers and command aircraft) to provide long-range bombing – and particularly nuclear bombing – capabilities across the globe. The command would become ubiquitous with keeping airborne command and control aircraft 24 hours a day in case of Soviet attack.

SAC would also oversee the ground launched Inter-Continental Ballistic Missile (ICBM) in the US, a system then based on the Atlas missile.

On 16 February 1948 USAAF Boeing B-29s took part in a joint air defense operation with the Royal Air Force Fighter Command for the first time. On 7 August 1948 Maj. Gen. Leon W. Johnson arrived at RAF Marham in Norfolk and assumed command of the 3rd Air Division (Provisional) and within 10 days he had moved to Bushy Park in Teddington, near London. On 23 August the 'Provisional' designation was dropped and 3rd Air Division was in business.

And on 8 September 1948 the entire 3rd Air Division HQ moved to Bushy Park – the same base from which General Eisenhower had set up his Supreme Headquarters Allied Expeditionary Force (SHAEF) during WWII.

It has been on 1 April 1948 that the then Soviet Union established its to "Iron Curtain" and blockaded Berlin. Then a new phase in the Anglo-American relationship, that would eventually include both the establishment of bases, personnel and even sharing (though somewhat one-sided) of nuclear weaponry, was underway. The USAAF became the USAF and, via the Strategic Air Command (SAC) and American aircraft and their crews flew over to undertake joint missions and the standby bases were beefed up as the Command ran round-the-clock operations. SAC bomber squadrons would be rotated from the US on *Reflex Operations* until about 1962 when *Project Clearwater* ended rotation deployments to overseas bases.

The personification of the new post-WWII special relationship also came in September 1948 when USAF and RAF aircraft flew together as part of Britain's first major air defense exercise. *Operation Dagger*, as it became known, took place from 3-7 September.

On 13 November 1948 USAF personnel and bases were accorded a permanent status in the UK. Britain and the US became part of the original North Atlantic Treaty Organization (NATO). New military groups and organizations and their associated acronyms started to spring up like fertilized grass.

The 11-month long Allied resolve to resupply Berlin by air transport on a 24 hour–basis without doubt further bonded US and UK military and government relations and also saw a rapid rise in US personnel arriving in the UK along with the B-50 aircraft (an upgraded B-29).

By the time the 3rd Air Division was released from the USAFE and assigned to Headquarters USAF on 3 January 1949 there were 374 officers and 4,637 airmen stationed in the UK.

The end of the Soviet Union's blockade on 12 May 1949 was, however, only the curtain opener to a further 40 years of Cold War operations and the build up of tensions and military posturing between the East and the West.

The USAF 3rd Air Force headquarters at the Victoria Park Estate was rebadged as USAF South Ruislip, in Middlesex, on 15 April 1949. It was a base some 15 miles west of London – but in reality it was a facility that comprised new factory buildings, one of which was apparently converted into barracks by the simple act of installing walls (and no ceilings) beds and toilets. There followed a further influx of personnel when the base also became HQ for the growing Strategic Air Command's 7th Air Division on 20 March 1951.

By then B-36D aircraft, known as the 'Peacemakers', were also being deployed to the UK (the B-47 Stratojets started coming into service in 1951) on Reflex Alert. The 7th AD Headquarters would stay at South Ruislip until 16 June 1952. It would then redeploy to USAF High Wycombe (aka High Wycombe Air Station/RAF Daws Hill) from 1 July 1958 – 30 June 1965.

Just over a month later, on 1 May 1951, the 3rd Air Division, became Third Air Force. Within a year it would have 25,573 personnel and there would be a total of 45,029 US military personnel in the UK including Strategic Air Command and US Army and US Navy along with 3,748 UK support personnel. There were also a further 28,436 US military dependents. The Third Air Force South Ruislip headquarters would grow (to 1,733 personnel including 1,246 US Military) along with some 2,339 dependents by 1962. The 3rd Air Force would remain there until 1971 when the command relocated to RAF Mildenhall in Suffolk.

As noted there was a US Army presence in the UK but on a relatively much smaller scale.

Records held by the US Army Centre of Military History show US Army personnel stationed in the UK from 1949 to 1969 as oscillating as fears over the former Soviet Union's 'iron curtain' materialized. A swelling in numbers from 1949 to 1957 was reduced as more US military ground troops were sent to the 'front line' US bases in Germany.

31 December 1949...94
31 December 1950...144
31 January 1952.....3,758
31 January 1953.....4,704
31 January 1954.....4,056
31 January 1955.....4,149
31 January 1956.....3,249
31 January 1957.....4,355
31 January 1958.....174
31 January 1959.....197
31 January 1960.....376
31 January 1961.....662
31 January 1962.....710
31 January 1963.....636
31 January 1964.....634
31 January 1965.....645
31 January 1966.....551
31 January 1967.....88
31 January 1968.....377
31 January 1969.....283

The Visiting Forces Act (VFA) – that Act which truly set out the framework for the US Military operations in the UK – was upgraded from its 1942 mandate in 1952 and covered everything from the exercise of powers and arrest and trials even to Coroner's

Inquests and the handling of deserters. The VFA appears to have migrated in more recent years into the Status of Forces Agreement (SOFA).

A year prior to this the British and US Governments agreed the Special Construction Program which set out plans for building or enhancing up to 26 bases in the UK.

Possibly in conjunction with that programme, several secret agreements were concluded in 1951-54 in which the British Government agreed to pay £125 million for the cost of the bases, despite British politicians being told that only £22.5 million was being contributed by the British.

The first combat unit, the 20th Fighter-Bomber Wing, which utilized F-84 aircraft, was stationed at RAF Wethersfield in Essex on 1 June 1952. A year later and the 3rd Air Force had 25 installations in the UK but had started to replace some non-key support personnel with British civilian workers. In 1957 and 1958 the USAF returned 10 of the bases back to Britain's Ministry of Defence and there were further draw-downs in the 1960s following a period when the Third Air Force came under control of the 17th Air Force.

In 1958, the US Naval Facility, London, was re-designated as the US Naval Support Activity London. As other US Navy operations in the UK and Northwestern Europe were established, the command was renamed US Naval Activities United Kingdom (NAVACTUK) and, in 1965, the US Naval Support Activities was disestablished and assigned to Commander NAVACTUK (COMNAVACTUK).

Other highlights saw the deployment of nuclear weapons to US bases in the mid-1950s and their shared operation between USAF and RAF units. The 1958 Mutual Defense Agreement became a cornerstone of the Anglo-American relationship on nuclear weapons with accords on training, tests (allowing the British to test weapons at the atomic test site in Nevada) and the exchange of information. In fact, so important was the Agreement that it continues to this day and will be reviewed again in 2014.

The presence of nuclear weapons and where they were stored was always something the US and UK declined to confirm as a matter of policy but among the USAF bases which had or had the capability to store the 'nukes' were: RAF Wethersfield in Essex (before 79th Squadron relocated to RAF Bentwaters in 1952), at RAF Shepherds Grove in Suffolk until it closed in 1966, at RAF Sculthorpe until about 1962, at RAF Manston in Kent until it was handed back to the MOD in the late 1950s, RAF Upper Heyford even until the 1990s and at RAF Lakenheath until 2008.

And then there were also bases at RAF Greenham Common and RAF Molesworth.

RAF Greenham Common in Newbury, Berkshire, was the former World War II base for the US Army's 101st Airborne Division who had taken part in D-Day and then later in Operation Market Garden.

Post-WWII the base had been put into inactive status for five years but ,in the wake of the Cold War, it was reactivated as part of the 7th Air Division and would, in 1954 become home to rotational deployments of nuclear bomb carrying Strategic Air Command B-47 Stratojets for the next decade.

Other units would then utilize the base, but Greenham would 'go nuclear' again in a high-profile way with the arrival of 96 nuclear capable Ground Launched Cruise Missiles (GLCMs) which were based there from November 1983 until March 1991. Likewise, RAF Molesworth in Cambridgeshire (from where the first USAAF Eight Air Force mission of WWII was flown), hosted 64 GLCMs from 1986 until October 1988.

The nuclear relationship was the focus of much attention by the US Third Air Force command, the British Government and peace protestors.

Notification of an accident at RAF Greenham Common in February 1958 would have gone through channels at Third Air Force HQ at USAF South Ruislip. A B-47E which had an engine failure on takeoff jettisoned 1,700 gallons of fuel which fell behind a B-47 loaded with missiles and set it alight. The aircraft burned for 16 hours (two personnel were killed and eight injured in the accident). There were claims that the bomber on the ground had nuclear missiles on board and the fire might have burned nuclear material but those claims seem to have been largely disproved following Gamma ray detector tests in the late 1990s.

In July 1956 there was a B-47 crash at RAF Lakenheath in Suffolk. The near-nuclear disaster that resulted when the plane, which had been practicing 'touch and go' landings, hit a nuclear bomb storage igloo and set it on fire probably resulted in what must have been one of the more heart-stopping communications that would be sent from the USAF Third Air Force HQ at South Ruislip to the Pentagon. The crash had killed the four crewmen set fire to the bunker housing three of the nuclear bombs. Indeed such was the apparent panic that a stream of people were said to have been seen driving hurriedly away from the base in the hours that followed.

A telex cable dated July 22, 1956, is from Lt. Gen. James H. Walsh, 7th Air Force, to Commander in Chief of Strategic Air Command Curtis E. LeMay and blatantly confirms the near disaster. The document refers to a Mark 6 nuclear weapon, which contained between 7,500 and 8,500 pounds of explosive.

FM COMAIRDIV 7 USAB SOUTH RUISLIP ENGLAND

TO CINCSAC OFFUT AFB NEBR

/TOPSECRET/CCMDR T-5262. PERSONAL FOR CINC LEMAY FROM WALSH. MORE TO MY PHONE CALL. HAVE JUST COME FROM WRECKAGE OF B-47 WHICH PLOUGHED INTO AN IGLOO IN LAKENHEATH ADS. B-47 TORE APART IGLOO AND KNOCKED ABOUT 3 MARK SIXES. A/C THEN EXPLODED SHOWERING BURNING FUEL OVERALL. CREW PERISHED. MOST OF A/C WRECKAGE PIVOTED ON IGLOO AND CAME TO REST JUST BEYOND IGLOO BANK WHICH KEPT MAIN FUEL FIRE OUTSIDE SMASHED IGLOO. PRELIMINARY EXAM BY BOMB DISPOSAL OFFICERS SAYS A MIRACLE THAT ONE MARK SIX WITH EXPOSED DETONATORS SHEERED DIDN'T GO. FIRE FIGHTERS EXTINGUISHED FIRE AROUND MARK SIXES FAST. PLAN INVESTIGATION TO WARRANT DECORATING FIREMEN.

(Source: Digital National Security Archive's declassified document set titled "U.S. Nuclear History: Nuclear Arms and Politics in the Missile Age, 1955-1968," Item number NH00842. The source of the document was the Library of Congress Manuscript Division, Curtis LeMay Papers. Box 206. File B-55745.)

Communications about another USAF death would have also been received by – and then sent on from – the Third Air Force HQ at South Ruislip just a couple of weeks later. General David C. Schilling, who was the Inspector General for SAC's Seventh Air Division and was on temporary assignment (TDY) from USAF South Ruislip to RAF Lakenheath in Suffolk, was killed in a car accident on the narrow country road from nearby RAF Mildenhall to the Lakenheath base on 14 August 1956. Gen. Schilling, who shared an interest in fast cars with SAC Commander General Curtis LeMay, had been on his way to discuss the sale of his Cadillac/Allard sports car when, at the start of an overtaking maneuver, he apparently lost his cap. In reaching for it he is believed to have lost control of the car which then skidded sideways into a stone bridge and broke in half killing him instantly.

The day before his death, Schilling – who had served in England during WWII as a P-47 Fighter Pilot, and become an 'Ace in a Day' pilot who ultimately scored 22.5 kills, making him the sixth leading ace in the 8th Air Force to that time – had made his last ever flight in a B-47. On March 15, 1957, Smoky Hill Air Force Base in Salinas, Kansas was renamed Schilling Air Force Base in his honor (the base was closed in 1965). The Air Force Association's Award for Outstanding Flight, which Schilling won in 1952, was named for him after his death.

There was one other higher-profile incident news of which would have travelled through the command corridors at USAF South Ruislip on its way to equally concerned officials on both sides of the Atlantic. In late October of 1958, Leander V. Cunningham, a USAF Sgt., armed himself with a .45 calibre pistol and locked himself in what Britain's then Secretary of State for Air described as a "base building" at RAF Sculthorpe in Norfolk and apparently threatened to commit suicide. The eight-hour standoff ended peacefully after the sergeant's wife, Nancy, and several chaplains and doctors were summoned. However, some reports seemed to indicate that the sergeant, who was reportedly being treated for depression, had held his pistol to the warhead of an atomic bomb.

Four years later in 1962 the incident was raised in the British House of Commons by the then Member of Parliament, the Right Hon. Michael Foot, who questioned the Britain's Prime Minister about whether, in the wake of the incident, he had held consultations with the US President about taking "fresh precautions" at such bases.

"The facts were given to the House at the time by the then Secretary of State of Air. The American Serviceman referred to locked himself in a shed for eight hours. The anxiety was not that he might explode an atomic bomb – there were none there – but that he might do something foolish with a pistol. I did not think it necessary to consult the President of the United States about this."

But with further assertions from another MP that it might not have been a bomb but it might have been "an atomic or nuclear device" that the sergeant threatened to destroy, the Prime Minister appeared to make a concession.

"It is quite true there was some explosive in the building and it is just conceivable that he might have caused it to explode had he carried out his threat, but this threat was to kill himself not to shoot the explosive. The was no fissionable material in the building and no possibility of a nuclear explosion of any kind."

From September 1954 when the USAF's first nuclear bombs arrived until sometime in the two years up to June 2008 (no clear date has been given as the removal was shrouded in secrecy) when the last of 110 B61 gravity bombs (the B61 is a variable yield delivery system ranging in destructive power of up to 340 kilotons) in the UK were said to have been removed from the 33 nuclear weapons vaults at RAF Lakenheath, the nuclear deterrent was set to become a cornerstone of US military operations in the UK. The B61 version of the nuclear weapons had been deployed in 2000 and probably included the 'bunker buster' version of the bomb.

In the later Cold War years, before the drawdown of US Forces in England, nuclear weapons known to have been stored at RAF Lakenheath in Suffolk, RAF Upper Heyford in Oxfordshire and RAF Bentwaters also in Suffolk) with the capability of them to be stored at RAF Fairford in Gloucestershire (which has a 9,993-foot long runway and was capable of handling both B-52 bomber aircraft and was also a designated Transoceanic Abort Landing alternative for the Space Shuttle – the base now remains in standby status). Most of these would have been B61 gravity bombs.

Beyond that a number of nuclear weapons (as well as nuclear propulsion systems) were thought to be on board US Navy submarines when they were calling at the US Navy's former Holy Loch base near Dunoon, Scotland, with some nuclear depth charges (thought to have been Mk57s (later B57s) fitted with hydrostatic fuses and providing five to 20 kilotonne yields each) stored not far away at the RAF Machrihanish base. The latter, located on the Mull of Kintyre (made famous by the eponymous Paul McCartney song), had also been home for a period to 20-strong US Navy detachment of the 'Special Warfare' Navy Seals and was like Fairford also, apparently, a designated "abort landing" option for NASA's Space Shuttle. The remoteness of the Machrihanish base and its 1.89-mile long runway have sparked many a rumor about its use for the Aurora, purported to be a top-secret hpypersonic spy plane successor to the retired SR-71 'Blackbird' (two of which had operated from RAF Mildenhall in Suffolk from 1976 – 1990).

More of those same nuclear depth charges, which were intended for use by NATO maritime patrol aircraft, were also stored under armed guard by a dedicated Marine Corps Security Force Company MCSFC (Det 2) at RAF St. Mawgan in Cornwall.

In July 1986 an apparent mistake by the US House Appropriations Sub-Committee on Military Construction revealed a $4 million spend for four nuclear weapon 'vaults' at RAF Bentwaters – now a former USAF base on Suffolk's East Coast.

The arrival of Ground Launched Cruise Missiles starting in November 1983 would also see both RAF Greenham Common (60 miles west of London near Newbury, Berkshire) and then RAF Molesworth (a USAF base in Cambridgeshire, 60 miles north of London) join – or in the case of Greenham Common, rejoin – that nuclear base list.

The formal policy – at least since about 1956 – seems to have been if at all possible to neither to confirm nor deny the presence of nuclear weapons or where they were stored. This policy even covered their removal from the UK.

However, a "US-UK joint release" agreement had been around since at least the early 1960s and was confirmed in a 5 August 1965 top-secret letter from then Prime Minister Harold Wilson to US President Lyndon Johnson and subsequent communications between lower-level officials. The Wilson letter referred to the storage of nuclear weapons at RAF St. Mawgan in Cornwall and went on to say:

"Release of these nuclear weapons in an emergency use by the United States and British forces based in the United Kingdom would be the subject of a joint decision taken by the President and the Prime Minister in accordance with the terms of the understandings, as amended, were confirmed by your letter of December 8, 1964."

There had also been complicated change of heart and relative rough patch in the Anglo-American special nuclear relationship over the delivery by the United States of the ballistic nuclear missile Skybolt GAM-87 system for Britain's Vulcan bombers. The delivery ran into difficulties with test failures and the US effectively pulled out of the agreement to provide a system upon which the British were basing their entire defense strategy. The cancellation, in December 1962, lead to problems between the US and UK often summarized by historians as the "Skybolt Crisis". A form of resolution came with the introduction of the US Polaris missile for use by Royal Navy submarines.

That 'tiff' aside, the Anglo-American relationship continued with occasional difficulties to be found in most alliance relationships.

But again the world was changing. On 30 June 1965 the 7th Air Division disbanded thereby formally ending the bulk of SAC operations in the UK. Within a year, however, more US Forces were deploying to the UK having been rapidly ousted from France following that country's decision to withdraw from NATO and promptly boot out the US military presence there.

There were probably quite a few coded messages and faster heartbeats at USAF Headquarters and bases in the UK during the Six-Day War between Israel & Egypt, Syria and Jordan (from 5 – 10 June 1967). The war saw the USAF Third Air Force go on full alert in the UK. It is believed that, for a brief period at least, the Third Air Force may have deployed jets armed with nuclear bombs either in the VA (Victor Alert – it would be later known as the 'QRA' (Quick Reaction Alert)) areas or at the end of the runways ready to immediately be launched should the order come.

While the bases were essentially USAF, most were 'badged' as RAF bases. An exception was the HQ at USAF South Ruislip or as it was also referred to: South Ruislip Air Station, USAF.

Another was High Wycombe Air Station USAF which became, for a brief while, RAF High Wycombe and then RAF Daws Hill. The last designation was to spare confusion with the RAF's own Strike Command bunker over at Naphill which is now known as RAF High Wycombe or RAF Strike Command, High Wycombe.

The RAF 'designator' wasn't just cosmetic – it alluded to a more formal 'buffer' between the Americans and the British and to the idea that the Americans were just tenants rather than an occupying force. Adding to that was, at most installations, the presence of a RAF Station Commander who could serve as community liaison. Additionally, it also meant that the bases could more easily utilize the British Ministry of Defense Police – another important and far more physical 'buffer' particularly at places like Greenham Common and Molesworth during the major protests over the deployment of Ground Launched Cruise Missiles. This possibly also helped reduce the potential for possible shooting incidents given the US military security police were armed.

One MOD official put it to the author thus:

"The explanation here is simple; the UK does not permit the basing of foreign forces in the UK, hence under the Status of Forces Act; they are on UK based and hence in the case of USAF units they are RAF bases and have an RAF Station Commander. Put simply the US are lodgers."

The year 1972 marked a watershed for the US military presence around London – at least in USAF terms. It was that year that Third Air Force Headquarters at USAF South Ruislip was relocated to RAF Mildenhall, a flight-line base near Newmarket (and just 5.3 miles from another major USAF base at RAF Lakenheath) in Suffolk.

The impact was to see the drawdown of several London area installations such as Bushy Park and consolidation of USAF operations largely to RAF West Ruislip, MOD Eastcote and High Wycombe Air Station the latter, which then came under the command responsibility of the USAF 20th Tactical Fighter Wing at RAF Upper Heyford 40 miles further west in Oxfordshire.

USAF matters, including the usual LSN (Local Salty Nation) training exercises, relatively sauntered along but with a couple of noteworthy events. First was the Falkland Islands War (2 April – 14 June 1982) during which there were short-term US deployments in quiet support of British operation. Then there was some return back-scratching tacit support by the British when USAF F-111s bombers flew from Britain – using RAF Upper Heyford and RAF Lakenheath – to carry out a raid on Tripoli along with aircraft from US Navy carriers operating in the Gulf of Sidra.

That mission, codenamed *Operation El Dorado Canyon*, saw the 'Aardvarks', as the F-111 jets were known, strike Tripoli on 15 April 1986 in revenge for Libyan agents bombing the La Belle discotheque in Berlin which had been packed with US service personnel two weeks earlier. The 1:45 a.m. bomb on 5 April killed three and injured 230 including 76 American military personnel. Intelligence agents had intercepted telex messages to Libya's embassy in East Berlin congratulating agents on a job well done. USAF F111-Es (the electronic jamming aircraft version) from 20th Tactical Fighter wing (TFW) at RAF Upper Heyford and F-111Fs from 48th TFW RAF Lakenheath. By chance 20th TFW crews had actually practiced such a long-range bombing mission (designated Operation Ghost Rider) in October 1985 using a simulated target near Goose Bay, Newfoundland to simulate the trip to Tripoli.

The 1980s brought the last major influx of US Military personnel to the UK for what would be two decades – namely through the deployment of the Cruise Missiles (GLCMs) to RAF Greenham Common and RAF Molesworth with some command and support elements located at High Wycombe Air Station.

It was revealed in 1985 that armed US soldiers were patrolling local residential areas around at least one of those bases to help protect US military personnel living in the civilian community under a proviso of the Visiting Forces Act of 1952. Indeed, another famous story of the time involving RAF Upper Heyford was that, during *Operation El Dorado Canyon* in 1986 base security police force went into combat posture and sealed off Camp Road, the main British public highway that bisected the base. It was understood that local British police had to actually negotiate with the deployed American troops to get the public road reopened.

At about the same time it was also revealed by the *New Statesman* magazine that the then existing 66 UK military establishments would be increased by an intended expansion of another nine air bases and potentially up to 30 civilian hospitals in the event of war.

Throughout the years the US Military presence in the UK made both news and occasional waves in other ways. Among them:

The mass peace protests in the early 1980s as Ground Launched Cruise Missiles were deployed at RAF Greenham Common. A US Airman's throwaway joke threat about killing Britain's Prince Charles in 1985 which made national headlines. The time in the summer of 1986 when a practice bomb fell off an F-111 from RAF Upper Heyford as it flew over Lincolnshire and a US Navy submarine ran aground off the coast of Ireland and had to be towed to Holy Loch. Concerns about US and UK submarines getting 'caught' in traveler nets and dragging them backwards. The murder of Tech. Sgt. Michael Sambogna, a young airman from RAF Benwaters, in the town of Saxmundham in 1986. The revelation that US Mass killer Patrick Henry Sherrill had come to Britain and taught servicemen and women shooting skills at RAF Mildenhall with other National Guard members just weeks before killing 14 postal workers in Edmond, Oklahoma.

Or the the time in 1987 that a US Air Force jet was reported to have been using Thirlestane Castle on the Scottish borders as a "simulated laster-designated toss target" when it crashed a mile away. Or when in 1987 the BBC Television attempted to deceive the USAF and then faced a backlash from politicians and others over its' controversial televised play "Airbase". Or the time in 1988 when telecommunications provider British Telecom (now BT) helped charge two airman with phone fraud on long-distance calls and issued a general warning to American airmen against using nickels rather than 20 pence (approximately. 30 cents) coins for pay phones. In just a few months the cost of the scam – well known to GIs – hit BT for more than $50,000 in long-distance calls. And then there was the revelation in the BBC's *Secret Society* television documentary that a British civilians could be evacuated to make way for US servicemen under British Government emergency laws.

Also making news was the base at High Wycombe which saw its WWII underground bunker, once home to General Doolittle and the 8th Air Force Bomber Command, receive between $13 million (some figures later indicated the spend to be about $19.5 million) for an upgrade as a programming centre for the cruise missiles and also as a fallback headquarters for the US European Command)in the apparent event command posts on mainland Europe – namely at Patch Barracks in Germany – were overrun by any invading Soviet forces).

Then came a series of watershed moments for the US military in general – with the repercussions strongly felt by the US military forces based in the UK.

- The signing of Intermediate Nuclear Forces (INF) Treaty in 1987 was an indicator of change ahead as it signalled the withdrawal of the (GLCMs) and their delivery platforms from the two UK bases (RAF Greenham Common and RAF Molesworth) and also from mainland Europe.

- The October 1988 agreement by the US Congress and the Senate to cut around 24 bases in Europe – something that hadn't happened for around 10 years. After that, the drawdown dam effectively burst in both the US and overseas with major implications for the US military presence in Europe.

- The dismantling of the Berlin Wall in November 1989, the reunion of East and West Germany and the break up of the Soviet Union in 1991 which essentially took away the principal Cold War enemy.

Things were now happening in rapid succession as the world order changed. Gulf War I kicked off in the Persian Gulf on 2 August 1990 and, in an echo of early Strategic Air Command days in the 1950s, B52s deployed to RAF Fairford in Gloucestershire to conduct long-range bombing missions as part of Operation Desert Storm (which commenced on 17 January 1991). The war was over just over a month later on 28 February.

For note, Fairford, with 1.89-mile-long runway, was used in 1969 to test fly Concorde from 1969-1977. The USAF also ran B-52s and B-1 bombers from there in Operation Allied Force targeting the former Yugoslavia in 1999 and B-52s for Operation Iraqi Freedom in 2003. In recent years that base has undergone a major refit (with more than $100 million spent on the runway and fueling infrastructure and two special hangers capable of housing B2 Bombers. It is now designated a 'standby' airfield with turnkey facilities kept ready to go with only a 48-hour notice.

On-going in the background, however, round after round of base closings were being announced by the Pentagon. And ever since then the US military presence in the United Kingdom could be categorized in a three-word term: *Reduction In Force*.

Base closings and the process for doing so had its entomology in the Federal Property and Administrative Services of Act of 1949 which was utilized in the 1960s to close some

560 US military bases in the US and around the world. "BRAC" (Base Realignments And Closures) entered the military lexicon with the first round of closures in 1989 and was formalized in 1990 with the Defense Base Realignment and Closure Act.

The intended plan, implemented during the President George H.W. Bush administration and continued by the Clinton Administration, was to cut the 300,000 American servicemen in Europe in 1989 in half by 1995. By the second round US base closures in Europe announced in 1991, meant that 381 European installations – 25% of the total bases there – would be closed. The majority of them were located in Germany. But it wasn't long before several major UK flightline bases – RAF Greenham Common, in Berkshire as well as RAF Bentwaters and RAF Woodbridge (known collectively as the "Twin Bases" because their base perimeters were no more than two kilometers apart) in Suffolk – were for the axe.

There followed a series of Cold War milestones for US Forces in the UK which, as of 1 September 1990, totalled around 22,675 military personnel largely stationed at seven major bases with a collective impact of $627,079,424 on the UK economy for FY 1990:

- The withdrawal of the 96 Ground Launched Cruise Missiles (GLCMs) from RAF Greenham Common and the 64 GLCMs from RAF Molesworth (a former USAAF WWII base in Cambridgeshire that is still in use by the USAF today as the US European Command's Joint Analysis Centre) in 1991. As the last of the missiles are Greenham were taken out in March 1991 watched by then British Armed Forces Minister Alan Clark,General Marcus Anderson, Commander of the USAF Third Air Force told the author: "It's the end of another chapter in both the Cold War and the US military presence in Europe."

- By May 1991 the number of US military personnel in the UK was down to 15,000 – about half what it had been five years previously and the lowest number since before WWII. In the previous 21 months 10 US military facilities in the UK had been closed. "We're outta here," one Third Air Force officer told the author at the time. "It's incredible to watch the speed at which this is all happening."

- In June 1992, the US Navy's nuclear submarine support facility (known as Site One) at Holy Loch in Dunoon, Scotland, also shut up shop when the submarine tender USS Simon Lake (AS-33) sailed back to the United States thereby ending a 31-year US Navy presence.

- In August 1992 the USAF partially withdrew from massive bomb storage dump at RAF Welford in Berkshire (a facility conveniently located adjacent – and with its own access road – to the M4 Motorway) which it had occupied since 1943 as an airfield and since 1955 for munitions storage; while the US Navy facilities at Glen Douglas in Strathclyde, RAF Machrihanish in Argyll and RAF St. Mawgan in Cornwall were handed back to the Ministry of Defence. At that point High Wycombe Air Station at Daws Hill was to be placed on "stand-by status".

- In May 1993, the USAF base at RAF Alconbury in Huntingdonshire and, the US Navy Base at Brawdy in Wales were announced for part and full closure respectively. At that same time it was announced that RAF Upper Heyford in Oxfordshire would withdraw all F-111 'swing wing' bomber aircraft and be closed by September 1994 and that High Wycombe Air Station, which had been due to go to a Standby Status (essentially mothballed) would actually transfer to the US Navy. It was the 11th round of base closing cuts and had followed consultation with the British Government.

The 7520th Air Base Squadron inactivation ceremony, effectively the end of the major USAF presence at bases close to London, took place at High Wycombe Air Station on 10 May 1993. The ceremony conducted by USAF Maj. Gen. Charles D. Link and Lt. Col Ben C. Williams was held on an old outdoor basketball court on the base's sports fields.

However the USAF drawdown of operations around London posed a logistical support problem for the US Navy as it still maintained its Naval Activities facilities in support of the London-based Commander US Naval Activities, United Kingdom. The loss of the USAF bases would have had a significant impact both on housing and privileges for US Navy personnel who had been able to use the USAF Commissary and PX (Post Exchange) as well as other aspects such as the schooling of dependent children.

The upshot was that the USN took on the responsibility for the USAF-managed bases at Eastcote, West Ruislip and Daws Hill, High Wycombe under the umbrella of Commander Naval Activities United Kingdom (COMNAVACTUK) in 1993.

However, COMNAVACTUK was a command now subordinate to the Commander Naval Forces Europe and, as such (and with emphasis on operations in other parts of the world), was starting to lose the gloss on its historical London link. Indeed it was viewed as an unnecessary duplication of effort and personnel. The Commander-in-Chief US Navy Europe (CINCUSNAVEUR) was also starting to look harder at logistics and savings.

CHANGING UK PERSPECTIVE

Aside from logistics in support of *Operation Allied Force*, the NATO bombing of Yugoslavia from March to June 1999 and the Second Gulf War that started in March 2003 against Iraq (known as *Operation Iraqi Freedom*), events did little to halt the drawdown of US military in the UK. If anything, perhaps, the Gulf in particular placed emphasis in the Navy's case, at least, that Headquarters operations had to be closer to where the bulk of the perceived action was – namely at that point in the Middle East.

As far back as November 2002 there were indications within the British Government in responses to questions raised in the House of Commons that the US Military –primarily the US Navy – might consolidate its bases around London to a single location. The British Government's "Major Investment Programme for London's Defence Estate", announced by then British Defence Minister, Dr. Lewis Moonie, on 29 October 2002, introduced the concept of "MoD Estate in London" or "MoDEL".

As part of the 2002 proposal it was first mooted that British forces would move from their famous and long-held command base at RAF Uxbridge in Middlesex and move to the Royal Artillery Barracks at Woolwich, with the Uxbridge base being redeveloped as the "core site for the USN" with US military units at Eastcote, Kennington Barracks, Daws Hill, West Ruislip, Blenheim Crescent and the US military's Defense Contract Management Agency (Northern Europe) office at Loudwater in Buckinghamshire, to be relocated there.

But no sooner than the Ministry of Defense had announced that plan than there seemed to be a sea-change within the US military in general and the US Navy in particular. For starters the US Navy Command in London changed its name from CINCUS-NAVEUR to Commander US Naval Forces Europe (COMUSNAVEUR) that same year. Crucially, however, the realization that US Navy really didn't need what was effectively a mirror command in London anymore percolated. In 2005 and following the activation of NATO's Joint Force Command in Naples, Italy, a year earlier, the COMUSNAVEUR relocated there with a significantly smaller 'co-located' command within the NATO unit at Britain's Permanent Joint Headquarters (PJHQ) at Northwood in Middlesex.

At that point NAVACTUK in London, the effective base command office, had 11 officers, 57 enlisted personnel and 257 civilian personnel supporting more than 760 military personnel and more than 1,200 of their dependents located largely in and around the London area.

The final stages of the main US drawdown came in four parts:

- Firstly, as part of a realignment of USAF operations the Third Air Force was inactivated on 1 November 2005 after a 50-year operational history in the UK. It would be reactivated exactly a year and a month later at USAFE Headquarters at Ramstein Air Force Base. The 3rd AF-UK remains headquartered at Mildenhall.

- Secondly, the Ministry of Defence revised project Ministry of Defence Estates London (MoDEL) – to include the sale of RAF Uxbridge as well as DOE Eastcote Complex (as that base was then formally known) and RAF West Ruislip to Vinci St. Modwen (VSM Estates), a joint venture between international construction company Vinci PLC and developer/regeneration specialist St. Modwen Properties PLC, with the monies from the sales of the land to be reinvested into upgrading and creating new facilities at RAF Northolt, the major RAF base with a flight-line located near West Ruislip. The contract was awarded to VSM Estates on 3 August 2006. In September 2006 the MOD said that Uxbridge, West Ruislip and Eastcote would be sold by 2012 with the intention of raising some £180 million to be spent on infrastructure at RAF Northolt.

- The Lend-Lease agreement (the Post-WWII one was known as The Anglo American Loan Agreement) with the British made in 1946 was paid off (a final installment of $83 million) on 29 December 2006. Part of the deal offset had been the UK bases – the lease of which had largely been free of charge (or sometimes on a 'peppercorn rent' – literally a peppercorn a year) for the last 60 years.

- The original Lend Lease deal loan made during the WWII with $650 million for the British to pay had been written off at the end of the War as part of the new post-War loan agreement. (It's worth noting also that the US also traded 50 destroyers in return for bases on UK territorial possessions in the Caribbean (Trinidad in Jamaica as well as on St. Lucia, Antigua and in the Bahamas and other parts of the world including Newfoundland (where the Naval Station Argentia was until 1994). The Destroyers for Bases agreement undertaken in 1940, was deemed paid off in 1995 when the US returned the last bases (which it had on a 99-year, rent-free basis) but with proviso for an anytime return to Bermuda and Newfoundland.

The final chapter – for Eastcote and West Ruislip in the first instance and for Daws

Hill High Wycombe in the second – had been sealed via the following two press releases (with the same title) issued by the United States European Command in Germany and the US Navy in London in 2006 and 2007 respectively:

DoD announces installation realignment in United Kingdom

Release Date: May 17, 2006

U.S. Department of Defense

WASHINGTON D.C. – The Department of Defense announced on May 15 that it would cease operations at RAF Eastcote and reduce operations at RAF West Ruislip, United Kingdom. Due to U.S. European Command force structure realignment and save $400,000 annually in facility costs, with additional personnel savings to be realized by 2009. It will affect 70 U.S. service members, 33 DoD civilian employees and 38 host nation employees. Navy operations at RAF West Ruislip will be reduced beginning June 30, 2006 with a scheduled turnover of affected facilities on Oct. 1, 2006. Facilities affected by the reduction will be included in the process of return to the host nation. This action is expected to save $1 million annually in facility costs, with additional savings to be realized in 2009. It will affect seven U.S. service members, 55 DoD civilian employees and 95 host nation employees. Ultimate total annual savings with these actions will be about $10 million. As with all stationing actions, the U.S. has coordinated with host nation officials before this announcement.

DoD announces installation realignment in United Kingdom

Release Date: March 30, 2007

Navy Region Europe Public Affairs

The Department of Defense announced today that the United States will cease operations at two military facilities in the United Kingdom.

With the U.S. European Command's force structure realignment and transformation, it was determined that RAF West Ruislip and RAF Daws Hill are no longer required.

Facilities affected by this decision will begin the process of return to the host nation. As with all stationing actions, the United States coordinated with host nation officials before making this announcement.

The US Navy formally ended its Commander Naval Activities United Kingdom (CNAUK presence in with a disestablishment ceremony on 14 September 2007 attended by US Ambassador Robert Holmes Tuttle, Rear Adm. Michael R. Groothousen, Commander, Navy Region Europe and Capt. David Dittmer, Commanding Officer of Naval Activities United Kingdom.

"Although we disestablish our command today, we will not disestablish our relations with our strongest ally," Groothousen said at the ceremony. "The commitment and pride our sailors and employees of CNAUK have demonstrated throughout the closure process is truly to be commended. Operational commitments of Naval Forces Europe have dictated that our troops need to be operating in other locations around the globe," Groothousen added in elaborating on the reasons behind closing CNAUK.

US Ambassador Robert Holmes Tuttle, who was the VIP Guest at the event, summed up the changing military posture but in doing so also underlined the Anglo-American relationship:

"Strangely, the closure of the U.S. Naval activities in the UK is a sign that we have been more successful than we might have dared hope when it opened – that our two nations could work together and face down the dangers of a changing world. We have proved over that half-century that our faith was justified and our loyalty to each other well-founded, as we have worked together to encourage and support country after country to move from totalitarianism to democracy.

"Many of the challenges of the 21st Century are different, but many remain the same. To face down these new dangers, we must find new ways of working together and new alliances – while remaining steadfast to those to whom we are most indebted."

Ambassador Tuttle continued:

"The relationship between the United Kingdom and the United States is not defined by the number of bases or types of services, but by our common spirit.

"I know that this base and the other U.S. naval facilities in the UK over the years – Holy Loch, St. Mawgan, Eastcote and in London – have carried that sense of spirit for many decades and we thank Captain Dittmer and his team for all that they and their predecessors have done in that effort. I have every confidence that our relationship is stronger than one command and larger than one base. I believe that the vision and values that join us are found wherever we serve together."

In October 2012, the Pentagon announced the return of the Medical Clinic at RAF Upwood in Cambridgeshire as part of the U.S. European Command's (USEUCOM) "...continued effort to remove non-enduring sites, bases and installations from its real-property inventory." The 57 military personnel were redeployed to nearby RAF Alconbury and the annual savings from the return were estimated at $554,000.

As of November 2012 the drawdown meant that there were just 13,528 USAF, civilian local national employees plus 5,721 dependents stationed across the UK. There are believed to be around 250 US Navy personnel also still stationed in the UK as part of the various elements including those at the Permanent Joint Headquarters at Northwood in Middlesex. One of the current prime focuses for the Allied Maritime Command (MARCOM) is Operation *Ocean Shield* which is the NATO operation combatting Somali pirates. Also run from the same HQ is Operation *Atalanta* which is the European Union-led mission with the same task.

Interestingly, Northwood's new role as the sole Allied Component Maritime Command for NATO, announced only in December 2012 may mean additional US Navy personnel coming back to the UK from Naples. NATO and US Navy from the base did not respond to author's requests for specifics in regards any increase in personnel but anecdotally it must be a percentage of the 925 US NATO military personnel and 1,250 who were stationed in Naples as part of the NATO component that had been the joint headquarters. By summer 2013, however, the Maritime Hub Command, as it is known was expected to grow from 220 to 300 personnel – a portion of which will likely be US Navy.

Also during summer of 2013 it was announced around 900 USAF personnel and their dependents would be assigned to RAF Mildenhall as part of the 352nd Special Operations Group). Arriving with them were the first two of 10 CV-22 Osprey tilt-rotor aircraft to support the unit's duties "...to infiltrate, exfiltrate and resupply special operations forces".

It's an interesting footnote perhaps that, in the hand-over of the bases (at least in cases that the author has witnessed), the process was clear-cut. There seemed to be a simple but effective rule worked out between the British MOD and the US Air Force, Army or Navy. It ran thus, as Captain Dave Dittmer explains in his own memories further on in the book: If you could imaging tipping a building upside down and shaking it, then everything that fell out would belong to the US Military. Everything that stayed fixed to the walls belonged to the British MOD.

At RAF High Wycombe/Daws Hill, like other closing bases surplus equipment was shipped to other bases in the UK and across the Europe.

The rule didn't just apply to the buildings apparently. It was a matter of minor annoyance for at least one RAF official that more than 80 pairs of shoes had been thrown into the branches of a tree behind the Mansfield boy's residence hall for military dependent children from distant bases attending London Central High School at RAF Daws Hill. The shoes, which had been tossed into the tree as a good luck tradition for many years, were left there when the US Navy departed despite a request from the RAF Commander that they be removed.

When it came to the public attention that any US military base was up for sale one of the facilities that had the greatest number of inquiries was usually the base bowling alley. Others inquired after garages/repair shops and or snack bar and club/bar dining equipment.

Meanwhile the general drawdown in Europe continued. On 8 April 2011 US President Barack Obama announced plans to reduce the US military presence in Europe by reducing the number of combat brigades stationed there from four to three by 2015. The total number of US military personnel stationed in Europe has dropped from 300,000 in the Cold War to just under 80,000 personnel as of 2011. In February 2012 it was announced a further 11,000 troops were being withdrawn from Italy and Germany.

The last BRAC Commission was held in 2005. In March 2012 it was reported that Pentagon officials were preparing proposals for further cuts in Europe with a further consolidation of some of the 300 sites ahead of proposed BRACs for 2013 and 2015. However Congress took a dim view of cutting more bases and refused to authorise the 2013 round though the request has returned in the Department of Defense's 2014 Budget request.

On 14 March 2012 The White House released the following statement on US and UK military relations:

Joint Fact Sheet: U.S. and UK Defense Cooperation

Today President Obama and Prime Minister Cameron reaffirmed their commitment to continue close cooperation on defense as the United States and United Kingdom build their Armed Forces for the future. The U.S. and UK share an unprecedented defense relationship that has helped secure our shared interests and values since the World Wars of the last century. We have developed unparalleled military interoperability and interconnectedness, working together to meet the challenges of the Cold War, leading in NATO and fighting side-by-side in defense of global interests. At every level of our defense establishments British and American service men and women train together, learn together, develop capability together and, when called upon, fight together.

Standing Together: British and American forces routinely operate side-by-side across a wide range of operations. A century of shared battlefield experience has led to a level of interoperability and familiarity that is unique in its breadth. This is exemplified in Afghanistan today where the U.S. and UK are the two largest contributors to ISAF and our Armed Forces are working together to degrade the insurgency and to train and mentor the Afghan Forces to provide security in Afghanistan. For example, in Helmand province the U.S. Marines' Task Force Leatherneck and the UK-led Task Force Helmand are working together to deliver stability.

British and American exchange personnel routinely deploy on operations with their host units. For example, British air transport pilots flew with the U.S. Air Force in Haiti earthquake relief operations and British F-18 pilots are currently flying operational missions from the USS Stennis. U.S. Marine Corp exchange officers have deployed on operational tours to Afghanistan with their host British units, in some cases in a command position and the U.S. Air Force has a long tradition of exchanging pilots on transport, aerial refueling and combat aircraft with Royal Air Force units.

Training, Learning and Developing Together: The ability of American and British forces to operate on the battlefield effectively is due in large part to the close-knit and constant training and exchange opportunities undertaken together. As close Allies, the U.S. and UK host each other's forces in order to conduct training, be prepared to forward-deploy when necessary and in many cases conduct current operations.

The U.S. currently has over 9,000 personnel permanently stationed in the UK, primarily on shared Royal Air Force (RAF) bases such as RAF Lakenheath and Mildenhall, where U.S. units conduct fighter, transport and aerial refueling operations. The Joint Analysis Center (JAC) at RAF Molesworth is a prime example of cooperation, where U.S. and British analysts monitor the world's trouble spots together. All four U.S. services send exchange officers to work with the British services and exchange both junior and senior military officers with British defense schools.

The UK currently stations over 800 British personnel in the U.S., conducting a wide variety of activities from conducting RPAS (Remotely Pilot Air Systems) operations in Afghanistan from Creech AFB, Nevada, to working side by side with American colleagues on major acquisition projects such as the Joint Strike Fighter and C-17 projects, to working with U.S. counterparts on cyber and space cooperation. Approximately 200 British officers are on exchange with the American services to develop joint approaches to develop capability and increase interoperability.

During the month of March, 2012, alone, 1,100 UK military personnel will take part in 10 training exercises with U.S. forces across the country, to include a detachment from the Royal Regiment of Artillery participating in an adventure training expedition near the Grand Canyon, a squadron of Royal Air Force (RAF) GR4 Tornadoes conducting live-fire heavy weapons training in Arizona and an RAF squadron participating in a Red Flag exercise at Nellis AFB, Nevada.

The U.S. and UK routinely entrust their best and brightest NCO's and officers to each other's academies, military schools and units to gain experience and insight into the other partner's way of doing business. Exchanging military personnel ensures a cadre of individuals in each military that understands their counterparts and cross-fertilizes the best each nation has to offer in ideas and doctrine.

Also during March a senior British officer, Gen Richard Shirreff, will conduct a speaking tour at West Point, Ft. Leavenworth and the Pentagon as part of the yearly Kermit Roosevelt Speaking seminar, a tradition between the British and American Armies that dates back to 1948. U.S. General Robert Cone is reciprocating later in the year at the British Ministry of Defence, the Land Forces headquarters and the Royal Military Academy at Sandhurst.

Today the UK also has a wide range of senior personnel serving in advisory or command positions in U.S. Headquarters, including the Deputy Commander of the 1st Infantry Division, senior planning staff at CENTCOM and senior liaison positions in NORTHCOM, CENTCOM, CYBERCOM, PACOM and STRATCOM. British officers also serve as faculty at West Point and the Naval War College. Similarly, U.S. officers serve with the British military in multiple advisory levels, attend British defense schools and are integrated into British combat units, sometimes in command positions.

Collaborating for the Future: The President and the Prime Minister agreed that both defense departments will continue to push for increased interoperability across the spectrum of military operations after today's operations come to an end. The U.S. Defense Strategic Guidance and UK Strategic Defence and Security Review reached many common conclusions, including the need for increased cooperation in dealing with the threats we face. We are committed to working together and with other close allies, wherever possible.

• Navy – Secretary Panetta and Secretary Hammond recently signed a Statement of Intent directing the U.S. and Royal Navies to seek ways to better develop aircraft carrier doctrine and maritime power projection capabilities. • Land – We will also seek to develop similar initiatives to enhance the already close ground force relationship though increased training opportunities in Europe and unit exchanges in the U.S.

- Air – The UK is a tier-one partner in the development of the Joint Strike Fighter – a unique program in which each country's defense industries are sharing in the development of a common platform that will ensure the U.S., UK and other partners own the cutting edge in air superiority for the next generation.

- Cyber – The U.S. and UK, along with other capable nations, are working together to protect vital information infrastructure from cyber attack. We are committed to building our interoperability in this vital new space, building on a Memorandum of Understanding signed in 2011.

- Space – The UK and U.S. will work with other partner nations to explore the potential for collaboration and information sharing in this expanding realm of activity.

- Management of Defense – Both countries are committed to ensuring that our Armed Forces have what they need for the future, are given the support they and their families deserve, while maximizing the value of the resources spent on defense. There are many areas where we can work together to make this happen, ranging from our Service Personnel Task Force, to work on future energy requirements, science and technology, to nuclear sharing.

Leading Together: The United Kingdom and United States stand shoulder to shoulder with each other to deter and, if required, defeat threats to our common way of life.

The President and Prime Minister agree that there are new opportunities to strengthen this relationship further. With new strategic circumstances come new reasons to cooperate. We cannot afford to miss these opportunities. Both countries recognize that many of the problems that we both face cannot be solved alone.

By working together more closely, we set an example to others and provide a basis for further collaboration with our Allies and partners, including through NATO, in the years to come.

In March 2011 It was announced that several US Senators were co-sponsoring S.402, a bill supporting the creation of a 'Cold War' Medal for anyone who served in the US Military from 2 September 1945 until 26 December 1991 and who had been honorably discharged. The bill was backed by Sen. John Kerry of Mass., Sen. Jim Webb of Virginia, Sen. Olympia Snowe and Sen. Susan Collins of Maine but appears to have gone nowhere. While there is a Cold War Medal it is not recognized as an official one that veterans can wear on their uniforms.

The CNAUK change of Command on RAF Daws Hill, High Wycombe in 2006. Center in the photo are Capt Steve Matts, USN (outgoing Commanding Officer (CO – on right) and Capt. Dave Dittmer (incoming CO – on left). In the photo is the then entire US Navy COMNAVACT military staff. Photo: USN

BASES, INSTALLATIONS & FACILITIES

So to the bases and other key locations. How did they come to be, what happened there and, of course, what has become of them?

A caveat. Not everything will, or can be, answered. Information about bases and installations is in more detail at some locations than others.

BLENHEIM CRESCENT, WEST RUISLIP

Blenheim Crescent is one of the few noteworthy surviving locations for the slender remaining US military presence in London. This base, which sits on a rough triangular wedge of land of 1.36 hectares (3.36 acres) close to the London Underground Central Line near West Ruislip station, comprises a long, three-storey main building with covered walkway entrances off to the sides as well as two older single-storey, sloped-roof buildings for the Base 'Shoppette', Post Office and other facilities.

The base was originally acquired by the Royal Air Force as part of the neighboring site West Ruislip in 1920 and the site and became home to the Royal Air Force Records Office. It was made its own entity, separate from the West Ruislip base in 1924. It is set in an old military housing area that was once occupied by RAF and then USAF families and, today, remains in part, at least, as a residential area comprising ex-forces homes.

Blenheim Crescent was first taken over by the US Air Force as part of the hand-over of West Ruislip from RAF to US control on 1 December 1955.

It was, for a time, home to the US Corps of Engineers which had enjoyed temporary residence at RAF Northolt before moving a staff of 75 to Blenheim Crescent in about 1969/70. From there they would be involved in most major base building projects across the UK including schools, hangers, command facilities and the NATO-standard Hardened Aircraft Shelters (HAS) – a familiar on most US flight-line bases.

Many of the latter were designed and contract managed in preparation for the arrival of the F-111 'swing-wing' aircraft in the 1970s. Their work also included designing the water-cooled 'hush houses' in which USAF ground crews mounted engines and ran them up for testing. Because they were water-cooled a distinctive plume of steam could often be seen 'venting' around 1,000 feet high above the unit and could be seen for miles around at bases like RAF Upper Heyford and RAF Lakenheath. In later years the hush house units would become air-cooled.

The Corps of Engineers at Blenheim Crescent were also involved in helping design the top-secret circa $150 million Cobra Mist project, located at Orford Ness on the Suffolk Coast. The marvelously code-named Cobra Mist was, in fact, an over-the-horizon backscatter radar station (AN/FPS-95) located in a in large rectangular box of a building resting on stilts for fear of the flood tides in an area known for its experimental radar testing over the years. RCA won the contract for the Anglo-American radar that was reportedly designed to provide early warning for missile launches from the Soviet Union.

Work started on it in 1967 and it became operational in 1970. Alas 'mist' seemed to be an apt code moniker. Depending on the version of the story you believe the system was either plagued by shifting-signal 'noise' problems or being successfully jammed by the

Russians. Whatever the cause, the facility, which included an 18-string 2,040 foot long radiated ground mast antenna, was shut down on 30 June 1973.

Like other agencies the Corps of Engineers would move to Edison House – then the name of a small office block between Edgware Road underground station and Marylebone mainline station in London before it was announced in December 2007 that they, among others, would be moving back to Blenheim Crescent.

By then the base was dominated by a long red-brick three-storey office-type building.

Blenheim Crescent was originally run by the USAF and then handed over to the US Navy in early 1975 and was under their control until 1 October 2007 when management and administrative responsibility reverted to the USAF Third Air Force and, as of 2012, to the USAFE-UK) via the 422nd Air Base Group at RAF Croughton, England, as the main leaseholder.

During at least part of the period the US Navy had control the American Red Cross also had offices there. A barbeque deck and covered patio were added in 1996.

Today Blenheim Crescent comprises around 130 personnel with around the bulk of those serving in the US Navy, USAF and US Army Research & Development Regional Headquarters. It is also home to the USAF's 422nd Communications Squadron. Post Office. The facility's front entrance has a blue linoleum-style floor inset with logos of many of the residing units.

In late 2007, probably following the closure of its namesake in Marylebone, London, the main building was also named Edison House – the name is etched into the glass entry doors of the building.

In an answer in the House of Commons on 16 October 2007 then Minister of State for the Armed Forces, Mr. Bob Ainsworth, said that Blenheim Crescent would be host to the following units: The European Office of Aerospace Research and Development (EOARD), the US Army International Technology Centre Atlantic, the Office of Naval Research Global, the Naval Criminal Investigative Service (NCIS), the Joint NATO National Support Element, the Defense Energy Support Centre and the Fleet Industrial Supply Centre.

The main role of the EOARD is to provide liaison with members of the European Scientific Community and facilitate contacts between Air Force Scientists and their European Counterparts. There are also other supply procurement and support organizations within the operational 'tree'.

At the end of 2003 the US Navy applied for permission to erect a guardhouse and sentry box at the entrance to the base. The US Navy had previously added a main building Conference Centre extension in May 2001. That extension was the first major works to the building since an entrance and entrance staircase addition to the building in 1991.

Around 2010 Edison House underwent a refurbishment of the building roof.

On January 10 2012, the British Government announced it was pushing through with plans for its HS2 high-speed rail system from London to Birmingham which it plans to have in operation by 2026. The proposed route would appear to go either right through or very close to the eastern portion of the Edison House and the base. The British Government has since indicated that the section of line in question could now run underground, however the future of the building and the base could still be in question pending actual disposition of the HS2 and, also the tunnel depths and any vibration issues.

The USAF is believed to pay a peppercorn rent on Blenheim Crescent to the Ministry of Defence.

The main entrance to Edison House at Blenheim Crescent in Middlesex. Photo: US Navy.

BUSHEY HALL, WATFORD (USAAF STATION 341 DURING WWII)

This former house-turned-hydrotherapy-facility-turned-hotel became Headquarters of the 8th Air Force Fighter Command which was set up on the 21-acre former golf course site by Gen Frank "Monk" Hunter in the summer of 1942 and was utilized by them until January 1945 when the command moved to Belgium.

It would also be the base where Lt. Gen. William E. Kepner would become the commander of 8th Air Force Fighter Command. The HQ 6TH Fighter Wing arrived at Bushey Hall on 16 August 1942. On 12 September the HQ 4th Fighter Group and its three (334th, 335th and 336th) Fighter Squadrons were activated and comprised Spitfires manned by US pilots from the RAF Eagle Squadrons. Some 17 days later the Group relocated to a base at Debden in Essex some 45 miles north of London. On 4 October both the 347th FS and the 350th Fighter Group left for the base at Snailwell in Cambridgeshire.

The 9th Air Force, which was established to support D-Day and the invasion of Europe, also established its headquarters at Bushey Hall in October 1943.

Perhaps one of the most significant tactical changes for the USAAF in WWII took place at Bushey Hall during a visit by General James Harold "Jimmy" Doolittle, who was by then commander of the 8th Air Force.

Whilst in Gen. Kepner's office after being present in the Operations Room and hearing radio chatter from US fighter pilots who were going back into an aerial fray to protect bombers even through they were low on fuel, he reportedly saw a sign reading "Our mission is to bring the bombers back" (Some versions of the story have it as "Our mission is to bring the bombers back alive").

Doolittle reportedly said: "From now on that no longer holds. Your mission is to destroy the German Air Force".

In a single sentence Doolittle mandated what Gen. Kepner had himself been strenuously arguing for in the months previously. Kepner himself recalled that Doolittle's final act before departing from Bushey Hall was, according to a detailed account by noted author Donald L Miller in his book *Masters of the Air: America's Bomber Boys Who Fought the Air War Against Germany*, to look at the sign and say: "Take that damned thing down!"

Of course bomber crews saw that new general order for their escorts in a somewhat different – and definitely unhappy – light.

During the later part of WWII Bushey Hall was to transition back to the RAF and post-War the base provided the command HQ role for the British in their support of the Berlin Airlift as the Cold War took hold.

As those tensions increased the US military also returned in the form of the Headquarters 32nd Anti-Aircraft Artillery (AAA) Brigade in on 1 October 1953 and used Bushey Hall headquarters as until the base was then deactivated again following the Brigade's departure in 1957. Key units, which also included chemical smoke generation units for defense against low-level attacks, were located at Lakenheath, Wyton and Sculthorpe.

The original Bushey Hall building from which the base took its name itself was demolished in 1955.

After a period of relative inactivity the USAF then utilized the base for London Central High School (a school for dependent sons and daughters of US military personnel sent to bases in the London area), which was being relocated from Bushy Park, Teddingdon. LCHS was the primary occupant from 1962-1971 when the school was relocated to USAF High Wycombe Air Station at Daws Hill (later RAF Daws Hill).

Bushey Hall was handed over to the MoD Defense Lands for disposal on 29 March 1972 as part of the drawdown of support installations after USAF Third Air Force relocated its headquarters from South Ruislip to RAF Mildenhall in Suffolk.

Today the 24-acre base is known as is the Lincolnsfields Children's Centre and the owners continue to try and preserve the military history with a special museum and exhibition as well as occasional military events. There are 13 remaining WW2 / Cold War period buildings on site with six being used for the Lincolnsfields Centre's Forties Experience program which is a mixture of museum, living history and educational programs designed to teach children and adults the realities of life in wartime Britain.

Some of the other buildings are used by non-profit making community activity groups including a kick-boxing club, children's dance school, model railway society, etc.

There are four other remaining WW2 buildings (three of which were Quonset huts used by U.S. troops) which the property owners have applied to the United Kingdom's Heritage Lottery Fund for a grant to re-instate these to add to war-time display areas.

The Lincolnsfields management team, previously headed by Francis McLennan and now by Phil Knight, remains keen to hear from U.S. veterans and their families about their memories of Bushey Hall for their archives and hopes to raise funds for a permanent memorial here on this historic site commemorating the contribution of the US Forces from 1942 – 56.

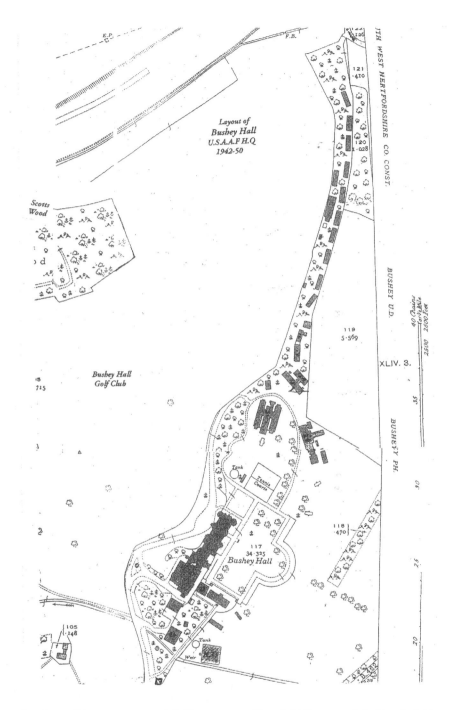

A map of Bushey Hall base as it was in 1942. Map courtesy of Francis McLennan and the Lincolnsfield Childrens Centre.

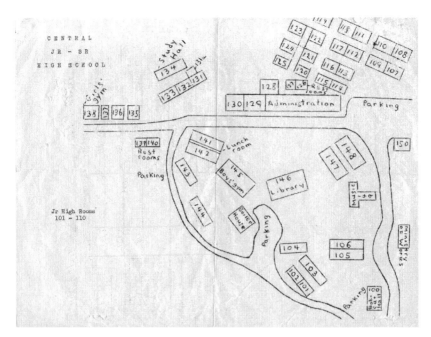

A map of the Central High dependent school campus (later to become London Central High School) at Bushey Hall Watford as it was in the 1960s. Courtesy of Cliff Williams.

This would have been a shocking surprise in WWII. What appear to be German soldiers man a checkpoint at Bushey Hall, Watford. But this is as a WWII reenactment weekend held at the former base in 2009. At the top left of the picture the bottom of the base water tank tower can just be made out.

The former school metalworking and woodworking building at Bushey Hall, Watford as it was in 2009.

BUSHY PARK (TEDDINGTON) – STATION AAF-586 IN WWII

Bushy Park, located next to Hampton Court Palace, enjoys considerable history (back at least as far as the Bronze Age) and royal patronage from when King Henry VIII when he took over the palace and parks from Cardinal Thomas Wolsey in 1529. In WWI large areas were turned into vegetable patches as part of the Britain's 'Dig for Victory' campaign, something that would be repeated in WWII.

The Americans also played a part in that history. Gen. Carl "Tooey" Spaatz established his wartime 8th Air Force Headquarters following his arrival in the UK on 24 April 1942. It formally operated as the Headquarters of 8th Air Force from 25 June 1942 – January 1944.

The base was given the codename "Widewing" but also known as Camp Griffiss –after Lt. Col. Townsend Griffiss – the first USAAF officer to die in line of duty in Europe during WWII. Griffiss was a special observer sent over in mid-1941 before the US formally entered WWII. He became the victim of 'friendly fire' after coming back from a mission to Russia via Egypt where he had been attempting to negotiate logistics routes for the delivery of aircraft to US Forces in England. Two Spitfires flown by Polish pilots from a squadron based in Exeter mistakenly shot the B-24 down killing him and the other crew members and passengers. Griffiss received a posthumous Distinguished Service Medal.

Gen. Spaatz set up the HQ in Building A, one of large four rectangular buildings on the parkland site. The 8th AF Service Command, which set up HQ on 6 July 1942, was located in Building B.

An Officers Club and Mess areas went into Buildings C and D respectively. Around them a vast sprawling base of other buildings, Quonset huts and facilities were created.

Before the end of the 1942 Gen. Spaatz was sent to the Mediterranean and was replaced by General Ira C. Eaker who had previously commanded the US Army Air Force's (USAAF) 8th Bomber Command at Daws Hill, High Wycombe.

Gen. Spaatz returned in 1943 as Camp Griffiss was reorganized to support the United States Strategic Air Force (USSTAF) which then combined the command for 15th Air Force in Italy and 8th Air Force in Britain. USSTAF remained in-situ until the end of WWII in 1945.

Bushy Park additionally became location of choice as the Supreme Headquarters for the Allied Expeditionary Forces (SHAEF) in March 1944 and was one of the key locations where General Eisenhower (who came back to the UK two months earlier) and his senior team laid plans for D-Day (from C & D Blocks).

Other US tenants at Bushy Park included HQ 8th Ground Air Support which relocated to RAF Membury (USAAF Station 466), pretty much where the Membury Services stop is on the M4 Motorway in Berkshire today, on 21 August 1942 and HQ 6th Fighter Wing Command moved to Atcham (USAAF Station 342) in Shropshire.

Post WWII the Royal Air Force also moved in and established RAF Transport Command, which played a key role in coordinated the UK efforts with the USAF in the 11-month Berlin Airlift starting in June 1948 to break the Soviet blockade of that city.

The USAF HQ, which by then comprised 400 officers, enlisted personnel and secretaries, remained at Bushy Park until April 1949 when it commenced a move to larger premises at USAF South Ruislip in Middlesex.

On 1 March 1949 the 28th Weather Squadron was reactivated at Bushy Park for a few months before relocating to USAF South Ruislip in June 1949 only to return to Bushy Park in 1951 when the base was handed back to the USAF on 21 April. However they were to move again – this time to RAF Northolt – in 1962.

In 1952 with the growth of the Cold War and the arrival of more US military personnel in and around London Bushy Park became the first home for London Central High School, the US Department of Defense Dependents School, which remained there until 1962 when it relocated to Bushey Hall in Watford. At that time it was determined that Bushy Park should be returned to parkland.

Demolition of the base buildings commenced in 1963.

Today Bushy Park is home to several memorials commemorating the US military presence including:

- The SHAEF Memorial which was dedicated in 1994 to commemorate Camp Griffiss (1942-1962) and General Eisenhower's D-Day preparations on the site. The plaque marks where his office was and there are other ground plaques to indicate the sites of key buildings.
- There is also the USAAF Memorial that commemorates the US 8th Army Air Force presence at the Park in 1942.
- A further memorial, unveiled by His Royal Highness Prince Charles in 1992, commemorates the 50th anniversary of the end of the Berlin Air Lift in which the Americans helped mastermind from the Bushy Park base.

There remains an intriguing story that a base command underground planning/map room was left intact and simply covered over during demolition in 1963, but this has never been confirmed.

Aerial shot of Bushy Park (aka Camp Griffiss) in 1947. Block C (middle of picture) was the location of the office of the Supreme Allied Commander Dwight D. Eisenhower' Office. It later became London Central High School. The site of a Glen Miller concert is also indicated. Photo courtesy of the Royal Parks /Bushy Park.

S/SGT Joseph Louis Barrow – better known as World Heavyweight Boxing champion Joe Louis – came to Bushy Park in 1944 to take part in a demonstration fight against Corporal Tommy Thompson. The 215 pound Louis, known as the "Brown Bomber", had held the title for eight years before joining the US Army on 10 January 1942. He first went into Calvary before being put into the Special Services Division and boxing 96 exhibition 'bouts' in front of something like 2 million soldiers. The Bushy Park bout took place in the ballpark for the Canadian School. Photo courtesy the Royal Parks Team at Bushy Park.

CAMP LYNN (STATION AF 101 IN WWII) / USAF HIGH WYCOMBE AIR STATION / RAF DAWS HILL

Codenamed 'Pinetree' but also known as Camp Lynn after USAAF VIII (as it was once known) Air Force crew member Lt. William Lynn, a pilot with the 15th Bomb Squadron who was became the first member of the USAAF to be killed in combat over Europe (on 4 July 1942), this was originally the Headquarters for USAAF VIII Air Force Bomber Command.

The base would go on from to play a key role in WWII and again in the Cold War and would then be transferred from the USAF to the US Navy and then play a part in the final chapter of US military operations around London.

The USAAF VIII Bomber Command, which would evolve into the 8th Air Force, saw its headquarters formally established in the spring of 1942 – just a day or two after Brigadier General Ira Eaker, who had been made Commanding General, arrived in England. A determination was made to be close to the RAF Bomber Command at South Down (the code name for a bunker located under a hill to the north of High Wycombe known locally as Nap Hill). Eyes (particularly those of Gen. Eaker and his Deputy Chief of Staff Col. Fred Castle) quickly fell on school grounds and hillside to the south and east of the town.

And so the Wycombe Abbey School with its extensive grounds sweeping up from the town of High Wycombe was requisitioned for use, despite the protestations of the then Headmistress, one Miss Winnifred Maitland Crosthwaite.

The requisitioning notice arrived on 30 March during the school holidays. By 15 April 1942 the students were to be gone. The upshot was a 16-day logistical feat extraordinaire with school staff having to relocate the entire female student body to around 40 different schools. During the US military occupation a small team from the school, lead by Miss Crosthwaite, remained on hand with a pastoral care brief that extended well beyond keeping an eye on the buildings. That brief included writing to each of the dispersed schoolgirls and keeping them updated about Wycombe Abbey. Miss Crosthwaite's efforts were so outstanding that when the school reopened in May 1946, the returning intake comprised 171 girls of whom all but a handful had never attended school at Wycombe Abbey before.

For the USAAF moving into a girl's school presented its own unique experiences. There remains to this day a legendary story, told more fully by a contributor later in this book, of the first night that junior USAAF officers spent on the upper dormitory floor of the school.

Gen. Eaker is understood to have taken up permanent residence on 15 April along with 20 staff and moved into the Headmistress's former office,with offices around for his adjutant and for briefing visitors and a hall for larger briefings. One of the first facilities that had been set up on the base had been a baseball diamond.

An Officers Club was set up in the former school Home Economics building while other buildings on the sloping campus grounds were parcelled off to various USAAF operations – Airlie House to Public Relations, Campbell House to Engineers. Famously – and in a sort of juxtaposition to what would happen some 15 years later at the USAF's Denham Air Base – the school Chapel was converted into a cinema, a meeting centre and a jury room for courts martial hearings. (Such was the apparent upset of some the student body when they returned after the end of the War about what they saw as the misuse of the Chapel that they had the building rededicated.)

Several of the buildings became the USAF VIII Bomber Command's telephone exchange, which, at its height, was reported to be handling 14,000 calls a day.

From 12-15 May 1942, some 387 officers and enlisted personnel arrived at the base to roll out bomber and interceptor command support and facilities with another cluster arriving in June. Eaker's original staff is believed to have included 12 officers and about a similar number of RAF liaison staff. Together they worked on the Combined Operations Planning Committee (COPC). It was on 19 May 1942 that the base took sole charge of all USAAF units in England for nearly a month until Gen. Spaatz took command of the entire 8th Air Force and moved his headquarters to Bushy Park.

There was a formal opening ceremony for the newly-created headquarters just a few weeks after arrival at which Gen. Eaker made his famous and incredibly brief speech quoted at the beginning of the book: "We won't do much talking until we do more fighting. We hope that when we leave you'll be glad we came. Thank you."

Soon after the Americans started arriving construction workers were set to work halfway up the hill above the main school and just above Daws Hill House which had been girls' boarding accommodation. There they started the construction of a top-secret bunker. British construction company McAlpine completed the bunker in 1943 at a then cost of £250,000 for the 11-month project. The 23,000 sq ft, three-storey bunker was 47 feet deep with a 10-foot reinforced concrete cover – all under a further 25 foot-deep cover of earth. It was built for use by the Combined Operations Planning Committee (COPC) and for a while, at least, an RAF Intelligence team.

The bunker and the base would be at the heart of operational coordination for the USAF precision daylight bombing offensives against Nazi Germany.

The High Wycombe base soon became festooned with tents, mess halls, a motor pool, photographic buildings and guard huts stretching up and across close to circa 200 acres with upwards of 12,000 personnel eventually stationed there.

Capt. Frances Cornick would lead the first party of 130 Women's Army Air Corps (WAAC) – arriving on 28 July 1942 to work as cooks, photo technicians, switchboard and teletype operators and a host of other jobs. Among them would be Juanita Folsom – just one of two women who served in the 8th Air Force Command Bunker at 'Pinetree' in High Wycombe who were thought to be left alive in 2013. Capt. Cornick would rise in rank to Major and become one of the first WACs (as they became) into Paris once the Allies had recaptured it.

Aside from the 'brass', film stars on active duty with the USAAF, including Jimmy Stewart and Clark Gable, passed through High Wycombe on their way to or from combat units. Another actor, Gene Raymond, husband of actress Jeanette MacDonald, was assigned to VIII Bomber Command at the base during 1942 as an intelligence officer.

Gen. Eaker, who served at High Wycombe from 20 February – 30 November 1942 was followed by Brig General Newton Longfellow (2 December 1942 – 30 June 1943) and then by Maj. Gen. Frederick Anderson (1 July 1943-5 January 1944). When the Headquarters of USAF 8th Air Force moved from Bushy Park to High Wycombe Lt General James H. Doolittle took command (from 6 January 1944 – 9 May 1945). He was briefly succeeded by Maj. General William A. Kepner (11 May 1945 – 31st May 1945) and on 16th July 1945 the 8th Air Force Flag was redeployed to Okinawa in the Pacific.

VIII Fighter Command had also established at High Wycombe briefly from 12 May 1942-26 July 1942 before relocating 17 miles closer to London at Bushey Hall in Watford. The Fighter Command returned to High Wycombe again at the end of the war from 17 July 1945 – 25 October 1945.

The biggest physical presence at High Wycombe was probably the 325th Photographic Wing (Reconnaissance). The wing had been activated on 9 August 1944 and was inactivated on 20 October 1945. A number of buildings in the lower grounds of the Wycombe Abbey school down by the Rye – the river at the bottom of the hill – were home to the 325th Wing Reconnaissance Base Laboratories as well as various camera and motion picture units and photo technical squadrons.

Between 9 August 1944 and 17 January 1945 Colonel. Elliott Roosevelt, the son of US President Franklin D. Roosevelt, served with the 325th as a photo reconnaissance pilot largely out of Mount Farm in Oxfordshire and commanded it from the base at High Wycombe where the photo processing took place.

Col. Roosevelt, who had flown more than 300 combat missions and earned the Distinguished Flying Cross and two Purple Hearts, would be promoted to Brigadier General and end his time there on 13 April 1945. He would later go onto face Congressional investigation into his recommendations made in 1943 in regards to a new generation of reconnaissance aircraft.

During 1944 Col. Roosevelt was a witness to the death of Joseph P. Kennedy, the eldest brother of the future President John F. Kennedy. Kennedy had been an anti-submarine PB4Y Liberator patrol pilot who had flown his required 25 missions, but who had stayed on and volunteered for *Operation Aphrodite*.

The operation, approved by Gen. Doolittle, involved taking off and flying a number of remotely controlled B-17 Flying Fortress to 2,000 feet and then bailing out. The aircraft, which had been completely stripped of non-essential gear, were stuffed with a British-made explosive called Torpex and would then be remotely guided by another control plane to crash into a hardened target such as submarine pens and V-2 launch sites in France.

Between August 1944 and January 1 1945 around 25 of the remote controlled aircraft were deployed but almost every mission ended in failure. It was on 12 August 1944 that the remotely controlled B-24 flown by Kennedy exploded in midair soon after take-off, killing him and the engineer Lt. Willford Willy. Roosevelt had been in a chase observation plane.

Doolittle was ordered by Spaatz to halt those missions on 27 January 1945.

Also based at High Wycombe was the Surgeon for the Eighth Air Force, then Col. Malcolm C. Grow. Col. Grow, who would eventually become a Maj. General, pioneered new thinking on aerial combat medicine with his "Care of Flyer" program that he launched in August 1942 at Daws Hill. It aimed to improve both the medical and psychiatric care of combat crews.

In 1942 – 43 only about 35% of bomber crews completed their required 25 missions. Grow's team would go on to work with metals specialist Wilkinson Sword (best known by consumers these days for its razors and garden tools) on body armor and to educate crews about anoxia (oxygen deprivation). They would also work with the British to roll out an Air-Sea Rescue service and procedures that would help save 650 aircrew. In 1987 General Eaker passed away at the Malcolm Grow Medical Centre at Andrews Air Force Base in Maryland.

Following the successful but largely propaganda-driven air raid on rail yard in Rouen, France, on 17 August 1942 in which Eaker flew on board a B-17 call-signed 'Yankee

Doodle', he ordered the turret guns be brought to High Wycombe and mounted on the wall of the Headquarters. *Yankee Doodle* was scrapped at the end of the war and what became of the mounted guns has not been resolved by the author.

One of the best-known WWII events to take place at the base was the performance of the Glen Miller Band on 29 July 1944 as part of a War Bond Rally and with General Doolittle and 4,000 others in attendance as they played *In the Mood* and *Stardust*. The orchestra performed on a flatbed truck marked US Army Air Forces on the side right in front of the main ivy covered Wycombe Abbey building. Staff from the WAAC formed the figure '8' inside a 'V' created by rows of enlisted men as part of the event.

"The war is going well," Gen. Doolittle said. "Substantial advances are being made on all three of the ground fronts — In Russia in Italy and in Normandy. The air battle is going well, too. We're able to go any place we want in Germany. The Hun air force is being gradually crushed. The Hun manufacturing capacity for munitions is being destroyed. Our losses are substantially a quarter of what they were a few months ago. But in the prize ring when the battle is going well, when a fighter has his opponent on the ropes he doesn't slow down. He gives him everything he has. He knocks him out. We've got the Hun groggy. We don't want to slow down now. We want to knock him out. One way to knock him out is to buy bonds," he concluded before stepping over to buy to the first inter-squadron war bond himself in cash.

British Prime Minister Winston Churchill visited the base on at least one occasion, thought to be in April 1943, when he was pictured in the main building.

British wartime chauffeur and later secretary Kay Summersby in her autobiography, *Eisenhower Was My Boss*, described the base as an "auto-motive airport" to and from which she had to drive Eisenhower and later Carl Spaatz – usually at speed.

The bunker was actually a building within a building and had five to six-foot thick walls. Furthermore, the inner building was cushioned by a five-to-six foot wide area of 'dead space' from the outer bunker. Inside was said to have been a classic map room (as seen in the movies) as well as communication links.

Parts of the void between the walls were filled – at least sometime after the 1980s – with banks of filters (thought to be 'air scrubbers') and air and water pumping equipment. The design feature of the internal building with the dead space and the foundations on which it rested was said to have been the architectural equivalent of a giant shock absorber to allow it to sway in the event of bombing.

There were several access points to the bunker. The main one was the one through which King George VI and Queen Elizabeth the Queen Mother travelled for a visit to the bunker on 11 May 1945 at the end of WWII as guests of the USAAF. A second entrance was about 40 feet further along.

And there was a third entrance. Set slightly further back and higher up on the site was a goods lift and rail designed for heavy-duty machinery to be lowered into the bunker. Even during a visit in recent years the concrete walls still had the imprinted grain of the wooden molding-formation boards.

Wycombe Abbey school received its main buildings back in 1945 but it took a further six months to make them fit for occupation by the girls and staff. However they did not get back the bunker site.

The bunker and most of the base was closed up with the bunker entrance itself believed to have been cemented up in 1946. For a while the base housing and other buildings were given over to help house homeless people in the post war years.

On 23 February 1946 High Wycombe Air Station was awarded the Freedom of the Borough of High Wycombe to recognize the base's wartime role.

However in May 1951 the Strategic Air Command agreed with British Ministry of Defense to re-occupy the underground bunker and the base. The base was re-activated on 1 October 1952 as part of the Strategic Air Command's (SAC's) 7th Air Division (7AD) along with the 3929th Air Base Squadron. As part of that activation was a SAC Joint Coordinator Centre – essentially an atomic warfare unit.

At that time the bunker was known as "The Block". It would also known variously over time as "The Bunker" and "The Hole". Water was pumped out of it and it was made ready for use by the 7th AD, which moved to the base as a HQ from USAF South Ruislip, where it has been since 1951. The 7th AD was in residence from July 1958 as the Cold War ramped up until its deactivation in June 1965.

Detachment 40, 28th Weather Squadron was also located at the base starting in 1952 when Hemispheric Weather Central (or European Weather Centre) as it was known, relocated from USAF South Ruislip to the bunker. The irony was, perhaps, that the weather forecasters were producing predictions from underground!

As part of the Cold War build-up, two barracks buildings were constructed in 1961. The matching and facing 210-room dormitories were each three storeys high. One of the enduring local legends about the base in the ensuing years was that the barracks were actually giant covers for missile silos and come any war they would slide aside down the hill towards the town of High Wycombe to allow the missiles to take flight!

In July 1965 when the Air Force Communication Service established the Automatic Weather Network (AWN) – It was said to be one of only three bases in the world acting as automated digital weather centers.

The 322nd Military Airlift Command was temporarily stationed at High Wycombe from 5 August 1966 – 24 December 1968 after it and other US military components had been kicked out of France. They had been based at Châteauroux-Déols Air Base. Following that country's withdrawal from NATO all US forces had to redeploy from the country by 1 April 1967. The 322nd was inactivated on 24 December 1968.

Perhaps the most important resident organization within the bunker during at least part of this time is believed to be what became known as the Joint Coordination Centre for the 7th AD. It included representatives from the Joint Strategic Target Planning Group or the Joint Strategic Target Planning Staff (JSTPS), a top secret multi-agency operation. The JSTPS had been created in August 1960 by then President Eisenhower and was responsible not only for the NST (National Strategic Target) list but the SIOP (Single Integrated Operational Plan) in event of nuclear war. Indeed, the top-secret US DoD War Plan was sent to the team in the bunker and updated around every 72 hours. JSTPS is understood to have operated organizationally until about June 1992.

At some point in the 1960s the NCO Open Mess at High Wycombe introduced brass tokens for use instead of American coinage for facilities on the base.

With the closure of SAC bases in 1965 the 3929th Air Base Squadron and Seventh Air Division were inactivated and responsibility for the base passed to the USAF Third Air Force and USAF Europe (USAFE). The 7563rd Air Base Squadron was subsequently activated and provided support for the London Area Department of Defense activities. At about that time the 322nd Airlift Division of the Military Air Transport Service, moved to HWAS and stayed there until inactivation in December 1968.

The three-storey barracks house up to 800 personnel who would work in 8-hour shifts in the bunker and also provide security and base services until 1969 when the numbers were heavily cut and the bunker was again closed down. The base, by then a part of the 20th Tactical Fighter Wing at RAF Upper Heyford in Oxfordshire. went to 'standby status' for a couple of years.

For a long time the base at Daws Hill also played an important communications role – with the 2719th Communication Squadron operating a major microwave relay station.

By 1971, when the USAF drew down operations at High Wycombe and closed up the bunker, the High Wycombe base sported four or five antenna towers (a 'sideways V'-shaped one was rumored to provide direct links to the White House for NATO via a microwave broadband communication network (sometimes referred to as ACE-HIGH) that also linked the bases at RAF Uxbridge, USAF South Ruislip and the US Navy Headquarters in London.

It was that year that the base was designated as the new home for London Central High School – the dependent school that had been located at Bushey Hall – and its circa 1,200 high school and junior high school students. For a year or two the school effectively became the sole occupier of the base.

In June 1975 in the wake of the closure USAF Third Headquarters at South Ruislip Air Station and its relocation to RAF Mildenhall the 7500th Air Base Squadron relocated to High Wycombe to form Det. 1, 20th Tactical Fighter Wing, effectively a subsidiary unit of the 20th TFW located at RAF Upper Heyford some 30 miles away in Oxfordshire.

The host unit was re-designated as the 7520th Air Base Squadron and came under direct control of USAF 3rd Air Force at Mildenhall. Its function was to provide administrative, engineering, maintenance, personnel and communications support to 38 USAF and DoD organizations on 16 sites throughout 10,000 sq miles of Southeast England and Greater London.

Among quaint local traditions the USAF Base Commander of Det. 1 at Daws Hill, High Wycombe was asked to attend the swearing in of the new mayor of High Wycombe. The town had an annual weighing-in dating back to medieval days – with all the local officials 'weighed' and their weight recorded.

In 1978 USAF the base commander Maj. Boyce Trout was pictured in the local press bravely taking part. As it was his first year he was spared any shout-out. 'And some less' or 'And some more' would be shouted out as officials were weighed to indicate if they had lost or put on weight). Local townspeople were known to take great delight in jeering if the "and some more" shout went out – it was said to indicate that officials had grown fat at the expense of the taxpayers!

Building 700, as the bunker became formally known at least by the 1980s, was reactivated around 1981 to create a Theatre Mission Planning Facility with a projected 20 additional personnel in support of the incoming Ground Launched Cruise Missile bases at RAF Greenham Common and RAF Molesworth.

The bunker would also become the US European Command Headquarters — a fallback/or replacement headquarters (formally known as the Headquarters United States European Command's Alternate Support Headquarters) for the US European Command then at Patch Barracks close to Stuggart in Germany. Congress reportedly set aside $13 million and possibly as much as $19 million in 1982 to fund the upgrade of the HQ.

It was rumored that if the High Wycombe Base Commander (usually a Lt. Colonel or Colonel) was invited to a meeting down at the bunker it would be he or she who was the lowest-ranking officer present!

The floors of the now revitalized bunker were reinforced to take heavy computer processors and cooling equipment and, at some point, Halon 1301 fire extinguishers were installed – certainly in the secure quarters of what was known as the J2/J3/J1C NCEUR WATCH (National Security Agency/Central Security Service, Europe). Among the computers thought to have been in the bunker certainly in the 1970s was the IBM 360 because the bunker was a key node within the WWMCCS (World Wide Military Command and Control System).

The entire bunker, at least under the floor space, was 'caged' with one-inch wide copper bands – apparently designed to turn the bunker like a giant lightening rod and to reduce the impact from electromagnetic pulse (EMP) from a nuclear bomb blast that might otherwise wipe out the electronics.

Sometime in the 1980s several buildings were constructed next to the bunker. Building 104 appears to have been a large storage shed of about 60 meters long that was large enough for several container trucks to pull into it. Building 701 and 705 which sat directly on top of the bunkers appeared to be office buildings for about 40/60 people each. Between them and the bunker entrance a raised metal air vent 'box'. Also with in the secure bunker area were Buildings 107 and 106. They were 30 meters up the hill across the road that lead to the bunker and were circa 30,000-gallon capacity fuel oil and water storage tanks.

It's unknown at this point if any the original USAF Bomber Command or later Cold War maps would still be on the walls of the bunker or if they were all ripped out during the 1980s upgrade or earlier, though at least one military serviceman who was stationed in the bunker believes there were Strategic Air Command wall maps still on the walls at least up to the time of the 1980s upgrade.

During a visit to the bunker by the author in 2003 the only WWII item the RAF Commander of the time knew to be inside the building was a metal ceiling light hook. It had been hidden away by modern suspended ceilings.

The modernized bunker had clearly been strengthened and optimized in parts specifically to house large computer mainframes and this was evidenced by the steel beam floor reinforcements and considerable cabling in the building. After the bunker closed down it is understood that a contingent of US Marines held hostage rescue training sessions throughout the facility and, as part of that, pulled down and cut whatever wires they found for practice.

One of the many interesting aspects of the abandoned bunker was that many of the rooms were fitted with a ceiling-mounted red 'emergency' beacon. It was said that this was turned on every time a non-military/British contractor was 'on the floor' to do work. It was also said that there were listening devices in the computer rooms allowing commanders to listen in to conversations on the floor.

The Alternative Support HQ was deactivated on 1 May 1993 as part of the overall inactivation of USAF operations. It was at that time the actual mission was declassified. In its nine years of operation the 7520th ABS (motto 'Understated Excellence'), won three USAFE Outstanding Unit Awards and three USAFE Base Appearance Awards.

Speaking during the Inactivation ceremony Lt. Col. Ben C. Williams, then base commander and 7520th Air Base Squadron Commander, said:

"Although inactivation of the 7520th Air Base Squadron closes a pivotal chapter in Air Force history and signals the end of 51 years of Air Force presence in the High Wycombe

Community, it will not terminate the unique bond that has evolved between the people of High Wycombe and the countless American men and women who were privileged to have served here."

After the US Navy took over management of the base in 1993, buildings were all re-numbered with a '1' prefixed in front of them and the bunker became Building 1700.

At some point the then provider of US Military banking facilities – there were a number including American Express (which operated the contract until 1 October 1980), Bank of Fort Sam Houston and Nations Bank) for the USAF (and then the US Navy) had its processing and office space in cabins next to the bunker entrance there – certainly during part of the 1980s – before a mysterious fire destroyed the offices. Part of an old metal stairway for the building could still be found in the undergrowth near the bunker entrance in recent years.

The High Wycombe base became home to the USAF's 7555th Theater Mission Planning Cell as early as 1981, prior to the arrival of the controversial Ground Launched Cruise Missiles (GLCMs) at RAF Greenham Common (in November 1983) and RAF Molesworth, The TMPC provided Ground Launched Cruise Missile (GLCM) planning and targeting software programming throughout the European Theatre including to-pography data that would allow the missiles to hug and weave around mountains and buildings.

For a period the 7555th also operated from a cluster of three above-ground office buildings known as the '500s' near the base Post Office and the three-storey dormi-tory buildings. Prior to their occupation the buildings had been linked and had their windows blocked up (with locked steel panels). Any surrounding ground access portals were sealed and the buildings were placed behind barbed wire. Inside one of the build-ings were offices and in one of the others was were special metal-panelled, raised-floor isolated spaces – effectively a room within a room – for the computer programming. The 7555th also operated from the underground bunker and probably concluded operations around 5 March 1991 when the last cruise missiles were removed from Greenham Com-mon following agreement on the INF Treaty in 1987.

By 1987 the base operations had fallen under the re-designated 7520th Air Base Group and included a Medical Aid Station which had been there in the 1970s.

For a top-secret major USAF command base High Wycombe was not the easiest of places to secure partially because a public footpath bisected it!

Campaign for Nuclear Disarmament (CND) peace protestors, who famously camped out at RAF Greenham Common but also had protest camps adjacent to most other USAF bases, managed to breach the fence-line for above-ground missile programming complex at least once during its approximate five-year operational period and daub it with symbols but not get in the building itself. Peace campers had established a small tented peace camp presence outside the main gate on the opposite side of the road from 1982 to 1984.

The public footpath (a surprisingly regular feature to be found at US bases in the UK) effectively cut the Daws Hill base off from the Eaker Estate military housing area and a former elementary/Junior High School that was there for many years before the build-ing was demolished. In the early 1970s the junior high school's boiler had blown up, thankfully on a weekend, and had blasted right up and out through the school roof. Staff returning to school after the weekend discovered it against the base fence line some 70 yards down the hill towards the town of High Wycombe.

In July 1989 a bronze bust of General Ira Eaker was unveiled in a small park next to the Base Commander's office at HWAS by Mrs. Ruth A. Eaker. The $8,000 sculpture was created by Lt. Colonel Jerry McKenna, a previous base commander at High Wycombe from 1975-1979. Three copies were made with one going to the Air Force Association HQ at Arlington VA and another at Eaker AFB. When the base closed the one presented at High Wycombe/Daws Hill was removed and due to be sent to the RAF's Strike Command HQ at RAF High Wycombe.

The Daws Hill bunker was eventually deactivated in June 1993 as part of the hand-over of the base to the US Navy at a ceremony presided over by the then 3rd Air Force Commander Maj. General Charles Link.

There is a story that makes the rounds of military personnel familiar with the base of peace campers breaking in to the bunker complex area and, after two hours of not being apprehended, calling the British police and requesting them to come and arrest them. By then the bunker had been closed down for several years!

As the USAF drew down its bases in the UK and Europe in the early 1990s, High Wycombe (which in 1992 had been home to about 260 USAF military and 130 civilian staff), was handed over to the US Navy on 10 May 1993, in support of Navy operations in and around London. The hand-over was reportedly a result of a decision to take responsibility for the bases made by the US Secretary of the Navy three weeks previously, primarily due to US Navy London personnel housing shortage issues. Otherwise High Wycombe would have likely been closed as part of the Pentagon's by then 11th round of base cuts.

"We are interested in it for the military housing," Capt. Donald Lachata, then Commander of US Naval Activities at the time, said. "We have a housing shortage in London and we also have some leased warehousing that may be moved here." The London Central High School, the military dependent school with its halls of residence converted from the military barracks in 1971, would also remain. At its height of activity the school had nearly 1,200 students; by 2003 were just 339 students and in 2006 when it was announced that the school would closed the following year the student population was 275 with 55 staff.

Starting in about 2001 or 2002 for several Octobers, the bunker found a new temporary role as 'The Haunted Bunker at Daws Hill' for an annual base Halloween event with a hayride from the school's music hall at a cost of £2 (or $3). The Halloween event was set up and staged by the US Navy Construction Battalion 'Seabees'.

The one apparent occasional military use that remained for the bunker was as an aforementioned training zone for the US Marines Corps Security Force (MCSF) Detachment 1 (Det. 1) who were stationed at MOD Eastcote for many years – and, also possibly, for Marines from the US Embassy in London for hostage rescue training and building clearance exercises.

A number of buildings on the base were upgraded around the turn of the millennium to handle and overspill of NATO support units from different nations who were part of RAF Strike Command/RAF High Wycombe, the RAF's main operational command located at Nap Hill several miles away. The Canadians, in particular, had a major presence there.

The formal disestablishment ceremony took place on 1 September 2007 for the US Navy COMNAVACTUK command and for the High Wycombe base. The 'Record of Transfer and Acceptance' covering RAF Daws Hill and another base, RAF West Ruislip) was signed on 27 September 2007 by Tammy Sadler, the contracting officer for the Naval Facilities Engineering Command, Atlantic and for the Government of the United Kingdom by Sqdn. Leader George Hannaford, then the RAF base commander at Daws Hill, the following day.

The transfer included 67 housing units at High Wycombe including 39 in the Doolittle Village and 28 in the Eaker Estate. The United States Navy, on behalf of the US Government, "…reserves any interest in residual value compensation from the United Kingdom Ministry of Defence which is related to the returned housing units. This issue is still being discussed between the U.S. Navy and the U.K. Ministry of Defence." the transfer document caveated.

The US Navy departed the High Wycombe base on 31 October 2007, with the London Central High School having closed in June 2007. On that day the base 'keys' were effectively handed over to the representative from Defence Estates, as the property wing of the Ministry of Defence used to be known (it has since been folded into the Defence Infrastructure Organization – DIO). In turn the MOD put freehold of the main base (but not the bunker) on sale in September 2008 through British estate agent GVA Grimley. At the time it was "for sale by informal tender" and listed as being approximately 24 hectares (60 acres) in size with "potential for a substantial residential and mixed-used development subject to the necessary consents".

The sale has also included the base's sports fields of approximately 20 – 30 more acres that belonged to the Carrington Estate, the historic property holding of former British Foreign Secretary Lord Carrington, Peter Alexander Rupert Baron Carington (unusually 'Carington' seems to be spelled with one 'r' in his name and two in his title) and his family. Part of that estate was actually sold to the USAF in the 1960s to enable the construction of the Eaker Estate housing area.

The only part not included in the sale was the Bunker site.

There were no acceptable offers for the base at the first marketing, so the base was re-marketed in February 2011.

The oldest parts of the present-day base, apart from the WWII built bunker, were the main base oil storage tanks (built in 1951), the former bowling alley building (built in 1957) as well as the antenna tower (built in 1978) and the gate sentry house (built in 1965).

After closing the base, the MoD continued to utilize the two main housing areas (the Doolittle and Eaker Estates) for British military personnel who were working at RAF Strike Command bunker three miles away.

Initially the MOD wanted to maintain a longer, rolling-term lease (five years per term), however in 2011 they changed this to warn housed British military personnel that they would be getting rid of the housing areas anytime within 18 months.

Following hand-over of the base back to the MOD the main 'technical site' including many of the base buildings and the school, was fenced off and patrolled by private security contractors. However in 2010 part of the technical site including the school campus was leased out to specialist British police units for training purposes including weapons training which took place among the empty hallways of the 900/1900s buildings. (This has been a common post-military use for some bases – for a long time the former base at RAF Upper Heyford in Oxfordshire was used by police for car pursuit driver training among other activities.) Some of the vehicle maintenance shed buildings at Daws Hill, High Wycombe, were also later let out to film/TV equipment companies in 2012.

Following the remarketing to the property development sector the base was sold to leading UK housebuilder Taylor Wimpey for £35,750,000 on 30 August 2011. It was anticipated that Taylor Wimpey could start development work on the first phase of the site in 2013. There has been no indication from them to date if any buildings or, indeed any of the existing roads or road names will be retained.

In November 2012 the local council had indicated that it would be publishing a development brief for the 24.5 hectare former RAF Daws Hill site including up to 550 homes, 4,340 sq m of business space, a neighborhood centre, local shops and a primary school. The council indicated it was not planning to release the adjoining land, owned by the Carrington Estate and known as Abbey Barn South "at the present time".

The bunker area remains a separate entity and is sited on land that ultimately belongs to Wycombe Abbey school. Under special British Government provisions for property appropriated in wartime – known as the Crichel Down Considerations – the school applied for the return of the land to which it had first rights. The bunker was formally handed back to the school in October 2012.

A 7 February 2012 letter from Louise Speding of the Defense Infrastructure Agency was sent to Wycombe District Council to ask them for a determination as to whether the MOD needed prior approval to demolish the bunker area with two options proposed and the ultimate intention to return the site (initially, at least) to grass. Set for demolition in the bunker area were:

Warehouse – 2,800 m3
Generator Building – 975 m3
Offices – 350m3
Offices – 350m3
Tanks Bldg – 150m3
Bunker Entrances – 30m3

On 15 March 2012 and probably based on that letter, the local *Bucks Free Press* newspaper reported that the bunker was set to be demolished before the summer was out. There was no mention of what the type of demolition below ground would be in the application.

The bunker site was referred to English Heritage sometime in the Fall of 2012. However it wasn't until August 2013, as this book was being finalized, that their report (Case Number 473300) was submitted to the Secretary of State for the Environment with recommendations about the bunker. The delay is seen as somewhat exceptional as most reports are concluded within about 3-4 months.

On 11 October 2013 English Heritage announced by letter to the author that the bunker, RAF Daws Hill, Bunker, RAF Daws Hill, High Wycombe, Bucks had been awarded Listed Building Status.

"I am writing to inform you that the above building has been added to the List of Buildings of Special Architectural or Historic Interest," wrote Gosia McCabe, Designation Co-Ordinator, South for English Heritage. "The building is now listed at Grade II*. Listing helps us to mark a building's significance and celebrate its special architectural and historic interest. It brings specific protection so that its special interest can be properly considered in managing its future. Please be aware that the listing of the building took effect on the day that the List entry was published on the National Heritage List for England."

The Grade II listing covers both the bunker and the generator building. There were several visits to the bunker by English Heritage's Assessment Team, and then a joint visit by English Heritage's Ancient Monuments Inspector and Historic Buildlings Inspector.

The main report backing the decision included key elements recorded or paraphrased as follows:

"Structural Interest

Although the bunker was quickly acquired by the USAAF, it was British built, being one of the largest of our purpose-built bunkers of the Second World War. While the method of construction is similar to others of its date, it stands out from its contemporaries for its unusual if not unique form, a double-skinned bunker comprising a cuboid set out on three levels, enclosed by a full-height perimeter void. It is built into the side of a hill and apart from the external steel vents and added entrance cages, which appear to date from the 1980s, it is totally concealed. Most unusually, an exit stair, flanked by a probably 1950s service rail for bringing in supplies and machinery, runs from the upper level entrance to the bunker to the lower level.

At lower level there are two entrances, secured by heavy blast doors, giving access via stairs to the lower levels and the perimeter void. Although each floor area was reordered in the 1950s and again in the 1980s, the overall structure and plan of the WWII bunker, and points of access and circulation within it, remain legible. It is structurally more complex and, significantly, more intact than the WWII bunker at Bentley Priory. There, despite the high resonance attached to the site as a whole the bunker was not recommended for designation because of the impact of alteration on its otherwise unquestioned historic significance during the Second World War. Daws Hill bunker however continued to be of very high significance in the Cold War and evidence in the structure to support its changing role contributes to its potential special interest.

Condition is not, in general, a consideration for listing and the deterioration of the Daws Hill bunker over the last year since it has been left open, whilst regrettable, does not undermine its structural interest.

Evidence of adaptation to military threat:

The building is both unusual and of considerable interest since it retains evidence of how it was adapted to meet changing levels of threat at periods of major international military and political importance over a period of some fifty years. Although all early plant and fittings have been removed, the structure and form of the 1940s bunker remains essentially unaltered, some lighter weight doors perhaps remaining from that initial period. Probably in the 1950s, heavier blast doors were added and the upper entrance was altered to accommodate the track parallel to the original stairs. Like other bunkers that continued in use, it was hardened against nuclear attack during the Cold War. Evidence of added protection in the 1980s is visible where it was applied to the most vulnerable faces of the perimeter void, creating a Faraday cage within the core of the bunker to protect electronic equipment against the effects of electromagnetic pulse (EMP), and in the very heavy blast doors at the entrances. The four, heavy, steel cuboid vents are evidence of the increased requirements and added protection specified during the Cold War.

Decontamination unit:

Attached to the bunker at ground level is a decontamination unit that is an extraordinarily intact and chilling reminder of NATO's policy of 'Flexible Response', to match any threat of Warsaw Pact aggression with a measured reaction and to harden key facilities against nuclear, biological and chemical attack. Small numbers – in the low tens – of decontamination units survive elsewhere in the country. Four are designated: the Magic Mountain, Alconbury, Cambridgeshire (Grade II*), the RAF Radar Bunker, Neatishead, Norfolk (Grade II); the Command Centre (Scheduled Monument) and Squadron HQ (Grade II) at Upper Heyford, Oxfordshire in each case attached to and included in the designation of the structures they served. The Daws Hill unit however stands out for its very high level of intactness, its stainless steel fittings and control room capturing in great detail the degree of precaution taken against the perceived military threat, and the procedure from the point of entry, through showers, provided with bins for the disposal of contaminated kit, before entering the bunker itself. Its size, smaller than the unit at Greenham Common, Berkshire (shortly to be assessed for designation under English Heritage's Cold War project), reflects the scale at which the bunker was manned.

Historic interest:

Historically highly significant, the bunker stands out from others in the UK for its exceptional role as a high level US command centre, confirming and relaying orders to US operational units in the UK and Europe in time of war, operating over three periods of highly significant military activity. The bunker demonstrates the close relationship between British and US forces in WWII; its British counterparts remain operational and are not subject to assessment. It was refurbished as the main centre of the USAF's 7th Air Division in the 1950s and was one of the main centres of NATO's Joint Strategic Targeting Planning Group in the 1980s. To date, no other high level command centres have been designated. Yet, the next stages in the chain of command during the Cold War, the Thor missile sites at Harrington, Northamptonshire and North Luffenham, Rutland and airbases at Greenham Common and Upper Heyford are represented by designation, some structures scheduled or listed at high grades. Although Daws Hill may lack the strong functional and physical relationship inherent in the group of Cold War structures at Upper Heyford or Greenham Common, the site provides a record of a longer period of use, as, for example at Neatishead. It is its isolation that is important. At this level within the command structure it was expedient that it was deeply hidden.

The unusual history of the site also contributes to the interest of the bunker. It was built on land requisitioned from the school in 1942, within the perimeter planting of 'Capability' Brown's C18 landscaped park. Most of the site was returned to the school after the war, but the importance of the bunker site was recognized, and it was retained for military use. In line with current MoD policy, the bunker site has now reverted to the school, of which the main school building is listed Grade II*, adding to the complexities of managing its heritage within a secure educational environment."

There was a 28-day appeal period but it is unclear if Wycombe Abbey was to contest this if there would be any appeal.

Even if there were to be any successful appeal against an order that effectively means the bunker must be kept intact, then, demolition – below ground, at least – appears unlikely. How the actual demolition a 23,000 sq ft bunker facility could be facilitated itself presents an interesting challenge. Simply digging out the bunker could be prohibitively costly. An alternative could be to fill it in. One expert consulted by the author suggested that the complexity of the building design and spaces might be overcome by using expanding foam to fill in the cavities. But for the moment it remains supposition.

A legal challenge to the overall development of the RAF Daws Hill base by residents who had formed a neighborhood forum and who were concerned that base site had been excluded from planning consultation by the local council, made it all the way to

Britain's High Court but on 13 March 2013 the judge found that Wycombe District Council had acted lawfully.

In October 2013 it was revealed that local residents were taking the case to the Court of Appeal. That was heard in February 2014 and the decision was upheld in favor of Wycombe District Council.

Aside from the bust of General Doolittle, there are three other plaques relating to the base. MoD officials have made arrangements to preserve one on the base dedicated to a Capt. Kennedy for whom the main road that ran through the base. Capt. Kennedy was a Royal Navy officer who was killed in the first naval engagement of WWII when his armed merchantman came up against the German pocket battleship Scharnhorst and prompted the British Prime Minster to say that he and his crew "fought until they could fight no more."

A smaller brass plaque that was placed in front of a pine tree planted by Gen. Eaker when he returned to visit the base is awaiting disposition. The third one is in the grounds of Wycombe Abbey school and is mounted on a stone near what was a farm house prior to the estate being turned into a school known as Daws Hill House. It reads "The Pinetree Memorial" and is "in memory of American Gallantry in World War II" and lists the commands and the commanding generals. It was a project of the WWII 8th Photographic Technical Squadron.

Above – High Wycombe Air Station in the mid-1980s. The base entrance is far left (*foreground*). Buildings (*near left*) among the trees comprise the Eaker Estate. The antenna was the only one left by the mid-1970s. Kennedy Avenue, the main road running vertically just to the right of the antenna, leads directly to the underground bunker (*set in the trees at top of picture*) with the town of High Wycombe in the valley (*very top*). Note: Baseball diamond (*foreground*) and a tiny segment of the M40 Motorway (freeway) at *bottom left*.

Opposite Page – RAF Daws Hill (as it was called by then) in 2006 just prior to the base closing. Building numbers are prefixed with a '1'. as appeared to have happened after the US Navy took control. The underground bunker is reached by a road running off from the bottom right. Two dormitories (1810 and 1814) – used originally for personnel working in the bunker and later for students.

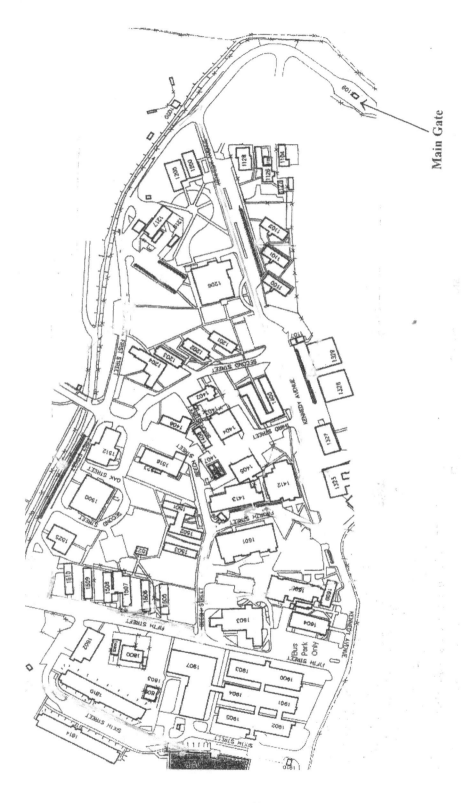

Main Gate

81

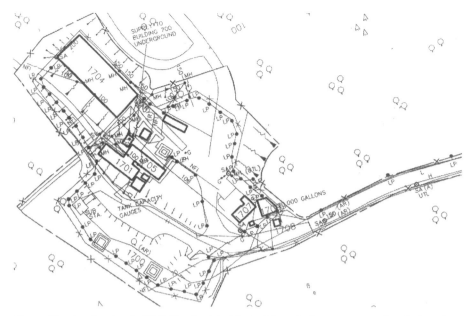

Above – The plan showing the HWAS bunker (aka 'The Hole') reached via a 200-meter long link road. The main bunker entrances (there were two) are either side of the black-edged square between above-ground drive-in storage Bldg. 1704 and office building 1701/05. The bunker ran from there and down towards the bottom of the map. The buildings to the right 1706/07/08 were oil and water storage tanks. The concentric square formations are bunker vents.

In May 1945 the Pinetree Headquarters received a royal visit from King George VI and the Queen who reviewed a formation of soldiers in front of the Headquarters Building down the hill at Wycombe Abbey and also visited the bunker further up the hill. Here they can be seen emerging from one of the bunker's two concrete entrance revetments which remained in-situ but were 'caged' in for additional security in later years. During their visit were accompanied by Gen. Doolittle and Gen. Kepner and an assortment of others including Maj. Mary Dixon, then Commanding Officer for the WAAC.

The High Wycombe base from WWII to 2007. Left – The 325th Photographic Wing (Reconnaissance) photo processing buildings once in the grounds of the Wycombe Abbey School. The road leading up the hill went to the Quonset huts, tents and the underground bunker. Right: The base main gate in 2006.

Above Left – The underground bunker area (Bldg. 700/1700) in the mid-1980s. At the top are the oil and water storage bunker. The main entrances to the bunker are at the lower right just under the tree. The stacked portable buildings were a one-time home for the US Military bank before it burned down. Right – The bunker entrance revetments seen in the WWII picture of the Queen Mother (opposite page) can just be seen behind the caging as it was in 2007. The yellow block building in the middle was the security post. In the far distance (centre left) is the door to the oil storage bunker.

Above Left – Inside a computer room space in the bunker on a 2003 visit. The warning beacon was reportedly switched on when a non-US 'foreign national' entered the room to work on equipment. Right: The base entrance and one of the antenna in the 1970s. 'US NAVY' was replanted in the topiary in later years.

CARPENDERS PARK (WATFORD)

The 70-home, 5.1-hectare (11-acre) Carpenders Park base housing estate was opened in June 1960 by Maj. General Ernest Moore, then Commander of the Third Air Force, on a former school (Highfields) site that had been compulsorily purchased by Britain's Ministry of Defense for £65,000. The entire site was sold after the base housing was closed down in 1994. It has been said that on the living room floor of one of those base houses singer/songwriter Gerry Beckley of the group America wrote the song *I Need You*. The entire site was sold in 1997 for £9.5 million to a house builder and demolished in about 1998 and replaced by a new residential estate.

The entrance to the Carpenders Park housing area just after it was closed down. At the core of Carpenders Park was the main Housing Office which included a workshop. The dishes on roof supplied satellite TV to all quarters. When first installed, rumors abounded locally that they were part of some a secret communications network! Photo courtesy of Neil Hamilton.

THE COLUMBIA CLUB

In 1954, just before losing its WWII Officers Club at Winfield House in Regents Park to the US Embassy for use as the US Ambassador's official residence, the USAF agreed a lease on what had been until then the Palace Hotel adjacent to London's Hyde Park in Bayswater. The hotel was owned by the Rose family who had purchased it in 1948. Under US military management it opened in February 1955 as the Columbia Officers Open Mess.

The Columbia Club, as it quickly became, garnered the reputation as one of the finest Officers' Clubs in Europe and, with an annex that would stay open until 2 a.m. on Saturdays and Sundays (far later than most British establishments at that time), one of the most popular.

The USAF took the hotel on a fixed lease that it had the right to extend (and did at least once) during its occupation that lasted until 4 January 1975. Major works to turn the hotel into a full-blown billeting and entertainment facility to USAF standards took until 7 February 1955 and even then a grand opening was only scheduled for late April of 1955.

During the US military occupation, the five-storey hotel had 83 rooms and 143 beds along with a ballroom, three banquet rooms and two games rooms. Its phone number in early days was known to all simply as *Paddington 8381*. The hotel also became home to the London Billeting Office which had previously been located at the nearby Kensington

Palace Hotel. That office was responsible for tickets to shows and sight-seeing trips.

The Columbia Club and nearby Douglas House, the enlisted personnel club which had itself had relocated from Mayfair, just 140 meters away, were pragmatic answers to a growing problem for the US Military command in the 1950s and 60s.

The problem was that American military personnel would head down to London from many of the big flightline bases in East Anglia and occupy every hotel bed around the area. Very quickly both facilities became a little bit of America and, as such, were staging posts from which to explore London.

There was American food and drink and all at American prices. It's understood that the local Westminster Council allowed USAF officials to have a maximum of 900 people in the building so when there were events on the place was heaving. The hotel was once said to have handled six different generals coming in for various events on the same night as well visits by the US Ambassador and by occasional celebrities.

The Columbia Club featured a bar known as the Tartan Bar – but few people knew it as that. It was informally referred to as Charlie's Bar – after one of the stalwart bartenders (Charlie Benfield) who had been serving there for years. It had a jukebox and a dartboard and it was a swinging place!

But it wasn't the only bar in the Columbia Club.

On the 2nd Floor (1st Floor for the British) could be found the Princess Caroline Suite which featured specially-flecked Air Force blue carpeting and a bar that overlooked the Lancaster Gate entrance to Hyde Park. Described as a "plush bar complete with a grand piano" the 'suite' actually comprised two rooms with an opening between them.

And if that wasn't enough then there was the Trafalgar Bar on the 1st (Ground) floor. This had pictures of the Battle of Trafalgar and heavy wooden armchairs – the same chairs that the hotel was still using in 2011!

On a busy night the staff would serve up to 500 people for dinner and drinks. But it was Sunday night that was the most popular – this was Smorgasbord night and the tables would be festooned with food – everything from suckling pigs to cream puff swans – attracting up to 700 covers.

Over the course of an average month the club would have some 3,000 guests to stay along with around 16 parties and four conferences. Even by 1958 it was calculated that the Columbia Club was a $750,000-a-year operation with three Americans and 135 local employees on the staff.

During the anti-Vietnam era of protests in London during 1968, a Molotov cocktail was thrown at the entrance of the hotel. *Stars & Stripes*, the US Military newspaper, reported that local security guard Thomas Moran – another Columbia Club stalwart – was on duty and helped put it out.

In one year of its operation (it's believed to be the late 1960s) the Columbia Club sold 19,473 tickets and put 17,795 people on tours around London in a single year to earn a profit of $26,170. During the year they provided 49,937 beds, charging $2 per bed for those on temporary duty (TDY), permanent change of station (PCS) assignments or leave. However those just on leave had to pay an additional service charge of either $1.00 for a room without a bath and $1.50 for a room with a bath.

For obvious reasons London and the military clubs also became popular for Officers and GIs stationed across Europe. By the early 1960s GI's stationed in Germany and the rest of Europe were being reminded to bring UK Sterling with them when they flew into RAF Northolt on what was the regular military route from Wiesbaden, Germany. As there was no currency exchange at the base they were recommended to have £3 and 10 shillings (in old English money) to cover their travel and a night as there was no foreign exchange facility at the RAF Northolt base.

Food, and its relative cheapness and quality compared to the fare available on the economy, was a particular attraction of the Columbia Club. By the early 1970s, the Chef's Evening Special included Roast Rib of Beef With Soup, Salad, Dessert and Beverage for $2.50. On a Friday the Chef's Evening Special was 'Fillet of Sole, Bonne Feme With Soup, Salad, Dessert and Beverage' which cost $2.25. Or you could have 'Lamb Chops in Wine Sauce with Soup, Salad, Dessert and Beverage' for $2.00.

A weekday menu of the day (Plat du Jour) included choice of soup or juice entrée, two vegetables and beverages. Dessert was pie, cake or ice cream. A fillet of plaice with lemon entrée was $1.75. At the bottom the menu promised: "Half Portions at Half Prices for Children".

When the USAF relocated the Third Headquarters from South Ruislip to RAF Mildenhall it spelled the beginning of end of the Columbia Club. Following an announcement the previous October 16 regarding the closing of both the Douglas House Enlisted Club and the Columbia Club the latter closed on 4 January 1975. The previous November 'Special Services', which arranged travel tickets and shows then moved in as part of the Keith Prowse Ticket Agency (a local economy business on Audley Street just down from the US Navy Headquarters).

The USAF formally handed the hotel back to the Rose family on 4 April 1975. When the family took the hotel back from the USAF they discovered a few of the menus in the kitchen which they have preserved in company archives until this day.

The Columbia Club subsequently became the Columbia Hotel with 103 bedrooms and four Conference Rooms along with a bar and a breakfast room. In its post-USAF 'civilian' life the hotel became a major music celebrity hangout particularly during in the 1980s. Jim Kerr, the lead singer of Simple Minds, recollected in a British magazine interview about his fond memories of the hotel — and being there at along with big '80s British bands ABC, The Human League and Frankie Goes to Hollywood – all staying or visiting the hotel bar on the same night!

In 2012 a single bedroom at the Columbia Hotel would cost around £76 ($118) per night while a double would be around £100 ($155).

At one point it's understood that a group of American military personnel got in touch with the Rose family to initiate a discussion about placing a bronze plaque on the site to commemorate the building's long-standing US Military status as part of a WWII commemoration program.

Hotel officials had to gently point out that the Americans weren't actually stationed at the hotel during WWII!

The Columbia Hotel as it is today. Externally the building is not much different than it was in its military days and there are internal features that would still be familiar to anyone who knew the Columbia Club, as it was, in detail.

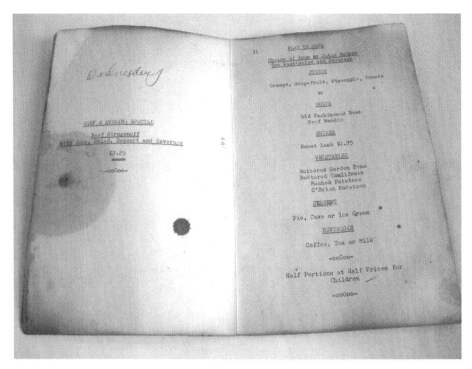

A menu from the Columbia Club – found in the kitchen area after the lease was surrendered back to the owners in the 1970s. (Menu courtesy of the Rose Family/The Columbia Hotel).

DENHAM STUDIOS / DENHAM AIR BASE, RAF DENHAM
BUCKINGHAMSHIRE

The US military had already been at the famous Denham Film Studios in Buckinghamshire long before they formally showed up to occupy the studio facilities in October of 1953.

By then the 165-acre Denham Studios, originally developed as a distinctive art deco-influenced film fiefdom of the legendary film maker Alex Korkda in 1935 (he occupied it for just three years), belonged to J. Arthur Rank organization which had operated it as a film studio until 1952.

The studios has been home to a host of classic movies: *The Thief of Baghdad, Henry V, Goodbye Mr. Chips, Brief Encounter, The Stars Look Down* and *A Yank At Oxford* had been made there before and during WWII.

Two films had particular portent to the Anglo-American relationship and future US Military use of the studios. The first was *A Canterbury Tale*, the remarkable wartime

(1943) film produced by Michael Powell & Emeric Pressburger which featured US Army Sgt. Bob Johnson played by a real-life US Army Sergeant John Sweet. In the plot, he, along with a British Land Girl and a British Army Sergeant, unintentionally become caught up in a mystery near Canterbury in Kent. Like the original eponymous book by Chaucer the story involves a bit of a pilgrimage as they set out to solve the mystery and learn something about themselves along the way.

In real life Sweet had been posted to General Dwight D. Eisenhower's staff in London in 1942 and had been quartered in Green Street, Mayfair. He had been into amateur dramatics and it was while he was performing in his off-duty time at London's Scala Theatre that he was spotted by Powell. He was released by the Army to make the film and then went on to make two documentaries and conduct post D-Day interviews with service personnel. The £500 he received for his starring role he donated to the National Association for the Advancement of Coloured People (NAACP).

Real-life American Sgt. John Sweet appears in a scene in the gentle wartime movie *A Canterbury Tale*, a film about an American serviceman who becomes caught up in wartime mystery set in and around Canterbury in England. It's very likely this scene was filmed at Denham Studios.

The second film, *A Matter of Life & Death* (aka in the United States as *Stairway to Heaven*), starred David Niven as a WWII RAF pilot who falls in love with an American military radio operator (played by actress Kim Hunter) with whom he is in contact as he tries to bring his badly damaged Lancaster bomber back across the Channel to England. The fantasy film love story sees Niven's character cheat death, fall in love and then have to prove his case in the heavens to stay on Earth. The film, which has still-acclaimed escalator and heavenly courtroom scenes, was part filmed at the studios.

The last film to be made at Denham was the Walt Disney film *Robin Hood* in 1952.

Eighteen months after Rank effectively mothballed the studios (all but one sound stage which continued to be used for music recording) the US military uniquely took a rental (rather than a leasehold) occupancy on 30 September 1953 with the idea of creating a logistical hub for its bases in the London region. The base was formally known as Denham Studios but was also called Denham Air Station. The USAF 494th Supply Group undertook the bulk of the logistics support for the USAF 3rd Air Force and the 7th Air Division and around 70 associated units.

Under the lease the US military was able to use four giant sound stages, up to 30,200 sq ft in size each, for the storage and logistics operation to supply the 3rd Air Division with everything from consumer items for the BX (Base Exchange) or PX (Post Exchange) to medical supplies as well as food and drink for the Columbia Club (the London hotel and club for the officers) and Douglas House (the London hotel and club for enlisted personnel) and for the mess halls at its London area bases. A US Postal Service was based there. Military equipment including spares for cars was also stored there.

Meanwhile, personnel were housed on the site in former actors' dressing rooms, some of which had their own bath or shower. And with a few more seats the Denham Studios 'preview' theatre was converted into a base theater for GIs, while another viewing or sound facility was turned into the Base Chapel. GIs stationed there report that, in some cases, you could walk from your barracks to eat and do other things without ever having to step outside or leave the complex.

In 1957 with the closure of the radar support facility at a base called RAF Shellingford near Faringdon, Oxfordshire, the USAF 7568th Materiel Squadron moved to the Denham facility and the command was rebadged as the 7500th Air Base Group.

While the base was primarily for storage and for housing personnel there were some other operational units including the 7501st Photo Film Library and the 53rd Medical Materiel (Depot) Flight. A number of the US Air Force sports programs were sponsored from there including the USAF American football team the London Rockets. The Base Commander's office was in the impressive former Art Deco-themed office once occupied by Alex Korda.

Among one of the focal events at Denham was a court-martial hearing in December 1958 for Master Sergeant Marcus Marymount then aged 37. MSgt. Marymount was charged with murdering his wife, Mary, on 9 June 1958, by poisoning her with arsenic trioxide at the USAF base at Sculthorpe in Norfolk.

It turned out MSgt. Marymount had fallen love with a Ms. Cynthia Taylor – described in reports of the time as "a shapely 23-year-old" British brunette who was separated from her husband and whom Marymount had first met at a club in Maidenhead.

The 14-officer court martial board held a four-day hearing and then took 117 minutes to reach a guilty verdict. MSgt. Marymount was, by secret written ballot, sentenced to hard labor for life. He apparently remained in the Stockade at the Bushy Park base as he went through an appeals process with the Staff Judge Advocate but was eventually transferred to Fort Leavenworth in Kansas.

A call to Fort Leavenworth in 2013 was not able to establish his fate other than that they had no record of him since the records had gone electronic.

At its height the Denham base is said to have had 562 military personnel along with 18 Department of the Air Force civilians and 75 'local hire' British personnel. At one point an Anglo-American fishing club was formed and the USAF even helped to stock a 1.5-mile stretch of the River Colne that ran through the studio grounds.

The Denham base was closed down in around 1962 with the bulk of operations relocated a few miles away to RAF West Ruislip and the site returned to the landlords on 30 Sept. 1963. Rank Xerox became the subsequent occupier but the site was demolished in the 1970s and has subsequently become the site of Broadwater Business Park.

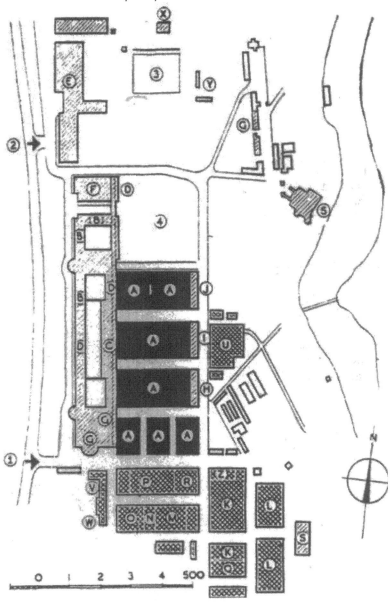

Denham Studios as it was prior to the USAF taking over. Two main entrances are seen from the road on the left. It's understood that the bulk of USAF operations was centered around buildings marked with an 'A'.

The main building at Denham Studios during the time they were used as a USAF base. Photo courtesy of Ted Clark.

DOUGLAS HOUSE - LONDON

The London club for enlisted US military personnel – this was *the* place to be and, judging by stories of the bevy of women usually found outside its double doors trying to get in, for most of the female population of London and the surrounds during the 1960s and 70s! How popular? There were even stories of officers 'dressing down' to try and get in!

Named after Lewis W. Douglas, the US Ambassador to the UK at the time, the original Douglas House was established in 1950 at 41 – 43 Brook Street in Mayfair next to the famous Claridge's Hotel.

The facility had been created from what had been the British military's Guards Club before becoming a US interrogation 'safe house' where US pilots who had been shot down and escaped back from Europe had been sequestered for questioning until their stories and identity could be confirmed.

Among its wartime residents had been Roderick L. Francis. The B-17 commander had been shot down over France on April 17, 1941 and been captured. He escaped from a train and spent several months hiding before the Free French helped repatriate him. Francis (who was shot down a second time later in WWII) had to spend two weeks in that building until his identity could be fully verified. He went on to become a Lt. Colonel and Commanding Officer at the main HQ base at USAF South Ruislip in Middlesex in the early 1950s.

In 1953 the hotel, which had space for 150 overnight guests, was charging $1 per night for a room (the equivalent of $8.62 per night in 2013) with no charge for children. It had two restaurants with a total of 130 covers and they were open from 7:30 a.m. until 10:00 p.m. There was a ballroom (four nights of dancing and entertainment and a Saturday night floorshow), laundry service, cocktail lounge and theatre booking service. The Military Police (or rather the 'Air Police' as they were known from 1948 to about 1966 after which they became Security Police) had an office there for their 'Town Patrol' and Douglas House also offered two free tours around London each day (with the touring parties apparently taken around by VW camper vans). Commander of Douglas House then was Capt. Winfred Starnes with club secretary being MSgt. Jack Crudduck. There were four other US Military staff heading up catering and supplies as well as a local British staff of 182. Monthly turnover was then at about $47,500 for the facility.

The facility also catered for outside events – in particular the 4th of July Celebrations – which, in 1953, took place in Battersea Park with 3,400 people in attendance.

On 3 December 1958 it was announced in the forces newspaper *Stars & Stripes* that the Douglas House's Mayfair location would shutter its doors on 7 December with the entire hotel vacated by 14 December when the building lease expired. It was also announced that there would be new facilities at 66 Lancaster Gate in Bayswater but that they would not be ready until early 1959. The new club opened in Spring 1959.

More than $500,000 was reportedly spent on turning what had been six terraced houses into the new Douglas House with around another $100,000 spent mainly on the 108 bedrooms (176-bed) facility making it what the *Stars & Stripes* newspaper referred to as the "most modern enlisted men's hotel in the world at that time".

As with the Mayfair facility the USAF Air Police had an office and from there they continued to run a 'town patrol' to pick up any personnel who got too drunk or in trouble around London.

By November 1959, six months after opening in Lancaster Gate, more than 100,000 visitors had passed through the 'D-House' as it was also known. So busy was the place that staff would also help to book up to 100 people a night into other hotels if the facility was full.

And it certainly was a showcase of entertainment. There were the house showgirls – the Douglas House Girls – were regulars (it's believed they may have actually come from the Astor Club after their show there) and also backed a host of entertainers including American comedian Phil Silvers who had made his name as Sgt. Ernest G. Bilko in the hit military comedy the *Phil Silvers Show*.

One young American recalls showing up at the club in his first week at work to see Ike & Tina Turner and then, a week or two later, Wayne Newton. The British TV star and singer Des O'Connor and the Irish group the Bachelors were among a myriad of UK star performers who showcased there.

"I saw Tom Jones there and then saw him 40 years later in Las Vegas and that was unusual," recalls Tom Miller, an airman who visited Douglas House in the 1960s: "I also heard Lulu sing and listened to Petula Clark. What a voice! It was a great place for someone of my generation."

At some point in the 1960s there was an audacious raid carried out by villains who obtained USAF NCO uniforms and held up the bar. They escaped with the money out the back fire door that led to the cobbled Leinster Mews behind where a getaway car had been idling.

In 1970 the lease was terminated by the landlord and Douglas House found itself on the move again – this time to 33-36 Princes Square near Queensway (a few blocks to the West) and into what had been four adjoining houses. The new location (which in the early days sported sign outside reading Columbia Hotel just to add to confusion) had a 67-bedroom capacity – only about half had private bathrooms – and had what the *Stars & Stripes* described at the time as a "burnt orange" decor.

There was a TV lounge and a bar but the bar was restricted to midnight closing under British licensing laws until the US military fully took over the quarters on 1 May.

The intention was to create a ballroom and a cocktail lounge with slot machines by August. Around 18 staff had transferred over from the old facility under director Ray Moser.

But the Queensway site turned out to be a temporary reprise. With the relocation of the USAF Headquarters to RAF Mildenhall in Suffolk, Douglas House was on borrowed time. The US Military closed Douglas House – along with the Columbia Club – on 4 January 1975.

The Douglas House site at Lancaster Gate became the Park Inn Hotel (now the Lanaster Gate Hotel), with the former Grade II-listed Thistle Hotel opposite (between the old Douglas House and Hyde Park) completely gutted and refurbished as The Lancasters. It proffers some of the finest apartments in London at prices of between about £995,000 and circa £35 million apiece! A four-bedroom apartment of about 251 sq m (2,705 sq ft) was on offer for a guide price of £7.75 million (circa $12 million) in late 2011.

The **Douglas House. Photo courtesy Tom Miller. Today the building is a hotel (right).**

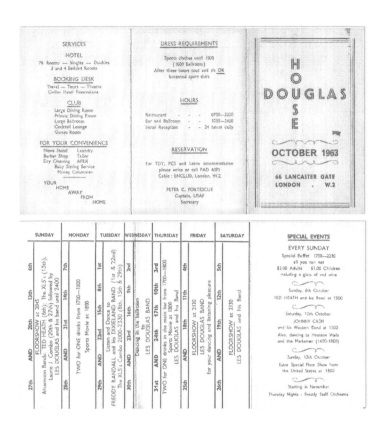

Douglas House hotel guide courtesy of the collection of Rich Silver.

EASTCOTE / MOD/RAF EASTCOTE / DOE EASTCOTE / OUTSTATION EASTCOTE / HMS PEMBROKE V, RAF LIME GROVE

When the phrase 'quiet backwater' was created, it might have been with this base in mind. Located in a leafy corner of Middlesex not far from other US military bases at West Ruislip, Blenheim Crescent and South Ruislip was, it could be argued, rather innocuous as bases go. In one sense it had lowest profile of any of the bases and despite occupying some 7.68-hectare (19 acres) of the Eastcote area even few locals or passers-by (certainly in later years) would even have been aware of its existence.

Or its historical importance.

It may be hard for those who live in Eastcote now to believe but the area was a tourist resort in the early 1900s and then became an exemplar of suburban living between WWI and WWII.

The base buildings on Eastcote were constructed sometime between 1942 and 1943. Thankfully the base never had to fulfill its apparent design brief as an intended hospital for the expected high number of casualties from the D-Day Landings. Its austere cement grey appearance gave it the luster of a prison (and may have started a few rumors along those lines – there was some discussion at one point of it being used for a women's prison.) Perhaps adding to the confusion there was a German prisoner of war camp called Eastcote however it was located in Towcester, England.

With that intention in mind the single-storey buildings were seemingly set deep in suburbia. Reached on one side (in later years this became the only side) by driving about a half-a mile down the Lime Grove residential street, at least half of the sloped base site primarily comprised of three of the single-storey buildings each with 'spurs' extending off a central hallway of about 120 yards long for Block 1 and about the same for the separate and perpendicular Block 2. Block 4 ran parallel to Block 1 and had an even longer hallway but spurs mostly on one side. Block 3 was on its own at the Lime Grove entrance to the base (near a canteen) and, in later years included more corrugated metal type storage buildings and a large boiler room and stack as well as several above-ground oil storage tanks.

The ever-busy public Field End Road flanked the lower end of the site and for many years until the 1970s, at least – a second gate was open there. In later years — due to security and increasing issues of traffic and possible accidents (the entrance, came out on what was actually a critical bend in the road) – it was shut and the fence-line flora had been allowed to grow to camouflage this already low-key base yet further.

Part of the land upon which the base was sited had originally been owned by Lady Anderson, the wife of a tea and rubber plantation owner, who had acquired it in 1909. The Government had requisitioned it at the outbreak of WWII to create the hospital and, somewhat unusually, in 1947 the Crown acquired the land.

As with the base at High Wycombe, the Eastcote base included a public footpath. However in this case the footpath, with origins back to at least 1565, effectively ran right through the operational middle of the base – cutting off Block 3 from Blocks, 1, 2 & 4. It necessitated security measures both in the early days and right to the end of base operations.

Sometime between 1947 and 1952 a wooden footbridge was constructed over the public footpath to link the two sides of the base and reduce security issues, but it was removed some years later.

Nearly 50 years later, with heightened security requirement of the US military in the wake of the 9/11 terrorist attacks in 2001, a permanent guard had to be put at the point where the footpath bisected the main base road and barrier gates had to be opened and shut on both sides to allow cars back and forth – there was also a turnstile security gate for pedestrians.

Given that Lime Grove was the only access road to the base in latter years, this must have been a costly and annoying logistical operation requiring a 24-hour manned security operation and the constant opening and closing of two sets of manually-operated gates.

But that was all to come. During WWII years at least part of Eastcote became Station AA 456 and was a base for the US 9th Troop Carrier Command. Local reports mention American GIs in Eastcote carrying out street-fighting exercises in the neighborhood possibly in training for D-Day.

More importantly, in the context of WWII and the Cold War, in 1943 Eastcote was given over to become what is believed to be the largest of five UK out-stations for Station X, the Government Communication Headquarters (GCHQ) at Bletchley Park – from where British cryptologists broke Germany's famous Enigma machine cipher communications codes.

Then, according to local historian Susan Toms, author of *Enigma and the Eastcote Connection* in the *Ruislip, Northwood & Eastcote Local History Society Journal (2005)*, Outstation Eastcote, also known as HMS Pembroke V, housed two of the all-important Colossus code-breaking machines and 110 of the 'Bombes' – early computer processors which were cased in bronze, weighed about 2,000 lbs each and sorted out the code-breaking permutations for the Enigma via a system of rotors and drums.

The staff largely comprised 800 British 'Wrens' and 100 RAF personnel.

The bombes were located in Block 2. At least one of the GCHQ Colossus machines, known as *Atlanta*, was being operated by US military personnel from the 6812th Signal Security Detachment. The 6812th was stationed at Eastcote from 1 February 1944 until 7 May 1945.

Harold Thomson, a member of the 6812[th,] recalled in a 2005 letter to local Eastcote resident Jean Dixon that he had arrived in England in one of three companies (the 6811th, 6812th and 6813th) in early December 1943 aboard the *Britannic* as part of the American integration into Bletchley Park's operations. For security purposes the unit only came back together months later in Harrow-on-the-Hill in England on grounds next to the famous Harrow School (one of the houses he lived in there was named 'Hollywood'). Thompson may not have initially known the name of his new duty station but he knew something far more important – that it was next to the Black Horse Public House (pub). That, he says, "…quickly became a place of importance!"

It was, of course, Eastcote. The US military had barracks built on an open field and, he said, the locals would watch them play baseball near the station.

"More importantly, until we arrived, it was manned by a group of English Wrens, the most beautiful collection of beautiful high-class girls I have seen," Thompson recalled. "They invited us to their dances and we had a lot of fun."

And a bit of adventure, too.

"On the far side of the Wren Barracks was an alley-way," he said. "I remember being chased down it after pushing a Wren, who missed her curfew, through a window. Some-where on the other side of the alley was an Air Force installation where they printed maps. They was a lot of activity in that alley after the invasion (D-Day)!"

Most of the machines were dismantled on-site in 1945 though two were said to have been left there untouched for a few years afterwards before being disposed.

What is not so well known and only really came to light only in 2003 was the Eastcote base's post-WWII role in continuing GCHQ's efforts to code break the cipher machines of the old Soviet Union.

GCHQ moved to Eastcote from Bletchley Park 1946 and set up what became known as the London Signals Intelligence Centre.

GCHQ's Administration offices and an organization possibly known as IV Intelligence Group were located in Block 1 along with a base canteen and snack bar. Government Offices were located in Block 4 while Block 3 up near the Lime Grove entrance housed either a Hospital area or an operation codenamed "H" along with Administration Of-fices, a Library and Communications/Teleprinters centre with links to an Aerial Mast right by the public footpath.

Sometime during or after 1946 British code-breakers working at the Eastcote base cracked or help crack the Soviet Union's main military cipher systems used across most of their military – giving the Allies access to some of the most highly prized intelligence they could seek. The British had acquired the ability to read material from a Soviet system known by the codename *Caviar*.

There is some indication that the US Navy may have kept an office at Eastcote between 1947 and 1952 at least as code-breakers started to really take on Soviet communications. It was in March 1946 that US Navy Commander Joseph Wenger brought a team to London to negotiate the BRUSA Agreement, a focal US-British agreement on exchanging intelligence.

The British shared intelligence gathered under what came to be the 'Venona Project' (it was also known by other less frequently used codenames of "Bride" or "Drug"). The codes were broken by unified teams of British and American personnel who wee apparently kept separate form other GCHQ teams.

It was, reportedly, a success until 29 October 1948 when all the Soviet ciphers and codes were suddenly changed.

The day became known in intelligence circles as "Black Friday", according to author Michael Smith, whose book, *The Spying Game*, originally revealed Eastcote's early Cold War role. Apparently a cryptologist in the US Army, who was in the pay of the Soviets, had informed his handlers about the Allied capabilities.

It's also understood that the British spy Kim Philby, who was the British Secret Intelligence Service liaison to Washington D.C at the time, may have been made aware of the Eastcote program, which, it is also claimed, helped in a wider way to uncover spying activities of German-born atomic spy Klaus Fuchs and English physicist Alan Nunn May, among others.

Following the Crown's aforementioned purchase of the site they essentially gave one of the blocks (Block 3) to the Royal Mail postal service in 1950 and then, over the years, for use by the Census Office (1956) and Department of Environment support offices before the Department of Trade and Industry (DTI) took occupancy in later years. With its arrive came the Export Intelligence Office – along with some of the best computers of the era including some of the most modern ICL computer mainframes of their time.

During this period Eastcote also established a reputation as a mini-crisis centre. On two occasions the UK Government established Petrol (Gasoline) Rationing Units at the base and, at one point, London's Flood Warning Centre was located there. They are believed to have operated there until the 1980s.

GCHQ had largely retreated from Eastcote as part of a move to Cheltenham by the early 1952, however a small cadre of about 50 GCHQ staff, including a number of members of a Joint Speech Research Unit, remained until 1978 at which time they had a celebratory group picture taken.

Eastcote as it was between 1947 and 1952 when GCHQ was involved in Operation Venona. Handwritten items on the map were provided to local historian Susan Toms by a person who was stationed at the base during those years. Map courtesy of Susan Toms.

O.S.E. FEVER
(with apologies to John Masefield)
I must go back to Eastcote again,
To the busy bays and the laughter
And all I ask is an EVENING WATCH
With thoughts of sleep soon after.
And the wheels click and the motors' song
And the typewriters chattering,
And the thankful look on everyone's face
When "Tea Boat" comes clattering.
I must go back to Eastcote again,
For the call of the running machine
It a wide call and a clear call,
That will remain evergreen.
And all I ask is a trouble free DAY
With the Relief Watch arriving,
And the evening free, the RAF hall full
    And everyone jiving.
I must go back to Eastcote again,
    To the communal way of life.
To the Wren's way and the R.A.F's way,
    Which "discip's" a disgarded wife,
And all I ask are pretty Wrens
        "Like the one that came from Dover."
    And quiet sleep and sweet dreams
        When the NIGHT WATCH is over.

    H. NEWTON
    C. CAMPBELL

Above – – Probably one of the few poems ever written about a base. In this case about the base's intelligence gathering operations either in WWII or in the 1950s/60s period. The 'tea boat' is the tea trolley and 'discips' is a believed to be short for "disappoints". The poem, attributed to H Newton and C Campbell, was printed in a one-off publication from the GCHQ staff and provided by Jean Dixon via Susan Toms.

In the early 1950s the USAF took over or took over more of the base as part of its expanding presence in support of the USAF Third Air Force Headquarters at South Ruislip. In 1952, when the majority of GCHQ had relocated, Block 1 became an US Department of Defense Dependents elementary and junior high school for children of US military personnel stationed in the area.

The parallel Block 4 building sat on a hill some 20 foot above and 60 feet back from Block 1. During the 1960s the USAF used at least part of this building as the central processing mortuary for any Third Air Force service member or family members who died at the any of the bases in the UK, along with a Dental Clinic and a Veterinarian Service, the latter being an agency whose duties included quality checks on food/kitchen facilities.

In later years, after the USAF had departed, Block 4 was understood to have been used by the Britain's Ministry of Defence photo processing and publication operations and comprised a series of large processing rooms with darkened windows and film developing tanks.

However, in the final few years under US Navy control it appeared that the US Marine Corps Det. 1 Security Force used Block 4 for hostage rescue and other training exercises certainly during the last few years of their 29-year deployment at Eastcote.

The Lime Grove entrance road to the base snaked around a general reception area and entrance building with a cafeteria to an equally long Block 3 which ran away to the right, again with spurs on either side. In later years this at least used by various departments of the British Government for activities such as publication storage and by Department of Trade and Industry's Export Intelligence Service (EIS) as its computer information processing centre. The EIS is understood to have used one of Britain's most sophisticated computers in the 1980s – believed to be a DRS20 computer system.

At its apogee the US dependent elementary school in Block 1 had 600 students in attendance and there were probably up to 500 other personnel working on the site for both the British government organizations and the US Third Air Force. However, when Colonel H.L Porterfield, Commander of the 7500th Air Base Squadron and Base Commander of the Ruislip Complex, announced the school's closure in 1972 as the result of the USAF relocating its Third Air Force Headquarters to RAF Mildenhall, there were just 224 students and 10 teachers with 14 directly or indirectly employed British staff. Many of the students transferred to the elementary schools at the nearby West Ruislip base and at the Daws Hill base in High Wycombe.

In 1980 at least half of Block 2 became the Headquarters for the United States Dependents Education Schools, European Area (USDESEA) which, at about that time, changed its name to Department of Defense Dependent Schools (DODDS) Atlantic Region. The school system administrators based there had control of some 44 schools in 10 different countries and would work with personnel in coordinating with construction experts located at the nearby Blenheim Crescent base for the building and refurbishment of schools – some of which were still using WWII buildings.

DODDS-A stayed there until it was closed in 1994 as part of the consolidation and the needed parts of the operation were transferred to Wiesbaden, Germany.

It was in 1993 that the base management was handed over by the USAF to the US Navy.

The MCSF Det. 1 US Marines, who had arrived at the base in 1973 and stayed until 2002 when they were redeployed to Rota, Spain, were a striking presence. But their flat-top haircuts weren't the only thing that marked them out as they came and went from the barracks in Block 1 (which would be designated Building 201 under the US Navy command). On occasion their cadence chants as they went on early morning training runs around Eastcote would get young children emulating them. And their rousing chants caused at least one letter of complaint to the local paper the *Uxbridge Gazette*!

The various wings off the main 120-yard long laminated wood sloping hallway (in earlier years it had been linoleum) that created the 'spine' for Block 1 were given their 'Spur' designations. But at some point etched white and black laminate signs reading Spur 1 to Spur 12 hung from link chains down the main hall at the door to each wing.

The Block was entered about halfway along the building. To the right for the most part (from Spurs 7 – 11) were the barracks and dining facility – mostly for the US Marine detachment but also with some rooms set aside for Ministry of Defense Police accommodation. There were washing rooms, a laundry room and at least one room with a bank of pay phones.

Resident British MoD police could, according to their Mess Accommodation Rules, not smoke anywhere in the building. They could consume alcohol in living quarters but "it should be in moderation". All barracks rooms had a 10-minute fire localized fire alarm delay before a general building alarm would sound.

Starting in the 1990s the US Navy worked up an upgrade of the Block 1 Spurs 7-12 primarily to accommodate the US Marines barracks and increase security. Design work was undertaken in 1991/92 reportedly by the firm of Barnett Briscoe Gotch, while Bell Fischer Landscape Architects took on the landscaping aspects including demolishment of a disused office and new block paving, lighting bollards, trees, road surfacing, drainage gullies and entrance doors of the entrance to Block 1.

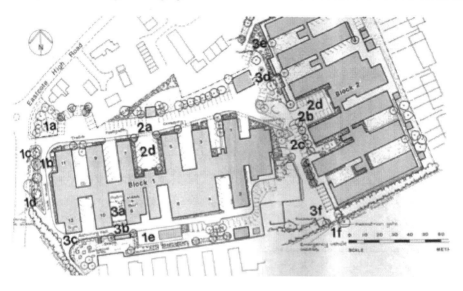

Above — An architect drawings showing Block 1 and Block 2 at RAF Eastcote –The plan indicates the layout of some of the key Blocks along with some of the planting undertaken by Bell Fischer. Block 1 is left and centre of image with the Shooting range being the narrow protrusion under the '1b' at the far left of the Block. The bordered white buildings below comprised Block 4 (one time home to the USAF Morgue and other veterinary facilities and then later as giant darkrooms for the RAF photo processing units. Block 2 is seen to the right. Map courtesy of Bell Fischer.

Bell Fisher Landscape Architects also worked on landscaping at a number of other US Bases including High Wycombe, West Ruislip and distant bases such as Chicksands in Hertfordshire and Brawdy, Wales.

(As an aside it's worth remarking that foliage and planting was a matter of pride at most of the bases around London – indeed in the UK – even prior to this the US Navy taking over. Indeed there was a large metal sign at Daws Hill, High Wycombe near the main entrance – and then after the US Navy took over – on the wall in the RAF Commander's Office – that read "You are on the best little base in the USAFE". The High Wycombe base won the 'Base Appearance Award' for Third Air Force in 1986, 88, 91 & 92 (the last year of USAF control) and for USAF Europe (USAFE) in 1988, 1991 and 1992).

A second upgrade to the base buildings at Eastcote followed a decade later when the US Navy revamped the bachelor quarters in the barracks on 25 March 2004. The upgrade project was started in 2003 and run by the US Navy's Seabees (the Construction Battalion) who were working for Naval Activities (NAVACTS). The upgraded barracks were given an official opening by then Rear Admiral Stanley Bozin, Cmdr of Navy Region Europe and Capt. Stephen Matts, then Commander of the US Naval Activities.

The $600,000 project upgraded the rooms throughout Eastcote with new furniture, carpet, window coverings, wall art, microwaves and bedding. Additional 'swipe' card door key access was installed to all rooms. The barracks development and upgrade may have taken a bit of pressure off the US Navy / USMC's Kennington Barracks in London which were vacated at about that time.

At least part of Spur 11 and all of 12 (adjacent to Field End Road) had a 'business' function. It was here in Spur 12 that the armory was located – an alarmed security room of about 12 x 10 x 8-foot with a weapons cage comprising inch-thick metal grills within the confines.

The armory was right next to a door leading to what had been an addendum to Spur 12 – the spur closest to Field End Road. There the US Navy had constructed an approximately 120-foot long pistol range. The range essentially comprised a breeze block construction externally clad in a light pistachio green corrugated-type metal. It was built on part of what had been a playground for the elementary school. Part of the playground remained – an area just large enough to accommodate an outdoor basketball court.

At the end of Spur 12 was another 'crucial' room particularly for military personnel; the Eastcote Fitness Facility. It operated 7 days a week. Here was an unstaffed weight room (complete with sign saying "Please Don't Bang the Weights. Thank you.") and wall-mounted (both internally and externally) pull-up bars positioned up about 7 feet high.

In Spur 8 was the Cafeteria and a recreation zone complete with pool table area; in latter days an overhead Miller Lite-branded light box was installed. There was also a ceiling-mounted large screen TV projection system that was left in-situ when the US Navy departed. To the left was a double lane darts zone. Over by the main bar area was a wood-lined alcove space with series of five or six standing bar tables set at chest height.

In the late 1990s the Enlisted Club was open for lunch each weekday, with dinner and movie nights on Monday and Thursday and a Happy Hour every Friday from 16:30 – 18:00.

Also left behind in the Eastcote base club (sometimes known as "The Foxhole") when the base closed was the cocktail list. The list, in white plastic lettering on a black perforated board, featured a B52 at $4.00, a Vodka Sunrise at $2.00, a Tequila Sunrise at $3.00 and a pitcher of beer at $5.00. A Brain Hemerage (sic) cost $2.00.

Along the main hallway Between Spurs 9 and 11 was a small trap door. The area between the two spur wings had been fenced off externally. This was the dog compound – primarily, it is believed, for British Ministry of Defense Dogs. One of the resident dogs in later years was a golden retriever named Tweed who specialized in explosives detection. Another dog, a Springer spaniel, was trained in drug detection and would, on occasion, be used not only through barracks but at London Central High School at High Wycombe. The two days of the year the dogs and their handlers definitely didn't have off were during the 4th of July celebrations held at the nearby RAF West Ruislip base that in the last decade of base operations, were opened to the British public.

There was at least one other dog associated with Eastcote in more recent years. The US

Marines had a English Bull Dog as a mascot but it is understood he was more of a pet.
✛ Between Spur 7 and 9 a sand-based court was established for beach volleyball.

If you turned left through the main entrance and walked up the long hallway you found Spurs 1 – 6 comprising both key offices and the operational end of the Block.

Spur 1 contained the COMNAVACT UK Training Room. It was reached down a couple of steps and was oddly shaped and was one of the largest rooms on the base. In here along one wall was a map located behind brownish-red or – pending your take –muddy purple curtains that could be drawn open or closed. The map centered on the Atlantic Ocean with bits of Europe, Africa and the USA. Different colored magnetic cut-out shapes of ships no more than about an inch and a half long were spread across the map and could be moved with the push of a hand.

The area and rooms around had steel-plated doors with Sargent & Greenleaf (S&G) combination locks featuring their distinctive white etched black dials. (There have may have been a second map/briefing area, but active demolition works had made it too dangerous for the author to enter.) Overlooking the room was a small hatch and a separate room. Nearby in the maze of rooms was what had been a large computer or communications room complete with a tiny two-inch square grey box set into a central pillar. The box had a red inset button. It was, it seems, literally the Panic Button and was marked so. In the side of it was the reset key.

Next door Spur 3 housed the offices for the Naval Criminal Investigation Service (NCIS) Resident Agent (sometimes referred to as the (NAVCRIMSERVRA). The NCIS, staffed by civilian special agents, provides law enforcement and counter intelligence services for the US Navy and the US Marines including investigating of computer crime, fraud, terrorism and espionage and other serious crimes. It had originally been known as the Naval Investigative Service (NIS). NCIS agents were based in the Eastcote office but the unit had links to other NCIS offices around Europe including Frankfurt (Germany), Keflavik (Iceland), Rota (Spain), Sigonella (Italy), Marseille (France) and Bahrain.

The NCIS office was one of the few – if not the only – spur in the block that had a separate external entrance from the main street. Next to that exterior entrance there was a large 5' x 3' sized blue acrylic NCIS sign. The Spur had barred window.

Entering NCIS from street side visitors would step into a small reception room complete with a polished steel-barred reception booth with a wood-framed, bank teller-style desk with a clear flip lid and pass tray into which items could be handed to the reception clerk. Visitors stepped through a punch code lock door and into a hallway off of which were various offices.

The facility included what appeared to be a small single-person holding cell that, perhaps unusually, was accessed by going up two steps. Accessing the NCIS spur from the main wing entrance required punching another code into a hood-covered code entry box.

In some of the NCIS rooms were small 'reminder' stickers exhorting: *"Have you Secured All Classified Papers? Your Files? Your Safe? You! Are Responsible."*

The two exceptions to the operational at this end of the building were, perhaps, the Spur 2 (where there was a laundry block) and, also, the large room at the very top end of Building 201 which was used largely for storage. The latter room was unusual because it encompassed what had once been an external wall complete with the original crittall steel-framed windows running within the building.

As with Block 1 visitors entered Block 2 at Eastcote approximately halfway along the length of the building. The building itself was about 50 meters away from the top of Block 1 and ran perpendicular to Block 1. Here past the blue double doors and the sign reading "Visitors Entrance" the two ends of the building had, at least in the latter years of occupancy, very different uses.

To the left were the aforementioned Department of Defense Dependent School (DODDS/DODDS-A) Offices. The offices were denoted by pair of substantial glass doors across the main hallway with the DODDS identity laminated into them. They had been left in-situ even though DODDS had long gone. The doors were notorious within the DODDS school system with much educator chatter that they had cost $5,000 to commission!

The spur sported an executive-style fit-out – with a number of wood-paneled rooms. After the departure of DODDS the executive wing was been home – for a period – to Naval Supply (NAVSUP) Systems Command and to the Defense Logistics Agency. East-cote had been the regional contracting detachment office for the US Navy.

Also noticeable further down the main hallway of this wing were two facing walls comprising brown velvet panels for mounting displays.

In latter years a number of the rooms further down were used for various activities including the Boy Scouts, Girl Scouts and Maryland College courses.

Prior to demolition Block 2 also sported a small 25-30 meter high microwave tower and feed-in transmission equipment for the likes of TV channels such as MTV, CNN and AFN (the latter being the US military news channel) with the 'feeder' room located immediately to the right of the main entrance. Also along one of the spurs was the MOD Security office.

Like Block 1, Block 2 had a large power plant area. Nearby was the block's most un-usual facility — a room with raised, almost rubber-like block flooring and perforated wall panels and a single grey prefab 'Ladies' toilet in the middle (complete with 'occu-pied' light).

It's doubtful that the room in that configuration, at least, was anything to do with the old GCHQ operation despite the unique flooring. At one point in the history of the base that part of the building had actually been part of a small USAF medical hospital facility. Aside from one or two spurs around the main entrance the other spurs in this building were flat-roofed.

Among other operational components that had been located in this Block over time had been the DFO UK, the MOD Police, the US Navy Recruiter's Office and the Naval Oceanographic/NRCC Detachment London. The Human Resource Management Cen-tre was stationed there for a while before being reassigned to Naples in 1986.

One of the last units to remain in the building before demolition was believed to be the COMNAVACTUK 'Emergency Management/Disaster Preparedness Office which had the duty of preparing, training and executing the US Navy's Disaster Preparedness Operation Plan (OP Plan) for Eastcote and for other US Naval bases and facilities in England.

This would have included setting up Field Command Posts (FCPs) and executing con-tingency plans for everything from simple local natural disasters to dealing with the aftermath of a WMD (weapon of mass destruction). The rooms contained the Disasters Preparedness Command / Emergency Control Centre Annex and it is from here person-nel practiced for disasters such as nuclear, chemical and biological (NBC) emergencies. The training also dealt with everything up to 'Condition Alpha' (a surprise attack) and

'Condition Bravo' (impending attack) as well as 'nuclear event' attack levels.

A small stockpile of CBW decontamination suits and certainly some Chemical Biological Masks (which were issued in sealed green tin cans) were also stored. Apparently, in case of emergency at Eastcote base staff had an on-base muster site and then three off-site locations (a primary, secondary and tertiary) at which to assemble.

Between and off to the side of the two main US blocks was a small red-bricked building with a large open bay and three or four 'Crittall' steel framed windows above. This had once been home for the base fire station in USAF days. In more recent times it appeared to have been the base rubbish collection point – a sign fixed to the building fascia read "Fine for Dumping $100". Inside the bay a small blue-painted door opened to reveal... a brick wall!

Security was always a consideration for US Military bases and this had already increased in the early 1990s, a decade before the attack on the World Trade Centre in New York resulted in a stringent Force Protection (FP) policy. At some point after the US Navy took over the bases in 1993 the command authorized around £144,000 ($223,200) for a swipe card pedestrian turnstile entrance system at Eastcote and at the nearby base at RAF West Ruislip and for better telecom links from the Security Post to Security Command at Eastcote. The swipe access turnstile entrances at both bases could still be seen in-situ in the summer of 2010 at Eastcote (it was demolished soon after) while at West Ruislip it was still in-situ as of May 2014.

The US Navy itself announced plans to cease operations at Eastcote by the end of June 2006 as part of a wider announcement to also cease operations at RAF West Ruislip and hand both bases back to the Ministry of Defense by October 2006. At that point the cutbacks were calculated to save $400,000 annually as costs for the Eastcote base.

As part of the Project MoDEL redevelopment for the Ministry of Defense, the MOD announced plans to sell the site for redevelopment in 2004 with some 300 people attending the local council planning meeting about the redevelopment in 2004. Subsequently the MOD through a developer, VSM Estates Ltd, applied for and received permission for 385 homes to be built in April 2005, despite 155 letters of objection and six petitions to speak at a planning meeting after a consultation of 400 local households.

In early 2007 VSM Estates, announced the sale of the site to Taylor Wimpey. VSM had obtained outline permission for 50 dwellings per hectare and housing developer Taylor Wimpey purchased the site in March 2007. At the time the sale price was a rumored £30 million.

The Eastcote base went on sale advertised with three road access points – even though for the 20 years previously only one of those roads had been practical to use (the other two came out on to different points of a dangerous bend at Field End Road (the B466).

The demolition of Eastcote started in early 2007 and largely finished at the end of the year, however Block 4 (the old RAF Photographic Block that was also part of the former mortuary for the USAF in the 1960s) could not be demolished immediately. A colony of Great Crested Newts was identified on the site and this meant that the landlord had to come up with a plan to remove them necessitating a six-month delay in demolition as they waited for the right time in newt hibernation cycle. This was because the legally required temperature and time of year conditions for newt redisposition did not occur in Autumn 2008. In mid-2009 the newts were moved and works started on demolishing the remainder of the site.

As this book was being completed the house builders had completed 213 residences on the Lime Grove side of the site (now known as Sandringham Mews) and were nearing completion of the lower half of the site (where Blocks 1, 2, & 4 were located) to be known as Sandringham Park.

In 2010 the New-England style apartments on the Sandringham Mews site were ranging in price from circa £200,000 for a one-bedroom; £250,000 for a two-bedroom and with a 4-Bedroom (known as 'The Hamptons' in a possible (but not confirmed by the author) nod to the site's US history) at £450,000. Those driving into the site from Lime Grove could see a sign saying *Welcome to New England HA5* – the latter part being a postcode for the area.

Another part of the site has been called "Pembroke Park" in honor of the GCHQ links. In 2013 Enigma Hall, a public hall, has been built but not yet fitted out. Residents had complained to their local Member of Parliament in April 2012 over various issues to do with the development but a number seem to have been resolved since then. The last remaining two-bedroom unit in the Sandringham Park part of the estate was on offer in December 2012 for £370,000 (circa $592,000).

Development of the entire site was due to be completed by early 2014.

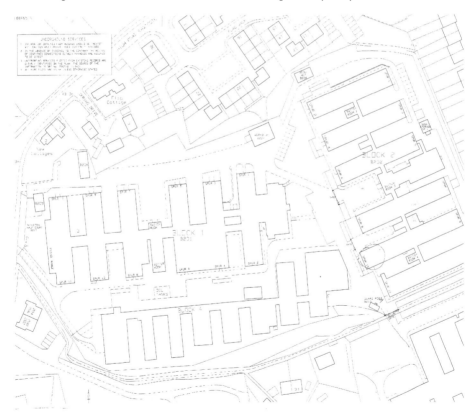

Eastcote with Block's 1, 2 & 4 primarily displayed with part of Block 3 in the bottom right corner. Near the bottom of Block 2 is a small security building located just above a footpath (running with a dotted line from left to right).

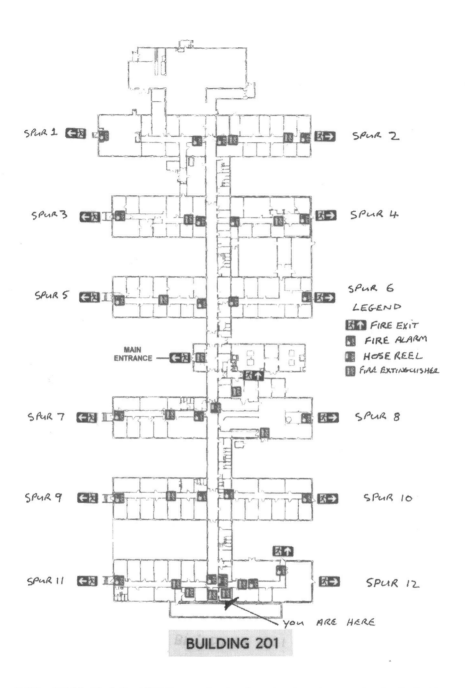

Building 201 (aka Block 1) at MOD Eastcote showing the various 'spurs' in more detail. What had been a Dependent School until the early 1970s was modified for operational use with Spurs 1-6 housing the bulk of the US Navy's operations including a computer room, a battle exercise planning room and the NCIS offices. Barracks for US Marines and MOD Police (were in 8, 9, 10 and 11). The long thin rectangle at the bottom is the shooting gallery. Spur 8 was the base club and dining space.

The walkway between Block 4 (up the hill on the left) and Block 1 (or Building 201) where the school was once located – as seen in 2006. For the view from opposite end it may be worth visiting Chapter 17.

The Eastcote base in early 2006 – Block 1 (later named Block 201) seen from the main base road running down to Field End Road. This block was used as an elementary school during the 1960s. In the 1980s the base came under US Navy domain. The Navy Criminal Investigation Service (NCIS) office was located in the second building down (where the sign is) while the block's main entrance was a little further on. To the far right edge of the picture is the corner of the red brick building was the Base Fire Station in the 1950s and 1960s while in later years it would be a rubbish storage area.

Changing scene – Block 1 (as seen from the Field End Road aspect – looked vastly different as of 14 March 2007 as demolition work on the Block neared its end.

Above – Block 2 (circa December 2005).

After the demolition was completed Tony Moore from the demolition company holds the sign up approximately where it had been.

Going. Going. Gone — The heating tower for Eastcote's Block 1 is toppled on the 27 March 2007 by the demolition crew from Syd Bishop &Sons. The crew had built a 'cushion' of debris to reduce the reverberation from the impact of the falling brickwork.

RAF MEDMENHAM (AAF-475 IN WWII)

RAF Medmenham was set in and around the grounds of Danesfield House, a former country house which sits above the Thames River on the road from Marlow to Henley, some 37 miles from London and just a few miles from RAF/Daws Hill / USAF High Wycombe Air Station.

RAF Medmenham was home to the RAF's Photographic Interpretation Unit from April 1941. As the United States became involved in WWII the USAAF Photo Interpretation Squadron and 2nd Photo Tech Squadron were established at the base along with the 19th AAF Photo Intelligence Detachment. Subsequently all the Allied forces were banded together under the re-titled Allied Central Interpretation Unit (ACIU) which eventually grew to 1,700 personnel.

Utilizing Spitfires armed only with camera guns the ACIU interpreted photography of potential targets and then, as the war progressed, photography of the bomb damage inflicted by the Allies.

Among the ACIU's key successes was Operation Crossbow (the identification of the V-1 & V-2 Rocket sites). Among those who served there was the noted British actor Dirk Bogarde.

The 2nd Photo Tech Squadron left the main building in August 1945. The house and land became an HQ for the RAF's 90 Group after the war and then was sold by the military in 1977 to become the HQ for the Carnation company. It was sold on as a country house hotel in 1991 and today is the 4-star Danesfield House Hotel & Spa.

Britain's Ministry of Defence kept its Police Training School on some of the land at Medmenham and trained up to 1,400 officers a year in the late 1980s and 1990s. Many of those British MoD police officers served security duty at US bases at Greenham Common and Molesworth during the deployment of Ground Launched Cruise Missiles.

The MoD Police facility was closed as a base by the mid-1990s and the last part of the base was sold on 15 December 1997. The Ministry of Defence retains a married quarters estate across the street from the former base that comprises around 190 homes. There is a small museum at RAF Chicksands in Bedfordshire (itself a former USAF base) dedicated to the work of the ACIU.

RAF NORTHOLT, MIDDLESEX (AAF-387 IN WWII)

RAF Northolt was not a permanent home for the USAAF's 302nd Air Transport Wing during the World War II but a good number of other American aircraft, pilots and crews dropped in there to pick up and drop off people and supplies.

Though historically and primarily an RAF base and one of the closest military flight-line bases to central London, there has been a small US presence at RAF Northolt over the years and the airfield is even used by visiting US military and political officials to the present day.

Perhaps best known as the home to the Polish Air Force's 303 Squadron during the Battle of Britain, in recent years the base is recognized by many for being the arrival point for the return of Princess Diana's body after her fatal car crash in France in 1997. It has been home to the Queen's Flight (now No. 32 (The Royal) Squadron) is based there. More recently it was home to four of the RAF's Typhoon fighter jets which were providing interception air cover for the 2012 Olympic Games.

The base, immediately adjacent to the main A40 road to London and within a mile-

and-a-half of the former USAF South Ruislip Base, has a single 5,525-foot long runway (now known as Runway 25).

In July of 1942 the USAAF 8th Air Force Service Command established an Air Transport Service with two aims: Managing the disposition of aircraft delivered to the UK by the USAAF's Air Transport Command (ATC) and the general movement of personal and equipment around the UK. It's thought that the 302nd ATW's 27 Transport Group and its 86th and 325 Ferry Squadrons were frequent visitors In the last two years of WWII (probably via their HQ base at Grove in Berkshire).

The subsequent US connections certainly may not be quite as major or direct as at other bases.

One such connection was the aircraft used by British Prime Minister Winston Churchill – a specially-equipped American Douglas C-54 Skymaster which was stationed at Northolt in 1944 as part of the Lend Lease Agreement. No expense was apparently spared in making the aircraft look luxurious complete with upholstered seating and what was said to be the world's first heated toilet seat! It would be the aircraft in which Churchill flew to the Yalta Conference and then to Potsdam Conference.. When Churchill lost the British election in 1945 his successor, Clement Atlee decided the plane should be returned to the Americans.

And post-War, as the USAF 3rd Air Force expanded at nearby bases including the Headquarters base at USAF South Ruislip, US military aircraft including Convairs (tail number 0-17899 in particular) were stationed at the base. They not only ferried the General staff and senior brass around their subordinate bases but also flew thrice-weekly services to the USAF Europe Headquarters in Weisbaden in Germany during the late 1950s and 1960s. One of the Convairs was subsequently presented for use as a static outdoor display at the Imperial War Museum at Duxford in the mid-1970s but was eventually scrapped about 2001 when museum officials became concerned that the undercarriage and main spar could fail completely.

"The restoration of this aircraft was, at that time, not within the capability of the museum's resources," a Duxford spokeswoman said. "A decision was made to safely dispose of the aircraft and the disposal was subsequently carried out by the Royal Air Force Joint Aircraft Recovery Unit with the approval of the United States Air Force."

The post-WWII period until about 1954 Northolt saw its heaviest use as a civilian airport while Heathrow was being constructed. Another quasi-American connection was probably made for all the wrong reasons when in 1946 a Douglas DC-3 Dakota aircraft operated by British European Airways famously crash-landed atop a local house in South Ruislip. The house was subsequently christened 'Dakota Rest'.

And the American connection on that front almost didn't end there. On 25 October 1960 a civilian Pan Am 707 (Tail Number N725PA) with 41 passengers on board landed at RAF Northolt having mistaken it for Heathrow. Passengers not only deplaned but the aircraft had a minimal fuel load to take off. British police also closed-off the A40. Local residents still recall the roar of the engines today as it barely made it back out. (*See picture.*)

Over the years US planes including the UC12M Huron, the MC-13)P Hercules, the VC-47 Skytrain, the C54E and the C-118A and C-131A have been amongst US aircraft calling at the base.

In early 2012 it was rumored that the base could be sold off by the Ministry of Defence to raise funds but this is probably unlikely given the amount of money recently poured into it through the sell-off of former American and British bases through Project MoD-EL starting in 2006. Among the changes, improvements and consolidations to Northolt has seen is a new British Forces Post Office (BFPO) Headquarters and an upgraded Officer's Mess as well as new personnel and sporting facilities. The ceremonial drill band the Queen's Colour Squadron was among units that have moved there).

While base officials say there are no US operations permanently stationed there, the base continues to receive at least one or two flights a day from US-registered aircraft. These could be military or civilian however as the base now offers private jet services to those who can afford the landing fees of up to £3,072.67 (circa $4,760) per time.

Above: The legendary Pan Am 707-321 whose plane crew mistook RAF Northolt for Heathrow Airport and landed at the military base in October 1960. It's seen here taking off from Northolt. George Trussell, who was stationed at the base, recalled that they off-loaded all the passengers and all the cargo (including some snakes). "Although we had a permanent Customs/Immigration presence at NLT (RAF Northolt), all the passengers had to be escorted and bussed over to London Airport (Heathrow) for clearance, much to the chagrin of a few who were involved with the USAF Unit then located at Northolt (they may have been architects), and who recognized, with alarm, where they were landing at the time." Photograph courtesy of George Trussell/Air-Britain.

ROMANY HOUSE, SURREY

Romany House in Portnall Rise on the Wentworth Estate in an area of Surrey known as Virginia Water was built in 1931 became part of the US Navy in World War II after it had been requisitioned by the Ministry of Defence for officers working at the nearby UK Defence Command.

After the D-Day landings the British Admiralty retained use of the house for US Navy Admiral Harold R. Stark, who had set up the US Naval Forces Europe. His successor

Admiral H. Kent negotiated a direct rent for Romany House in 1946 before it became an annual leasehold. It was in June 1947 that the US Government, acceding to a request from Admiral Richard Conolly, purchased the house through the US State Department's Foreign Building Office for the princely sum of £16,250 or $25,025 (or about $280,000 in 2012 money back then) for use as the formal residence of the US Navy Commander – in-Chief in the UK.

The 8,990 sq ft house, which also had included a further 2.09 acres of grounds, was located between the acclaimed Wentworth Golf course and Windsor Great Park – the grounds of Windsor Castle. The house is right next to the first fairway of the East Golf Course. Wentworth is a residential golfing estate in Virginia Water and abuts Windsor Great Park – the grounds of Windsor Castle (and the Virginia Water Lake).

Romany House had eight bedrooms including four for guests and two separate staff apartments for the chauffeur and gardener. The main reception rooms were oak-panelled.

It seems a matter of debate about whether or not General Eisenhower was a guest there on occasions. What is known is that there was a large back-up bunker for Britain's Home Defence Command under the main car park of the golf club with a passageway leading to the golf club. The bunker was said by some to have 44 rooms (though some who have been inside in more recent years say the number is far less). There was often rumored to have to be a network of tunnels that linked various houses/buildings on the estate but local experts claim that is a 'tale'.

In recent years the house had been used by the US Embassy's Military Attaché to London but as the drawdown took effect the decision was made to dispose of it. Given there were no US Government agency takers for the property the US Navy put Romany House on the market through local estate agent Barton Wyatt and sold it on 20 May 2009 for £4,554,000 (circa $7 million).

"Romany House still stands today, but will inevitably be demolished and replaced by a much larger property as favoured by wealthy Eastern Europeans these days," believes James Wyatt, Partner at Barton Wyatt.

Romany House as it was when it went on sale in 2009. Photo courtesy of Barton Wyatt.

USAF SOUTH RUISLIP / SOUTH RUISLIP AIR STATION, MIDDLESEX

What can be loosely described best as a large triangular plot of land wedged between the Central Line train tracks and Victoria Road in South Ruislip, was originally owned by Stonefield Estates and then developed by Percy Bilton of Bilton Estates with 'modern' factory spaces on it.

The site became the headquarters for the 3rd Air Division from 15 April 1949 – 1 May 1951 when the 3rd AD was deactivated and replaced by the Third Air Force. The base would grow to become the headquarters and the hub of activity the next 22 years of the Cold War before closing when the command was relocated to RAF Mildenhall in Suffolk.

USAF South Ruislip, as the base was known, was where all the action was – both militarily and socially.

The base ran along much of the Victoria Road which also served as the conduit for the main entrance. American 'yank tank' cars were frequently parked out front. From time to time stories of an occasional late evening drag races down Victoria Road make the rounds. One unconfirmed story has it that the British police occasionally raced against the American cars.

In February 1951 the USAF formed the 9th Rescue Squadron at South Ruislip. Among other units and operations at the headquarters were the 1933rd Airways and Air Communications Squadron (AACS) and the 11th COMMS Squadron.

While South Ruislip wasn't large a number of social activities including a Rod & Gun (hunting and fishing to the uninitiated) Club functioned from the base. South Ruislip also had a major Commissary, a PX and Hospital as well as a major Gym along with the retinue of command offices.

In December 1955 the base came under the general UK spotlight when the BBC Television Show *Saturday Night Out* featured the life of the US personnel stationed there and also focused on the American Boy Scout Troop 194.

The base was not just a headquarters for military operations. It was also a leading entertainment venue for the Command, something actor Larry Hagman refers to in detail in his memories further on. The Harlem Globetrotters performed in the base gym. Singer and cowboy actor Roy Rogers came there. People stationed there mention a series of 60s and early 70s chart headliners who came through and performed. The NCO Club was said to be even more popular than that officer's Oak Room and there were reports, as with Douglas House, of officers trying to feign excuses to get in for some of the shows.

A USAF South Ruislip nurse, Major Jewell Patterson, is said to have opened the Silver Dollar Pizza Parlour in nearby Rayners Lane/Harrow. Patterson was said to be a sister or relative of Floyd Patterson, the undisputed heavyweight boxing champion of the world. She reportedly refused to serve people if she didn't like the look of them and was also said to keep a shotgun under the counter of the popular pizza parlour – in the 1960s it even attracted the likes of Frank Sinatra and The Who. It is said that Mick Jagger once claimed that it was about the only restaurant he was ever thrown out of.

Patterson sold out in 1970 but the Silver Dollar remains in operation to this day run by the family of the London chef who purchased it. Boxer Floyd Patterson did have two sisters – Carolyn and Deanna – but no further information was discovered about them or other relatives at the time of book publication. In latter years, certainly in the 1970s,

Bruno's, another pizza restaurant this time thought to have been started by two GIs, opened near South Ruislip Tube Station.

A modern headquarters building for the Third Air Force was constructed in 1963 and opened in 1964. It became known as Building Number 1.

In January 1972 the USAF announced that the base would close and impact 1,300 military and civilian personnel as operations were centralized to RAF Mildenhall in Suffolk and to Wiesbaden in Germany. In February 1972 just before the base closed, The Flirtations, who frequently appeared with British pop star Cliff Richard, played one night at the NCO club and were followed the next night by the Drifters. Through the 1960s it had been a regular smorgasbord of Hollywood and Motown entertainment.

In 1952 the South Ruislip Base Hospital, a key part of the South Ruislip headquarters operation, was opened and operated by the 494th Medical Group. The 7520th took over the hospital operations in 1954 and came under the unit umbrella for the 59th Veterinary Inspection Flight.

It wasn't long after the hospital opened that USAF personnel from the base took part in rescuing people in one of the biggest accidents in the history of British Railways. The Harrow & Wealdstone rail crash took place at 8:19 a.m. (with the accident causing the station clock to stop) on 8 October 1952. It involved three trains that collided into each other in the station killing 112 people and injuring a further 340.

It is clear that the numbers of dead and injured would have been much worse had it not been for up to 150 US Military and civilian personnel from the South Ruilsip base who became a critical part of the rescue and care process. Some of those personnel from the 494th Medical Group had actually been on one of the trains heading to work. They were able to get police to contact their colleagues at the South Ruislip base and summon more help. What would today be known as a ERT (Emergency Response Team) arrived from the base soon after. The team, including seven doctors and one nurse, were commanded by Lt. Col. Weideman, USAF.

Responding to the emergency they descended upon a scene of utter chaos and tragedy. Among the team was Lt Abbie Sweetwine, a USAF nurse from Florida. In fact, she was one of first black nurses at the hospital and the only black person on the team.

It was the team's action on that fateful October day that is widely credited with changing the process by which major accident emergency care would be carried out by Britain's National Health Service thereafter. Sweeetwine and the US military doctors effectively treated the area as a combat zone and established a triage area on Platforms 5 & 6.

It was Lt. Sweetwine, then 31, who went out among the injured and assessed them for care in a move that would subsequently be incorporated by the British medical profession. Most memorably she used her own lipstick to circle the injuries and or mark on the foreheads of the victims – an 'X' if they had been treated by Doctors already and an 'MS' if they had received Morphine Sulfate already (thereby preventing those 'receiving hospitals' from 'double-dosing' them). Her battlefield coding was explained to the British ambulance personnel so that they in turn could explain to the bemused British hospital teams receiving the injured.

Her presence and her actions on that day earned her the moniker 'The Angel of Platform Six' in Britain's *Daily Mirror* newspaper as well as considerable media coverage including mentions in leader columns of other papers.

Lt. Sweetwine and some of her USAF colleagues can be spotted in newsreel footage of the day.

"She took it upon herself to carry out those actions," Sweetwine's niece, Burgandi Alexander (a nurse herself) explained in an interview with the author. "And she kept going. She actually received an injury on her right arm from a jagged piece of metal. But she didn't get it dressed for two days."

Lt. Sweetwine made international news and she was would be feted by many. She was celebrated by the British Variety Club who gave her an engraved silver cigarette case.

For all her pride Sweetwine was a modest woman. In the wake of her actions she was commended by USAF Gen. Hoyt Vandenberg (Chief of Staff of the US Air Force) and later promoted to Capt and would eventually rise to the rank of Major. She was invited, along with Gen. Vandenberg, as a guest of Her Majesty Queen Elizabeth II to attend the Royal Coronation ceremony on 2 June 1953.

"She was not one to boast – she wasn't that way," Ms. Alexander said of her aunt. "Her life was of service. One of her fondest memories of her time in England however was that she had been able to go to the Coronation of Queen Elizabeth II."

Among those involved with the South Ruilsip base was one George Markstein. He was for a time, one of the reporters for *The Overseas Weekly* (an newspaper based in Germany and published from 1950 to 1975 that was also known as the *Oversexed Weekly* for its often more salacious content), before he would go on to devise the background and stories for the cult TV series *The Prisoner* starring Patrick McGoohan. Markstein's earlier involvement covering the military would also see him write the WWII-related book *The Cooler* and then co-adapt *The Odessa File* for the silver screen as well as scripting the famous car chase scene in the Steve McQueen film *Bullitt*.

Markstein's work at South Ruislip probably provided good background for another of his books, *Ultimate Issue,* a fictional mystery/thriller about a USAF Captain from a base in England awaiting a court martial trial for adultery because the authorities believe he knew too much about "a forthcoming event."

Speaking of courts martials and USAF officers an event that would have crossed the desks of senior 3rd Air Force officers at USAF South Ruislip was that of the very real and public protest and subsequent court martial of Capt. Thomas Culver. Capt. Culver, a USAF lawyer himself, became the first US serviceman in Europe to be court martialled following his involvement in an alleged demonstration against the Vietnam War.

This arose from the opposition to the war in Vietnam from a surprising number of American servicemen in the UK. A group of enlisted men at RAF Mildenhall had begun a newspaper entitled P.E.A.C.E. (People Emerging Against Corrupt Establishments.) This was a paper against the war and was distributed on all the USAF bases in the UK. It was financed by the British film actress Vanessa Redgrave and apparently had wide support among airmen and their families.

Captain Culver had become their unofficial legal advisor. He also wrote for the newspaper and helped distribute it (in civilian clothes) outside RAF Lakenheath.

Air Force Regulation 35-15 forbid demonstrations in a foreign country but allowed petitioning Congress or the President. The group circulated petitions on all the bases and obtained 19,000 signatures to a petition against the war "...so mildly stated that President Nixon might have signed it," says Culver. The group wished to take the piles of forms to the US Embassy in London, deliver them and then go for a party in a London park organized by Ms. Redgrave.

The Metropolitan Police would not permit so many servicemen to be that close to the embassy. They suggested the busses unload at nearby Hyde Park Corner and then parade to the Embassy. Some leaders of the group negotiated with the police. Captain Culver was not one of them but as their legal advisor had been on the phone advising. A parade might be thought to be a demonstration and so was rejected. Finally it was agreed that small groups of 5 or 6 would each carry a bundle of petitions and deliver them to the embassy and then return to the busses to go to the park.

"The police insisted that each group have an identifiable marshal and so each group contained one wearing a white armband with a helmet and a clenched fist stenciled on it," Culver recalled to the author. "That is what occurred. There were no placards or banners and the public probably barely noticed them."

Captain Culver was one of more than 200 military personnel attending and was identified as being there and was quoted in a newspaper regarding what was occurring. It was the tradition in the military that in such circumstances it is the senior person who should first be subjected to a court martial and so he was.

An eight-officer court martial held at RAF Lakenheath in Suffolk in 1971 found the Vietnam veteran guilty of conduct unbecoming an officer and gentleman for distributing the newspaper outside the base and violation of a lawful General Regulation. The issue in the case was whether what had occurred was a demonstration.

"Once the military judge defined a demonstration as "Any gathering of 3 or more people to express a common purpose" the outcome was inevitable," Culver said.

Before the case Culver had already made a name for himself as a civil libertarian by writing a series of letters to the *Stars & Stripes* in 1970 to promote women's rights. After leaving the USAF following the courts martial he returned to the UK and became a lawyer for US servicemen and worked on representing them in a number of cases including two charges of murder while representing others in American Civil Liberties Union (ACLU) cases. He became a naturalized British Citizen in 1979 and went on to become a Recorder (a part-time Crown Court judge) and, later, an Immigration Appeal Judge hearing cases against Britain's Home Office. Today he lives in retirement here in England with his wife.

Among those born at the South Ruislip base hospital was a child who became Col. Gergory H. Johnson, USAF (Ret.) in May 1962. In 2007 Johnson was selected to pilot the space shuttle Endeavour on the STS-123 mission that launched in March 2008. After he returned from the flight, Johnson served as a CAPCOM for STS-126, STS-119, STS-125 and STS-127. Johnson piloted Endeavour's final flight, STS-134 that missioned in May 2011.

The 100-bed USAF hospital, which had boasted some the most modern equipment of the time, was offered to the local authorities when the base was closed in the 1970s. However the offer was turned down and major parts of the base were demolished in the late 1970s through to 1994.

Several buildings from the original base still remain on the site which has largely been given over to a retail park, trade buildings and a major sorting office of the Royal Mail.

Property agents Fuller Peiser advertised parts of the base for 'free standing storage units of 23,375 sq ft to 36,635 Sq ft (total 209,725 Sq ft) for short-term let in 1993. Their particulars at the time read: "The premises comprised of seven mainly single –storey freestanding buildings with on-site loading and parking facilities. The initial rent at that point was £3 ($4.65) per sq foot per year.

In 2012 it was announced that part of the site, occupied for many years by Express Dairies after the USAF left, could now be redeveloped by Arla Foods and London-based developer Citygrove as The Old Dairy – a 10-acre site to incorporate a Cineworld 11-screen cinema several restaurants, an ASDA supermarket and residential apartments with an anticipated 650 jobs. The site, empty since 2008 was mooted to become a mixed-use scheme with supermarket, a 10-screen cinema, homes, a hotel and a pub.

An enduring mystery and still a source of discomfort about the base for older locals in the area was the murder of Jean Mary Townsend on 14 September 1954. The 21-year-old local girl's body was found on what was then waste land next to the base (now the site of St. Gregory the Great Catholic Church) after a night out and there was always speculation (though never proved) that one or more US servicemen stationed there might have been responsible. There were claims that an at least one American voice was heard in the aftermath of what might have been a cry for help from her.

The idea that the killer might have come from the base was seized upon by some when it was learned that the South Ruislip base commander declined requests by the police to interview service personnel. There has been further speculation that the murderer might have been a GI who had previously been stationed in Germany.

In 1982 there was a 'cold case' review after what were described as anonymous phone calls to the police who, after their review, seemed of the opinion that no American military personnel were involved. The names of several possible suspects (now apparently deceased themselves) have also been proffered over the years.

In 2005 a request to see the files held by National Archives by a school friend of the victim was refused. The decision was backed by an Information Tribunal in 2007 and pending any successful reapplication the files remain sealed until 2031. The Government closure of the Forensic Science Service will likely reduce any chances of any resolution to the case though.

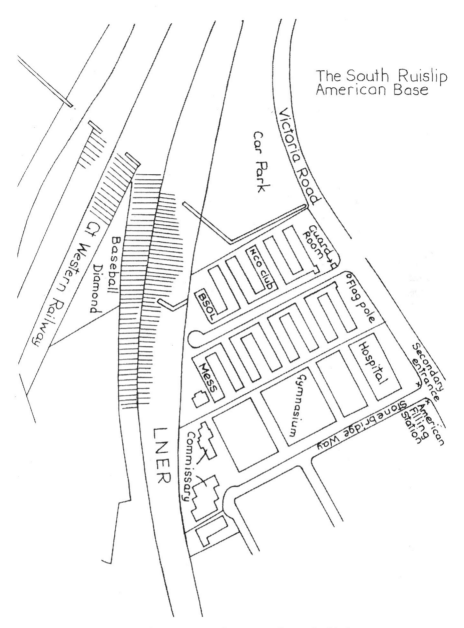

The layout of USAF South Ruislip base as it was in the 1960s. Today much of the base is given over to retail park units and other industry/light engineering businesses as well as a major Royal Mail sorting centre. Victoria Road was the main road for the base. Map courtesy of Eileen Bowlt of the Ruislip & District Natural History Society.

USAF South Ruislip as it was in the 1960s. These were the Airman's Barracks – in what was a converted factory building. Photo courtesy of Rich Silver.

USAF South Ruislip in the early 1960s. At the time the office on the left was USAF Assistant Deputy Chief of Staff for Personnel – among its occupants the DCS/Personnel for 3rd AF, Colonel Richard M. Douglas and one of his subordinates, the actor Larry Hagman. Right: Col. Douglas. Photo courtesy of Bill Douglas.

The USAF South Ruislip Base (date unknown but thought to be mid 1960s). Note the 'Yank Tanks' parked on the street. On the left side of picture (front to back): Building 1 — HQ Third AF, AG, ACS Comptroller, ACS/P&O. Building 3 – HQ 7th AD, Building 5 – 7th AD Materiel, 3rd AF Surgeon, Officers Club. Building 7 – Commissary, PX, Airmen's Mess, Snack Bar. On the right side of the picture (front to back): Building 2 – Third Air Force ACS/Materiel; ACS/Installation, Inspector General, SJA. Building 4 – Communications, Dental Lab and NCO Club. Building 6 – ACS/Personnel, Post Office. Photo courtesy USAF AETC/AFHRA/AAC.

The USAF Third Air Force HQ Building being constructed in 1963. It became known as Building No. 1. Photo courtesy USAF AETC/AFHRA/AAC.

The USAF Third Air Force Hospital as it was the day it opened on 16 February 1953. The opening was attended by General Griswold and Colonel Braswell, USAF. Photo courtesy USAF AETC/AFHRA/AAC.

The former South Ruislip base as it was in 1994 as a major part of it was put up for sale. For comparison look at the previous page picture.

One of the few original buildings left over from the USAF South Ruislip days as seen in 2007.

RAF WEST DRAYTON, MIDDLESEX

Given its original history as an athlete's village for the 1948 Olympics in London RAF West Drayton was variously used by the USAF and the RAF but is primarily known as the former home of London Air Traffic Control (until that operation moved to Swanick, Hampshire in 2008).

RAF West Drayton also had a military air traffic control centre for liaison with civilian counterparts and a RAF School of Fighter Control.

The US presence is believed to have commenced in the early 1950s with the 3911th Air Base Group taking residence up in 1951 to 1962). The base was also a key operations centre for the Linesman civil air defense radar system — the Main Operation building was known as L1 and base personnel plotted all flights (military and civilian). In the mid-1950s at least the base had around 900 military personnel and dependents. Among units stationed there was the USAF's 3921st Reconnaissance Squadron. There was also base housing. USAF courts martials would also be convened there for a time and there was a medical facility for personnel on base.

While not necessarily high profile in later years, the base was recognized within the USAF for a series of innovations and firsts:

In November 1952 the USAF set up a Non-Commissioned Office (NCO) academy at West Drayton. It would get Strategic Air Command (SAC) accreditation by 1954 and would offer 60 students a six-week course running eight times a year. That program led to it becoming a role model for other USAF NCO Academies.

Two years later the 3911th faced a mini security crisis when a rotation saw base Air Police (later known as Security Police) numbers drop dramatically. It was Floyd B. Cressman, Commander of the 3911th who reportedly launched a pioneering project to incorporate guard dogs as part of the security force. In March 1954 three dogs called Mark, Prince and Rex were the first recruits.

West Drayton also enjoyed another first. When the Base Exchange opened to shoppers on 12 November 1955 it introduced the concept of self-service shopping. It may have also been one of the first US Military Exchanges to incorporate mirrors on the building support beams within the shopping facility (it was designed by Capt. Humber L. LeMirande and Lennard Sampson, General Manager of the AFEX military retailing system). Mirrored columns in the shopping areas seem to have become a staple fixture of US military store design thereafter.

Staff from West Drayton's 7th Air Division Honor Guard provided escort duties for actor Jimmy Stewart and fellow cast members when they arrived for the June 1955 premier of the film *Strategic Air Command* in Piccadilly, London.

The base was home to a number of US military operations including the 7513rd Air Support Squadron which had moved to West Drayton from RAF Marham in the early 1950s. The USAF's 3911th Air Base Group was stationed there from 1951 to 1962. The 3911th ABG provided some administration and oversight function not only for West Drayton but for the USAF base at High Wycombe. The 3921st Reconnaissance Squadron was also stationed there – certainly during the 1950s.

Strategic Air Command operated at West Drayton until 1 July 1959 when it ended operations and turned the base over to the USAFE.

It's not generally reported when the last Americans departed but the bulk would have departed during the general command movements in the early 1970s. British military operations ended in 2008 and the base was demolished and set to be redeveloped as a 773-home residential site with an additional nursing home, shops as part of two areas known as Drayton Garden Village and Parkwest being created by separate developers. Parkwest has been a phased project of 574 units started in 2006 and will finish in about 2014. Drayton Garden Village opened in 2011.

A number of residential units adjacent to the former base were occupied by the US military over the years but they were sold off with the main site in January 2009.

(Note: It remains unclear to the author just how much of a role or liaison, if any, the USAF might have had at RAF West Drayton with RAF base personnel who, among chores, examined Unidentified Flying Object (UFO) reports.

For a while the RAF had a unit at West Drayton as part of Military Air Traffic Operations (MATO) with its HQ at RAF Uxbridge. MATO was tasked with receiving and analyzing UFO reports and it's likely that there will have been some discussion between the two nations – even if at an informal level. Certainly it's possible in at least one incident in February 1962 – where London Control tasked a USAF aircraft on refueling operations from the 420th Air Refueling Squadron at RAF Sculthorpe (116 miles north of London) to look at a reported UFO between Stonehenge and Oxford. That tasking certainly started the USAF pilot, who would rise through the ranks to become Major George Filer, off on a lifetime of UFO research.

And there is probably still more to yet come about many aspects of the Anglo-American military relationship in future years. After all, it was only in August of 2013 that it was revealed that several RAF pilots had flown top-secret spy missions in the U-2 aircraft over the Soviet Union in 1959 and 1960 and subsequently received the Air Force Cross for their efforts.)

US NAVY HEADQUARTERS, LONDON (STATION 576 IN WWII)

The US Navy had actually been operating in force Britain from World War I and so already had a considerable London legacy. By the end of World War I the Headquarters for the Commander of US Navy Vessels Operating in European Waters, located at Grosvenor Gardens, had 1,200 military and civilian personnel attached to the command.

During the inter-war years the numbers subsided but US Navy personnel were on hand to support the 1930 London Naval Agreement signed between Japan, France, Italy, the UK and the US to control submarine warfare and shipbuilding.

US Naval Forces Headquarters, London, was set up in 1942 by Admiral Harold F. Stark (COMNAVEUR) and became the primary command for the US Navy in Europe during WWII and beyond. Stark had been Chief of Naval Operations but was assigned to London in an administrative role following what is generally agreed by historians as a failure by him to pass on information about the Japanese attack on Pearl Harbor.

The location of 7,8,9,10 and 42 North Audley Street was originally the premises for the Jack Olding Motor Car Agents – with a car show room on the East Side of North Audley Street. The premises also incorporated Providence Court.

In the 1930s Jack Olding & Co Ltd, which sold Bentleys, became Aston Martin's sole dealership in London. Later the company would become sole agent for the American company Caterpillar and its bulldozer output. The company would establish a headquarters known as Caterpillar Island in Hatfield, Herts. As a side note it was at the company's Hatfield plant that it helped to prepare the US tanks such as the Sherman and the

Grant for use in WWII).

Next door to the dealership Audley House, a block of flats was erected in 1927-29. This would become the US Navy building.

The US Navy and its London links could be actually said to go back to 1861 and the American Civil War when, reportedly, the US First Minister (there were not yet Ambassadors) in London, one Charles Francis Adams, commissioned pictures of ships thought to be breaking the Union blockade of the South and also had agents keeping an eye on Confederate ships being built in the Liverpool docks.

Come World War I the Commander-In-Chief, US Naval Forces, Europe was sent to London. Admiral S. Sims duly came with an American flag, which was blessed at St. Paul's Cathedral and then flown alongside the British flag at Parliament. The US destroyer division followed with destroyers and additional vessels on operation in Europe and administered from London. By the end of World War I the US Navy had 37 destroyers and the five battleships (from the 6th Battle Squadron) in theatre.

Between the wars the US Navy presence was accordingly diminished but with the advent of WWII the Navy came back – and in force.

The US Navy's own official fact sheet entitled: *Biography of a Building: US Navy London Headquarters*, provides one of the best descriptions of what would be the major operational headquarters for the US military in London.

Located in an formidable red brick building, the headquarters of the Commander in Chief, US Naval Forces, Europe (CINCUSNAVEUR) (as the organization would become) is situated at No. 20 Grosvenor Square in London's fashionable Mayfair District. The exterior of the Neo-Georgian building resembles a block of flats, for which it was originally constructed. It later years it was distinguishable only by the U.S. and admirals flags, the communications antennae on the roof and a plaque commemorating General Eisenhower's use of the building during WWII next to the Grosvenor Square entrance.

Grosvenor Square itself was laid out in 1725 by Sir Richard Grosvenor. The buildings on the Square have housed dukes, generals, ambassadors and other dignitaries. One of the first "Yankee" residents of the square was John Adams, the first US Ambassador to the Court of St. James. He maintained a combination residence and embassy at No. 9 Grosvenor Square from 1785 to 1788.

The US Ambassador moved his embassy to No. 1 Grosvenor Square in 1938 and Americans have remained a permanent fixture in the square ever since. The US Embassy has been located in a modern building opened in the 1960s by Eero Saarinen that dominates the west side of the square. In October 2008 it was confirmed that the embassy would be relocated to Nine Elms near Battersea in 2017. Some believe it will be closer to 2019 pending actual building development and security processes.

In July 1940 Rear Admiral Robert L. Ghormley, then assistant chief of Naval Operations, was ordered to London to become Special Naval Observer and to establish "exploratory conversations" with the British. On June 1, 1941, he and his staff of 12 officers and six enlisted men moved into Numbers 18-20 Grosvenor Square.

During the London Blitz, Grosvenor Square served as a placement for anti-aircraft guns and barrage balloons. The US Navy building was not damaged apart from a few broken windows. A reinforced concrete wall protected the Grosvenor Square entrance. Fire watches patrolled the roof to extinguish any blazes caused by incendiary bombs.

In March of 1942, the command of US Naval Forces, Europe (COMNAVEUR) was created. Its war mission was to report Allied intelligence and research data to the Navy

Department. COMNAVEUR shared the building with the Army and the State Department Public Relations Division.

General Dwight D. Eisenhower, the Commander in Chief, Allied Forces, maintained his headquarters at the US Navy Headquarters building from June to November of 1942. In November the Allies staged the initial invasion of North Africa. The invasion's success was credited in good part to the constant stream of hydrographic and weather intelligence the Allied Forces received from commands in the London headquarters.

Eisenhower in his role as the Supreme Allied Commander, Allied Expeditionary Force, returned to the Second Deck (US Navy speak for 'floor. Note: Even in a landlocked US NAvy buildings 'deck' is the favored term) of the Navy building in January 1944 and left in March when his headquarters moved to the London suburb of Bushy Park.

Generals George S. Patton and Omar Bradley also worked in the building prior to the June invasion of France. Much of the planning and preparation for the cross-channel invasion of Normandy originated from commands headquartered at 20 Grosvenor Square. Among these commands was Task Force 122 headed by Rear Admiral Alan G. Kirk, a former Naval Attaché to London. Kirk's task force was responsible for landing troops on the Normandy Beach on D-Day.

In 1949, nine years after Admiral Ghormley had moved in, the US Navy established its European headquarters at 20 Grosvenor Square. Construction crews revamped the building for the command's needs and added the 7 North Audley entrance with is small flanking cement eagles above the doorway .

Rear Admiral Robert D. Carney, who would later become Chief of Naval Operations, was named Commander in Chief, US Naval Forces, Eastern Atlantic and Mediterranean (CINCNELM) in December of 1950. Six months later he assumed additional duties as Commander in Chief, Allied Forces Southern Europe (CINCSOUTH) and his headquarters moved to Naples, Italy. In June 1952 the two commands were separated again. Admiral Carney stayed in Italy as CINCSOUTH while Admiral Jerauld Wright took over as CINCNELM in London. In February 1960 the present command of Commander-in-Chief, US Naval Forces Europe (CINCUSNAVEUR) was established.

As indicated post-WWII the US Navy continued to maintain a presence and then, with the Cold War, expand its London and UK-based operations. US Naval Activities, London, was established in 1951 and then became US Naval Support Activity in 1958 with an eventual redesignation as US Naval Activities United Kingdom (NAVACTUK). The entity came under COMAVACT UK (Commander, Naval Activities United Kingdom) in 1965. The command would eventually go on to become subordinate to Commander-in-Chief US Naval Forces Europe, which would also operate from the building.

In a minor bit of historical trivia it is understood that some of the equipment for Glen Miller's band was stored in rooms on the top floor at the US Navy Headquarters during WWII.

During WWII the building had been a temporary duty station for, among others, actor Douglas Fairbanks Jr. who served as a special naval observer from July to November 1942.

The entire Grosvenor Square was nicknamed 'EisenhowerPlatz' in joking reference to the presence of Supreme Allied Commander Europe and the sheer number of senior American officers in the area. As a US Navy release noted the military headquarters main VIP entrance was at 20 Grosvenor Square with a 'tradesman's entrance' at 7th North Audley Street, became the Headquarters for United States Naval Forces Europe during the Cold War. Over the years the entrance would become the main operational entrance for all visitor bar VVIPs. Unsurprisingly, the building would become known in local parlance as "7NA".

The US Navy's presence in the UK grew during between 1960-1967 with a US Naval Security Group stationed at Edzell, then the creation of a submarine base at Holy Loch in 1961, a new Naval Radio Station in Thurso and a Naval Aviation Weapons base at Machrihanish (all four of the aforementioned bases are located in Scotland) and a US Navy installation at St. Mawgan in Cornwall.

CINCUSNAVEUR's landlord is the Duke of Westminster, who owns the freehold (ground) of most of the property in the surrounding Mayfair area.

The lease for the main building at 7 North Audley Street was taken from the Duke of Westminster and cost the US Navy a so-called 'peppercorn rent' of £100 ($150) a year (*see lease image further on*). However, the lease, dated April 16, 1952, actually ran from November 28 1947 for 999 years. At the time the lease was signed the US made a "cash down payment" of £164,150 or about $460,000 in 1952 money.

At its post-WWII height there were approximately 1,140 US Navy activity staff and 361 civilian employees (1998) working in the UK with about 200 military personnel working at 7 North Audley Street facility with some of them also working in offices surrounding buildings. Commander of Naval Activities at one point was located at 17 Great Cumberland Street. By 1991 they had 836 personnel.

One of the key operations for the HQ was communications and, in 1991, the Command was picked up another command acryonym – the U.S. Naval Computer and Telecommunications Station (NAVCOMTELSTA), London.

NAVCOMTELSTA London was an Echelon 3 Command reporting directly to U.S. Naval Computer and Telecommunications Command (COMNAVCOMTELCOM), Washington, D.C. It's understood that by 1991 some 20,000 messages were passing through NAVCOMTELSAT each day. The operation was renamed as U.S. Naval Computer and Telecommunications Area Master Station (NCTAMS) on 1 January 1995 and then disestablished in London less than six months later only to become submissive to the new MasterStation in Naples, Italy, in its new role as Detachment London (NCTAMS MED DET). Two years later another realignment saw it become NCTAMS EURCENT DET London).

Over the years the US Navy expanded operations and took over the running and occupation of a number of former USAF bases including West Ruislip (in 1975), Eastcote DOE Complex and Daws Hill, High Wycombe as well as Blenheim Crescent (just across the railway tracks from the old West Ruislip base) in support of their activities.

The 7 North Audley (7NA) Headquarters building (with associated offices at Providence Court) remained as the jewel in the crown in the US Navy presence around London until the Reduction in Force (RIF) program that started in 2002.

Also falling under the US Command from about 1980 was the Headquarters Fleet Marine Force, Europe, which was a designate command. A vanguard staff of 40 US Marines was established in London and grew by 1990 to 180 Marines in 19 locations in Europe working in support of NATO or joint local national projects. The Command moved from London to Stuggart, Germany, in 1993.

For a period at least until the 1980s the US Navy also utilized offices in Keysign House (as it was) on Balderton Street perpendicular to the Selfridges department store building and its famous clock. US Navy personnel also worked in Providence Court. The 7NA building underwent a series of changes and improvements over the years with its communications role having a particular impact on infrastructure (with raised floors and sub floor cabling). Several dish antenna were erected on the building in April 1986 and in May 1994 the USN added in a several roof-mounted satellite communication dishes with cover domes.

At the same time the US Navy changed its front entrance configuration with three new high security bullet and blast-resistant entrance doors. Additional refrigeration and cooling was installed in 1999 and 2000 in the basement and first floors of the building possibly to help cool communications equipment but also as part of the installation of cold storage for a Navy Exchange shopping facility in the basement.

In March 2002 an application was made for security netting, cameras and grills for the windows along North Audley Street and additional cameras in the next door offices at Providence Court). An emergency access staircase was also installed in the rear of the main building in 2003.

A tour of the vacated building by the author in 2009 found a fascinating mishmash of a structure. Past the security entrance on the ground floor the reception opened out to a main hallway running perpendicular. Along it the Barber Shop had been charging $6.75 for a regular hair cut, $5.00 for a shampoo and $4.50 for a beard trim with a mustache trim at $1.75. A Ladies Hair Cut was $11.50 and a Ladies Neck Trim was $6.50.

Nearby a countered area was identified with a sign in blue and red lettering reading *US Navy Post Office*. Behind the counter the heavy-duty, combination-locked door to a safe that previously stored Registered Mail and stamps (and was framed in black and yellow safety tape) was left wide open but was empty.

The small Naval Exchange (NEX) downstairs was kitted out in de-rigueur mirrored columns with a bank of refrigerated doors. It had closed on 2 July 2005 with operations moving to the US Navy base at West Ruislip.

Aside from the entrance, that deck included a Class VI store (for liquor sales) and the ITT (Information Tourism & Travel) office.

The Second and Third Decks combined both offices and operational facilities. In recent years the Fourth Deck was host to a bastion of acronymed organizations including NCTAMS MED DET. London, AIS Maintenance, COMM Centre, the Naval Criminal Investigation Service (NCIS), NLSO BRANCH, GCCS-JMCIS, CMC-DAPA and CNE.

Also on the Fourth Deck was, perhaps, the most interesting and, in later years, possibly the most historical part of the building. Part of the floor was given over to 'Admiral Country'. This was VIP space and included Admiral's Quarters featuring several reception rooms, a large kitchen, bedrooms and bathroom suite.

The rounded wood hallway frame alcoves may have been the oldest part of any original US Navy structure. In the Admiral's reception room an old fireplace surround had already been removed in preparation for demolition. Everything – even the toilets done in white floor tiles and pink-tinged sandstone with a pink plastic curtain and the country-style kitchen area — seemed to be frozen in some sort of 1960s design time-warp.

Fifth Deck apparently included PSD London, SATO Travel, N1 Manpower, Personnel & Training, Fleet Morale Recreation and Welfare (MWR), the Navy Passenger Transportation Office (NAVPTO) and Passenger Control Point.

Sixth Deck at the top of the building was home to organizations and units with designations such as: N4, N9, N17 and N19 as well as the O22 Fleet Medical Centre and the Education Centre. N4 was thought to be Support Logistics located in Room 6S03.

Intriguingly some of the windows on the Sixth Deck had internal shutter panels – literally sheets of metal that could be opened or shut with turning catch locks.

The NCTAM Eurcentre Detachement London room with its special London Bridge over an Anchor logo and its 'explosion hazard – charging batteries present' warning sign, lead to a room that had obviously once been humming with computers and communications equipment. By the time of the visit it just featured a crop of cut wires sprouting from holes in the raised floor and a cassette-paneled switching station and an aquamarine-colored executive chair.

NCTAMS had been badged on 1 January 1995. Additionally, 7 North Audley also hosted the NCTAMS Mediterranean Detachment NCTAMS MED DET London was created on 1 July 1995. Both operations had morphed into the NCTAMS Euro Centre (NCTAMS EURCENT).

Another room on one of the lower decks was occupied by the US Marines complete with a white board filled that, on the 2009 visit by the author, was still covered with commands and exhortations in red and black dry marker:

SITUATION REPORT: SALUTE

S – SIZE

A – ACTIVITY

L – LOCATION

U – UNIT

T – TIME

E – EQUIPMENT

Another room – possibly one called the DCOS Intelligence N2 space, contained an entire wall map centered on the Mediterranean. This appeared to be a briefing room – but there was also said to be a briefing 'theater' within the building that the author did not see.

Here in N2 also was space for what might have been at one time the WWDMS, possibly a modified Honeywell computer system that formed part of the US Military's worldwide Military Command & Control System (WWMCCS).

The whole building was a bit of a rabbit warren with in some cases, stairways that went up a few steps to dead end with brick walls! The basement area seemed to include a complex heating system and various service pumps that seemed to service not just the US Navy HQ but several adjacent buildings around Grosvenor Square.

One occupant of the building in the latter US Navy days recalls hearing repeated rumors that the Navy was shutting down its London Headquarters.

"None of us believe they were true," he said. "There had been a large model of a US Navy ship on the Ground Floor and I was convinced that as long as that model stayed the headquarters was staying. So the day it went..."

(For the record the model was either the *USS Constitution* or the *USS Constellation* — both of which were moved to – and still are in – the US Navy headquarters buildings in Naples, Italy. The *Constitution* had been made by Commander R.F. Stokes-Rees, Royal Navy and presented to the US Naval Attache in London on 2 April 1953 as "as a symbol of respect and affection of The Royal Navy for their comrades in arms of The United States Navy during two World Wars."

Following the military drawdown the 179,000 sq ft building was sold in May 2007 for £250 million to a consortium, GSL, headed by noted restaurant operator and property developer Richard Caring for expected conversion into 41 luxury apartments under various codenames including 'Project Navy' and 'Project Eagle'.

By 2012 after a series of funding renegotiations (including with the Irish NAMA bank which had take on some of the loan funding of Allied Irish Bank to the project following the collapse of the Irish economy) the consortium had reduced the number of proposed apartments to 31 with basement parking. It had been suggested that the reduced number of apartments might allow for more American-style horizontal living with apartments possibly ranging in price from £5 million to an expected £75+ million for a 12,000 sq ft penthouse. The lower floor of the building had been covered in hoarding.

In December 2012 Caring's consortium put the building back on the market. In April 2013 it was confirmed by *Property Week* that the Abu Dhabi Investment Council in partnership with Finchatton had purchased it to create 27 apartments. In April 2014 a construction worker was killed when a digger plunged through the second floor.

It was interesting to note that as of February 2013 the former residents of 7NA had revisited their old building and left graffiti – 'USN 1993-95' but respectfully done so via finger-in-the-dust writing on the lower building windows.

The US Naval Headquarters in London as seen from its Grosvenor Square Entrance (above). The VIP entrance with its US Navy logo floor was closed up sometime in the 1980s and the 'foyer' was used as a conference room. The original floor can be seen in the entrance from Grosvenor Square (below left). The entrance on North Audley Street was substantially enhanced over the years and included a large mobile x-ray machine (left behind when the US Navy departed) and heavy-duty doors. Even before 9/11 heavy duty blast net curtains had been fitted to the lower floors of the building.

The USNAVEUR tiled floor in the main North Audley Street Entrance (centre) and the US Navy's Barber Shop space (right) as they were in Sept. 2009. By the time the facility closed a regular haircut cost $6.75, a Shampoo was $5.00, a Beard Trim was $4.50 and a Mustache Trim was $1.75. A Ladies Hair Cut was $11.50 and a Ladies Neck Trim was $6.50.

Like most bases the US Navy building has its own US Post Office facility (above left) complete with walk-in safe. In the basement of the building was a small Navy Exchange & Commissary facility (above right) known as NEX London. It closed on 2 July 2005.

Above Left – the heart of US Navy London operations appeared to be in the 2nd and 3rd 'decks' where much of the mainframe computer and computer facility seemed to be located on as part of NCTAMS EURCENT. Above Right – severed cable bushels poke through the raised floors where on which computers and telecoms equipment were once placed. Below: Admiral Country – a hallway and the kitchen.

Admiral Country was extensive – the Fourth 'Deck' of the US Navy Headquarters included the quarters for the senior officer – usually an Admiral, though the Admiral would seldom live there except during possible crises. The quarters included reception room, bathroom and a kitchen – but the whole thing seemed caught in a bit of a time-warp that, visually, seemed to transported a visitor to the 1960s.

View from the top – looking down to the back of 7 North Audley Street (left) and satellite dishes on the building rooftop (right).

Above: The US Navy Headquarters from its North Audley Street aspect in 2013. Above Right: The main entrance for 7NA as the HQ was known.

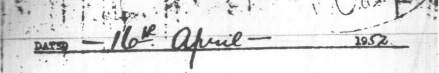

DATED — 16th April — 1952

THE MOST NOBLE THE DUKE OF WESTMINSTER D.S.O.

- to -

THE UNITED STATES OF AMERICA

Lease

- of -

No. 20 GROSVENOR SQUARE
Nos. 1, 2, 3, 4, 5, 6 and 7 NORTH AUDLEY STREET and
No. 20 PROVIDENCE COURT, LONDON.

FROM :   20th November, 1947
TERM OF YEARS :   999

EXPIRES :   28th November, 2946

R E N T: £100 per annum
CAPITAL CASH PAYMENT: £164,150.

(N.B. Notice is to be given of every assignment of this
Lease, whether by deed or not. See within).

The main page of the 999-year lease for the US Navy Headquarters at 20 Grosvenor Square 1-7 South
Audley Street and Providence Court. The lease commenced from 1947 and would have run until 2946
for a capital cash payment of £164,150 and rent of £100 per annum.

RAF UXBRIDGE (AAF - 409 IN WWII)

Few Americans – even servicemen and women – may ever have heard of a base called RAF Uxbridge.

And that's just way everyone – particularly those US Military personnel stationed there – seemed to like it. No wonder it earned the moniker: "The best-kept secret in the Air Force".

The 110-acre base, no more than few miles away from West Ruislip, South Ruislip and Eastcote and standing between them and Heathrow Airport was primarily home to No. 11 Group of Britain's Royal Air Force and, as such, was the one of the prime RAF Battle of Britain defense coordination bases during WWII with responsibility for defending the entire Southeast of England from the German Luftwaffe.

However, the base had history going back to WWI (indeed the base was purchased in 1915 – nearly three years before the Royal Air Force was formed). It would be the original training base for T.E. Lawrence (aka Lawrence of Arabia) in 1922 and would be the base at which acclaimed British fighter ace Douglas Bader would have a hospital stay to be fitted for artificial legs in 1932 after losing both legs in a plane crash in 1931. Bader would go on to participate in the Battle of Britain.

And it was that same battle that cemented the base's history due to the crucial command role it would play.

RAF Uxbridge's famed Battle of Britain bunker, built in 1938, has been preserved and is able to be visited to this day. It was on 16 August 1940 where Prime Minister Winston Churchill watched the plotting table as Spitfire and other pilots took on wave after wave of attacking German aircraft at the apogee of the Battle of Britain.

As Churchill emerged from the bunker he commanded his chief military assistant, Major General Hasting Ismay: "Don't speak to me – I have never been so moved." After a silence said to have lasted several minutes he then spoke the immortal words: "Never in the field of human conflict has so much been owed by so many to so few." He would repeat that comment four days later in the House of Commons and it would go down as one of the great quotes of all time.

US General (and later US President) Dwight D. Eisenhower also made several visits to the bunker in his rule as Supreme Commander of the Allied Forces Europe.

Other American links to the base also go back to WWII when RAF Uxbridge was home to Det. A HQ and HQ Sq. 9th USAAF. Additionally, a detachment of the 417th Night Fighter Squadron staff (there was no airfield there) was briefly deployed at RAF Uxbridge from May 14 – June 10 1943. For a period in the 1950s and 1960s it was home – literally – to the commander of the USAF Third Air Force. From his officer's residence one commander, USAF General John H. Bell, used to regularly run around the base keeping fit. For many years until the late 1980s a small contingent of homes located up near Vine Lane side of the base continued to provide officers quality housing for US military personnel.

What few people know about RAF Uxbridge is that there was a second command bunker located nearby the Battle of Britain one and that, in post-WWII years, the Americans occupied it.

The responsibility of doing so primarily fell to the USAF 2119th Communications Squadron activated on 2 June 1979 at RAF Uxbridge. Prior to that it had just been a Det. 1 of the USAF Communications Squadron located at RAF Croughton in Oxfordshire.

For a while the 2119th had its administrative offices about a half-mile across base on the first floor (2nd floor US) of the former RAF base hospital building. The USAF's Office of Special Investigations (OSI) unit – effectively the USAF's crime investigators – was also located for a period above the hospital.

Buck Pennington, Chief of Plans and Programs for the 2119th Comms Squadron from June 1980 until June 1983, recalls spending the "best three years of my 22-year career at RAF Uxbridge".

The 'business end' of the 2119th' was located in a largely unknown operations bunker on the base. However the operation necessitated the installation of a large microwave radio antenna. The curved roof bunker building housed the military communications switch and the 2119th's technical control facility.

"In those days the 2119th also had an adjacent site lounge called, strangely enough, 'The Site Lounge,'" Pennington says. "It was in a Quonset hut and could hold maybe 25 people. But I did see the odd occasion where there were at least 50 people in there crammed in cheek-to-jowl. It was really a great place."

In about 1985 the USAF named a single-storey off duty social building next to the antenna and bunker as "The Pub".

The 2119th became an independent communications squadron in 1980. Pennington recalls the struggle the unit had to get a proper logo with his office helping manage the process...or attempting to.

"The original design the unit submitted in mid-1981 was disapproved by the Air Force Office of Heraldry," he says. "We (The Squadron) were sent back to the drawing board sometime around late 1981 or early 1982. The squadron was still without an emblem when I was reassigned in 1983."

Eventually a unit logo with a griffin at the centre was agreed and a USAF commissioned a brown painted and silver-highlighted three-foot high wooden outline griffin was placed high up on the side of the bunker and flanked by silver painted wood stars. Today that griffin-shaped sign, which was rescued by the author and one of his daughters, forms part of the Battle of Britain Museum exhibition at a building above the bunker where they now have a special room dedicated to the RAF Uxbridge base.

The 2119th had two primary missions: Operating one of two 'switching stations' for AUTOVON (the military telephone network) in the UK and managing long-haul communications (microwave) for all USAF bases in southeast England.

The 2119th also had responsibility for remote manned operating locations and ensuring the were operable or sending out crews to maintain them.

Originally there were eight antenna in the chain in South Eastern England, however by early 1980 there were just two: OL-A at Botley Hill Farm and OL-C at Swingate, near Dover. OL-B (believed to be at RAF Christmas Common – an antennae site in Oxfordshire) was decommissioned in late 1979.

By then the base formed part of the Defense Information Systems Network (DISN) to provide both non-secure and secure voice and data, video teleconferencing and data facilities for the Department of Defense including long-distance communications.

The network was given primary delivery status during peacetime and wartime operations with the exception of actual nuclear attack) and could offer 'flash' overrides for its users. The operation also supported the Defense RED Switch Network (DRSN) and Secret IP Router Network (SIPRNet) among other communications systems.

In later years the US military buildings at RAF Uxbridge comprised:

Bldg. No.

4 . . . . USAF Control Centre / ITT Northcomm
Federal Services International Corp.

6 . . . . Antenna Tower

68 . . . . Standby Generator.

83 . . . . Uniter Building

Building 4, or "The Forgotten Bunker" as it has sometimes been called as it was over-shadowed by its more famous Battle of Britain predecessor, was also located next to Vine Lane on the northeast side of the base and almost opposite the entrance to the ACS International School on the other side of the fence and across the street

The bunker was built during the 1950s when the British military was growing increas-ingly concerned about enemy aircraft armed with atomic bombs and commenced an upgrade of the Anti Aircraft (Ack-Ack) guns as they were known.

"In each gun-defended area, purpose-built two-level bunkers were constructed to a standard design to control gunfire and plot enemy aircraft," says Sgt, Stew Thorpe who, as one of the RAF's base closure staff, researched the building history.

Each of the UK's The Anti Aircraft Operations Rooms (AAOR), as they became known, were linked to the RAF Sector operations rooms that also contained Army liaison staff.

"The RAF Uxbridge AAOR was completed around 1954-55 to control the London West gun-defended area," Thorpe said. "It was designed to withstand the most power-ful conventional bombs but not against a nuclear strike. It was very much a traditional World War II-designed operations room with a gallery overlooking the main plotting table."

The long-term plan was to replace guns on many sites with surface-guided missiles such as the Bloodhound. However in 1955, before all the AAORs had been fully com-pleted, missile technology brought about the abolishment of Anti-Aircraft Command.

Shortly after, in 1957-58, all the RAF Sector operations centres were shut with many becoming Government and Civil Defense control centres. During 1959 both bunkers at Uxbridge were considered as candidate locations for the launch coordination centre for the Blue Streak intercontinental ballistic missile then under development. The idea would have been to link the bunker to other RAF sites at Duxford (Cambridgeshire) and Odiham (Hampshire) where there would be underground silos. The cancellation of the entire Blue Streak program (largely due to its prohibitive cost) ended the plan.

So in 1962 the Uxbridge AAOR bunker was first taken over by the USAF with the USAF 2119th Communications Squadron becoming the primary resident.

The 2119th evolved into the 703rd Communications Squadron. The 703rd which had about 175 personnel, ran communications and communication systems for seven head-quarters and other groups at 17 sites around the South of England. By the time it inacti-vated on 25 June 1996 there were just 50 personnel on hand for the ceremony.

'The best kept secret' tag-line for the 703rd appeared not only in a headline about in the *Stars & Stripes* newspaper but also in a limited edition commemorative poster for the 703rd specially commissioned by the unit from leading UK-based military artist Keith Hill. (*See image at the end of the section.*)

The Air Force Office of Special Investigations (AFOSI) UK District Headquarters was known as District 62. There had been either a separate or co-located OSI unit (AFOSI/FldInvS.51/OL-C) at West Ruislip for some portion of time, too.

Interestingly an April 2010 article by Anthony Kimery in *HS Today,* a Homeland Security Insight & Analysis publication (www.hstoday.us), indicated that AFOSI Uxbridge had provided intelligence in relation to possible Iranian suspects in relation to the 1995 Oklahoma City bombing and cited recollected information from 1987 as having formed part of that information. It's unclear just what that was.

Today an OSI office, believed to be 51 FIS OL-C, still apparently operates in the London area along with Det. 514 at RAF Alconbury (Cambridgeshire), Det 514 OL-A at Fairford (Gloucestershire) and Det. 512 at RAF Lakenheath (Suffolk) and Det. 51S OI-A at Menwith Hill in (Yorkshire).

Under the American tenure the forgotten bunker was extensively developed including the construction of offices built above and next to it.

After the 703rd disbanded and the USAF departed the base on 21 June 1996 at which time the bunker and the general facility was declared to be surplus to needs.

A raised power line rack, covered by concrete capstones ran from the bunker to an antenna (one of two major ones at RAF Uxbridge) about 80 feet away. The American antenna, which was festooned with *WARNING – Controlled Area* signs and was probably installed first in the early 1960s as part of the USAF's microwave network.

Actually the US military could be said to have had two bunkers on the base. About 150 meters away (and even closer to the Battle of Britain bunker) was what was Bldg 83, better known as the Uniter Building. It was, outwardly, the most 'combat-looking' of the American (and, indeed, any) of the buildings at RAF Uxbridge. Sitting above ground in a fenced-in compound just 60 meters from its famous RAF underground relative, the windowless Uniter building, built in the late 1980s by the British, essentially comprised great surface areas of reinforced concrete walls about 2 feet thick.

In two locations around the building there were heavy-duty blast walls in front of the internally locking steel doors. The entire facility was designed so that inhabitants could reportedly continue operating for three months without having to come out.

The Uniter (essentially meaning 'united system') building is one of about 14 similar bunkers found around the UK that were meant to create a 'survivable fixed communication network for the RAF as part of the UK's air defense system.

RAF officials closing the base didn't know what the USAF used the facility for but it was thought to have been handed back in 1996. On a tour of the building in 2010 just before the RAF base closed the author found the main room devoid of everything bar a file drawer safe and some other furniture. However the side rooms were packed with power generation equipment – a substantial amount probably worth in the hundreds of thousands of dollars. The Uniter Building at RAF Uxbridge also had two fuel storage tanks of 55,000 and 2,000 liters respectively.

In late 2011 or early 2012, long after the building had been closed, part of the Uniter Building was part covered in camouflage paint, apparently in order to create a set background for use in a film.

As with matters at RAF West Drayton (*see separate entry*) not confirmed by this author there were believed to have been some USAF personnel attached to a joint unit from the Military Air Traffic Office (MATO) headquartered at RAF Uxbridge.

Among the high-profile investigations the unit may have been discreetly involved itself in either investigating, monitoring or exchanging information was the famous Rendlesham Forest UFO incident in December 1980 – a sighting of something unusual in the woods close to the then USAF Base at RAF Bentwaters in Suffolk that was enough to entice both USAF Security Police and the then Deputy Base Commander out to search the area. It has become the UK's most celebrated UFO incident and has remained the source of endless speculation, debate and newspaper headlines ever since.

By the time RAF Uxbridge was closed on 31 March 2010, Building 4, the 'Forgotten Bunker' that had become a key part of USAF operation, was fulfilling its moniker. During a tour the author found flooding in the lower level of the bunker. It had been visited periodically as part of a care and maintenance regime managed by Defence Estates/ Defence Infrastructure Organization.

As is the case with abandoned buildings – particularly those underground – visits were made harder and harder as Health & Safety concerns would come into play with concerns over either the air quality, asbestos, damp and even structural safety.

RAF Uxbridge was turned it over to VSM Estates for housing and other civilian developments as part of project MoDEL. VSM announced plans to develop the property through parceled deals to property developers as follows:

- 1,340 new homes, including affordable housing.

- 20,000 sq m of mixed-use commercial development, including new retail and offices.

- The inclusion of a site for a new 1,200-seat theatre with ancillary café.

- The inclusion of sites for warden-assisted elderly accommodation, a GP surgery, a primary school and a hotel.

- 45 acres of open space, including a public park alongside the River Pinn, part of which is in the Green Belt.

The former base is now being residentially rebranded as St. Andrew's Park. Demolition work had started in 2012. By November 2012 demolition company Cuddy started demolishing the American part of the site and by 24 December they had finished demolishing the 'forgotten bunker' to ground level and the antenna almost to it base. The Uniter Building remained as of Summer 2013.

It's interesting to note that in late 2012 a local visitor to the Battle of Britain Bunker took a picture that appeared to have an apparition in it – possibly an American solider. The matter was covered by the *Uxbridge Gazette* on 2 November 2013.

RAF Uxbridge had also been home to the RAF's Music Services and the Queen's Colour Squadron, The RAF has preserved the famous Battle of Britain Bunker which was also used to control air support for the evacuation of Dunkirk, France by the Allies and then later in the war, the D-Day landings and it is available for public visits. In early 2014 four new four/five-bedroom family homes, which had been built on the site of the former bunker by residential developer Charles Church, went on sale. They formed part of the developer's larger Crescent Collection dwellings built on part of the former base. Prices for those homes started at £599,950 (about $994,000).

Above: The Uniter Building (Bldg. 83) at RAF Uxbridge just before the MOD handed the base over to developers in 2010. Below: The bunker from the side in 2010 prior to it being 'part camouflaged' for a film or television show setting. In inside comprised one large but now mostly empty room along with other rooms packed with generators and electrical control equipment. Other Uniter buildings around the UK have been sold off – this one faces demolition.

Demolition of the American part of RAF Uxbridge took place starting in late 2012. Top: The key US operational area with the bunker and the antenna behind. Between then and March 2013 the substantial bunker and the antenna were demolished. The last picture shows a demolition machine looking more like a dragon completing work on the basement of the bunker (with below ground foundations just able to be seen to the right).

The limited edition poster commemorating the USAF operation at RAF Uxbridge was created by British aviation artist Keith Hill. Courtesy of Keith Hill.

RAF WEST RUISLIP, MIDDLESEX

RAF West Ruislip, the 21-acre support base for the USAF Headquarters at South Ruislip originally opened in 1955 on a former RAF supply and maintenance base just a couple of miles away from that HQ.

However West Ruislip had been a base long before that thanks to American involvement – in the form of US Army Civil Engineers.

It was built from 1914 to 1917 with some of the original massive 'stores' sheds being built in a contract carried out for Britain's Air Ministry by the Americans for a storage shed-dominated base with railway sidings servicing 13 sheds offering at least 400,000 sq ft of storage for everything from more than 2,000 aircraft engines to propellers to the wing fabric for aircraft. The base also had to accommodate up to 1,600, mostly female, personnel.

The sheds were located to one side of the railway lines (then for the Great Western Railways, the Great Central Railways and the Metropolitan Line Underground). On the other side went housing, support and what would become the RAF Records Offices.

In about 1939 the RAF ramped up operations in equipment, supply and repair with just over 1,800 personnel working on the site by the end of 1944. In the period immediately following WWII, the site turned out radio and radar vehicles. The USAF would also stake a partial claim to WWII base history by utilizing one of the massive sheds for its own logistics supply storage.

It was in 1948 that London Underground's Central Line was extended from Greenford to Northolt, South Ruislip, Ruilslip Gardens and West Ruislip (For Ickenham) as it the end of the line station was once named.

West Ruislip was handed over to the USAF Third Air Force in a formal ceremony on 1 December 1955 that was covered by British television. The USAF took full control of the base in December 1957 as the 7553rd Air Base Squadron. Within five years Third Air Force Headquarters at USAF South Ruislip (just a couple of miles away) would relocate personnel from the 7500 Air Base Group to the site as part of a consolidation of operations at nearby Denham, as well as at Bushy Park in Teddington and Bovingdon in Oxfordshire.

West Ruislip was effectively a supply and transport operations base and, despite the arrival of the Americans, was known locally for a long time as 4MU (Maintenance Unit) after the previous occupying RAF operation.

The increased USAF presence manifested itself through the construction of a Chapel of Faith and an elementary school to support personnel and families.

The West Ruislip base effectively comprised five separate elements – some separated by British public roads. In later years included the main 'technical base' with its warehouses and PX/Commissary, Post Office, Bank Autoport, Theatre and Club and Childcare facility and snack bar. Second was the base's Chapel of Faith and a large parking area.

The third area was West Ruislip Elementary School site and several tennis courts. The fourth 'zone' was the sports field with single level Morale Recreation & Welfare (MWR) gym building (featuring a weight and training room and other facilities that had opened in 1962). The MWR facility also had a barbeque pit out in front and the baseball field and dugout (refurbished in the mid-to-late 1990s) beyond that. Next to it was a racketball court that had been built in the late 1980s.

The fifth element was a base housing area of some 80 homes that would be incorporated into in what would later become known as Brackenberry Village (a much larger civilian residential site built on former RAF base land divested years before). The remaining US military homes had been renovated in a project starting in September 2003 and as the base closed down one of the houses was given over to the Child Development Centre operations around 2004/5 to support the remaining military community.

In its later years, aside from the one large remaining storage shed, perhaps the most eye-catching base building was the Chapel of Faith. Designed by Brandt O'Dell in the late 1950s it was built by the USAF and opened in the early 1960s. The 350-seat chapel also had adjacent flanking (and linked) function and chapel office wings.

The Chapel building's internal laminated wood arches were a distinctive feature. They were designed by Rainham Timber Engineering Company Ltd. The campanile (bell tower) appeared as if it were topped by a candle flame – a large 11-foot high parabolic aluminum-frame mast. There were no real bells in it the tower but speakers from a stereo system controlled from the Chaplain's office simulated them.

The US Military Vicar General's office says that its Catholic baptismal records for 'RAF Ruislip' (probably meaning USAF South Ruislip as well) start on 4 February' 1951 when US military Priests recorded for the year:

Baptismal Records 127
Marriage 16
Confirmation 10
First Communion 7
Civilian Baptismal 2
Profession of Faith 2

Over the years the figures will have waxed and then certainly waned as the US military community around London grew and then declined. With the relocation of Third Air Force Headquarters to RAF Mildenhall in Suffolk in 1972 operations at RAF West Ruislip were downsized until the base was transferred to the US Navy in support of its US Navy Headquarters in London in 1975. The Chapel is thought to have closed briefly in about the mid-1970s before reopening to serve the US Navy community.

For many years (certainly until 2001) London Transport, which ran the Underground Lines produced a continuing journey map with the Words "R.A.F. West Ruislip (U.S. Airforce Barracks)" printed over the location of the site – even long after the US Navy was in charge.

And even until around 2007 the nearby London Underground Central Line station at West Ruislip offered maps showing an area marked "Barracks" and the "West Ruislip Elementary School" with an outline of the school buildings.

It's believed that the main entrance to the RAF West Ruislip base was re-sited from Ickenham High Road to Aylsham Drive in the 1980s to reduce the problem of cars queuing to get into the base blocking the main road. Aylsham Drive was a side road that effectively bisected the base chapel and main car park from the main base.

The old entrance was gated and, at about the same time, a 'short-cut' turnstile entrance for base pedestrians with a swipe pad was installed between it and the West Ruislip Underground station to save people walking a further three minutes around to the new main entrance. As of May 2013 that forest green-colored turnstile was still in place though chain locked.

Many of the 420-foot long x 120-foot wide sheds with their eye-catching truss supports were demolished in the 1970s and what had essentially been a massive storage base reduced in size over the years until just one of the aforementioned 13 large sheds remained (the one nearest Ickenham High Road).

When the US Navy took operational management of the base in the 1970s a small Naval Exchange (NEX) shop and Commissary was created within the shed units along with a snack area.

In the 1988 a contemporary Children's Welfare & Family Development Centre was created at a reported cost of $5 million.

In June 2004 the US Navy completed works to create a new car park and play area within the main base.

The base club went through a number of names changes over the years from The Crown & Anchor to the Liberty Club. Some personnel may also remember it as Club Chameleon or the Lizard Lounge and before that Centre Stage and before that The Flagship. At some point it had been also named the Constellation Room and then the Mavericks Bar (in the mid-1990s at least). Whatever the name it was at the West Ruilsip club that the rock group America was thought to have played some gigs about the time they were forming.

In the final few years the Liberty Centre, as the encompassing umbrella building was known, was run by the US Navy's Morale Welfare & Recreation (MWR) had a bar and a 48-seat cinema (it was known as Centre Stage Cinema when opened in about 1996 – and Chameleon by 2002 and The Liberty Theatre at the end.

The cinema was closed on Mondays but operated from about 11:00 until 20:00 hours on weekend and slightly later on Fridays and Saturdays with a seven-hour opening (noon to 1900) on Sundays.

A US Post Office operated from the buildings nearest Ickenham High Road fence-line as did a US Military contract bank and the Morale Recreation and Welfare (MWR) office. In a small building once occupied by the guardhouse at the original main gate and which dated from the early days of the RAF base, a barber shop was later located along with a souvenir shop. The main gas station with its two pumps was located almost immediately ahead on the left as one drove in. The pumps were removed by early 2007 one of the first bits of base infrastructure to be removed.

To the other side of the old main gate was a U-shaped complex of buildings comprising of the Navy Exchange Vehicle Repair and Service shop (originally the NAVEX and later the NEX (Navy Exchange)) Autoport for use by GIs who wanted to have their cars serviced and repaired (front end alignments ran from $50 in 1996) or do it themselves. One unusual feature of the base was a single 'normal' civilian home named Fairlight House. This was at the corner of Aylsham Road and Ickenham High Road and dated from 1914. The house, once home to the RAF Station Commander (most US bases had an RAF officer who liaised or oversaw property practicalities), was demolished as part of the clearance for the site that was purchased by retirement home developer McCarthy & Stone. As of late October 2013 the site remained empty.

The West Ruislip Chapel – and the one at High Wycombe – were also home for the USAF Protestant Choir which, encouraged by a US military Data Technician and his wife (Jim and Sharon Scott) became known as 'Reflections'. The choir, eventually comprising some 40 teenagers, cut its own album and produced 500 copies in the mid 1970s. The group was also credited with performing at least one track on *Essential Hymns* "O come, O Come, Emmanuel.* Choir members also performed several times at the Tower of

London and Ely Cathedral.

The US Navy held a courts martial of a female US Navy Petty Officer at the West Ruislip base in January 2006 – thought to have been the last one held there.

The final Chapel service, conducted by the US Navy Catholic priest Father (CDR.) Mike Zuffoletto, took place on 4 June 2006 (services for the Protestant community had ended in September 2005 – in earlier years the Chapel had at times enjoyed the services of four Protestant chaplains, a Rabbi and a Catholic chaplain all at the same time) with an End of Era Celebration dinner taking place the night before. Zuffoletto had been preceded over the years by among others Father Patrick Sheeran, Father Carroll Wheatley and Father Pete McGeory and Father John M. Gubbins, among numerous others.

"You can chose to look at this as an ending or as a new beginning," Cmdr. Zuffoletto said at the closing service attended by then British Bishop George Stack of the Diocese of Westminster (now Archbishop of Cardiff). "We have come a long way together as a community and as two nations. I hope you all look at this as a new beginning with new experiences be shared and new friends to be made as we all move on."

The Chapel was demolished in November 2007.

The Chapel was one of the more unique and regular outreach programs to the host nation local community in Ickenham and West Ruislip by the US military with British national able to attend Catholic services at least. After the 9/11 terrorist attacks non-military parishioners had to bring their passports with them for about 18 months).

But the West Ruislip base also reached out to community in other ways. Many locals will recall the loud 'booms' usually preceded the start of the annual 4th of July weekend celebrations at the base – with one and sometimes both days opened up to the local British community. In later years there would be fairly stringent search procedures and plenty of US military security police along with some British civilian police before visitors could pass through and experience a slice of American life with hot dogs, hamburgers and, in later years, Pan-Pacific/Asian fusion foods and drinks, music and entertainment including motorcycle stunt displays, bands, and of course the concluding fireworks show – often with a finale featuring a US flag created in fireworks.

In March 1994 the West Ruislip base was subject to an arson attack with fires started in five areas of the MWR gym building and surrounds. There was speculation at the time it was related to a stabbing incident earlier in the year involving an 18-year-old US dependent and a local teenager.

With the general drawdown of the US Military the closure of the base, which played a support role for what was in the 1980 and 1990s a peak of about 3,000 service personnel and their family members stationed around London and was, by mid-2005 down to just 1,800 Navy personnel, civilian workers and dependents, was announced for 30 June 2006.

On 6 September 2007, following a phased hand-over that saw the US Navy retain the gym and fields for use for a final 4th of July Party and also retain housing, West Ruislip base was handed back to the MOD in and became part of Project MoDEL (the joint venture between VSM Estates partners Vinci and St. Modwen) which aimed to take money raised from the sale of West Ruilsip, Eastcote and the RAF Headquarters base at Uxbridge to underwrite development of RAF Northolt, flightline base just a couple of miles away from the aforementioned bases.

For a while part of the base was rented. A large portion of the remaining shed that had been the commissary and NEX became a temporary two-year home to the London

Motor Museum owned by model-turned-car customizing specialist Xavier Elo with cars (some belonging to celebrity soccer players) parked where the shopping aisles had once been. Another part was rented out to a television company for storage of stage sets. Elo, who has starred in the TV show *MTV Slips*, relocated his museum and car customizing operation to nearby Hayes when his temporary lease ended.

In 2007 Cala Homes, a Scottish residential property developer, purchased the site to develop 415 homes and an 80-bed retirement home. It was guesstimated (as neither vendor or purchaser would confirm the price) the land was then worth approximately £3.5 – £4 million an acre (making a unconfirmed project value of circa £70-80 million). Cala Homes did confirm to the authoritative *Property Week* magazine at about that time that the "…gross development value, excluding the land price, will be £180 million".

Demolition was carried out by Careys with bulldozers moving in from 19 November 2007. CALA Homes started major construction works on the cleared site in early 2010 with plans to build 40 detached house, 85 townhouses, 7 linked detached houses, 143 flats and 140 affordable housing units, over five years, with work planned to start on site in Autumn 2008. Construction actually started in September 2009 with a projected completion date of 2013. As of September 2013 those homes appear complete.

Ickenham Park, as it has been called, is "a mixture of four-, five and six-bedroom family houses, mansion-style apartments and two-bedroom affordable apartments, all set in an established landscaped setting with many mature trees". It has been CALA's "Fastest-selling development in the country throughout most of 2011."

As part of the development there is supposed to be an 80-unit care home to be developed by McCarthy & Stone but work on this has yet to start at time of writing. Another are for Explore Living next to the main road has also yet to be developed as of October 2013.

The only untouched part of the base still left is the former elementary school. The building, laid out in the shape of an 'H', remains as does a bit of the playground with a map of the American states in it and, for a time, at least, a 'tetherball' pole. At least two time capsules planted by American students over the years remain buried in the grounds.

In 2008 there was a proposal to house students from the local Glebe primary school there and redevelop their site but it fell away following intense local opposition. Part of the school has been rented out to a child-care scheme since by owners London Borough of Hillingdon. The school building is occupied but there is little sign of infrastructure car around the building judging by the growth going on.

Privately purchased homes at Ickenham Park range in price from just under £250,000 for a two-bedroom apartment to more than £1 million ($1.55 million) for a 3,418 sq ft family home known as The Elgin in a part of the development called "The Chase".

Around 10 of the original Air Ministry marker stones which denoted the original base perimeter from 1914 (but had not been inset until 1928) can still be seen around some of the outer edges of the base to this day.

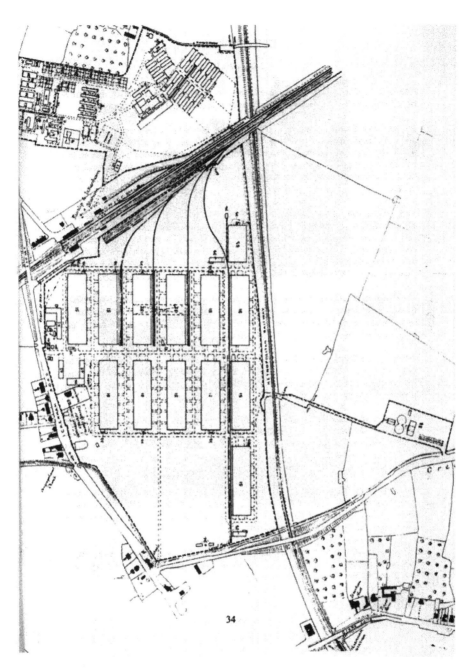

34

A map of RAF West Ruislip as it was in the immediate post-WWII years – with the giant storage sheds linked by railway spurs. By the 1980s all but one shed (the one to the far left) had been demolished. In later years the last remaining shed was used as by the US Navy for its Navy Exchange and, after, disestablishment, as a temporary home of the London Motor Museum. The buildings at the top (above the main railway lines) comprised the former RAF Records office – today part of it remains as the Blenheim Crescent base.

Above: A map of RAF West Ruislip effectively showing the base layout in parcels as it ran along the High Road, Ickenham. Note the baseball field at the top with the 'H' shaped elementary school just below it with two tennis courts denoted just to the right. The smaller units running around it were largely US military housing. To the right of the baseball pitch was the Chapel of Faith complex. Much of the area below is parking. Then below that and running down to the railway tracks is the main base with its massive former shed (listed as B101).

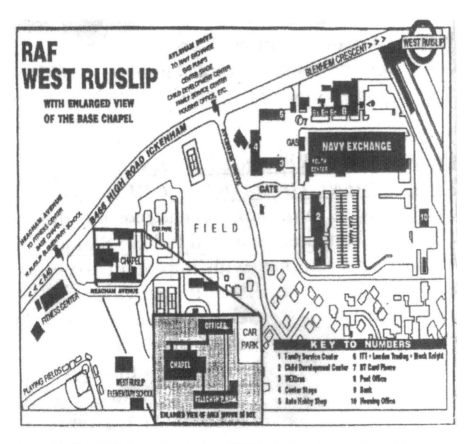

A map of the West Ruislip Base as it was in about 2005/6 just before it closed. Only one of the massive storage sheds remained. It is marked "Navy Exchange".

Above: The last remaining storage shed (see on the right of the road) just after the base closed in 2007. The buildings to the left held MWR offices, the Bank and Library in the last few years. At the very end of the road is the Central Line – one of the tube trains can be seen.

The Chapel of Faith at RAF West Ruislip on the day before the base was handed back to the MOD in September 2007. Note the concrete barriers at the front – put in as Force Protection measures in the wake of 9/11.

The London Motor Museum, run by Xavier Elo (pictured above left and right) became a unique but temporary letting at the base after it shut in 2006. The Museum with around 70 amazing cars (including a Batmobile and a Flintstones car) occupied the former NEX Commissary building until the Museum relocated to nearby Hayes when the base was sold.

Above, Left & Right: The candle-shaped campanile at the Chapel of Faith at West Ruislip with the moon directly atop and contrails (left) was always a striking architectural feature. Shots taken during afternoon of 3 June 2006 – the evening before the last service. Demolition work started on the base during the week of 21 November 2007.

Above Left and Right: November 2007 – Demolition experts Scudders in the midst of taking down the Chapel of Faith building.

By 2008 most of the West Ruislip base had been demolished. The site as it was in April 2009.

Some of the luxury homes on the former West Ruislip site now completed by CALA and as seen in the winter snow of January 2013. These are located approximately where the Morale Welfare & Recreation Gym Facility and the Chapel car park were located.

OTHER BASES, INSTALLATIONS & FACILITIES

There are numerous other bases, installations, offices and clubs in and around London that were populated or frequented by American forces over the years from WWI to the present day.

Only the best would do when, in WWI, **General John J. Pershing**, Commander of the United States Expeditionary Force to Europe, stopped off in London. Arriving in Liverpool, England in June 1917 aboard the steamer *Baltic,* Pershing and his cadre of staff travelled to London by train to be met Vice Admiral Sims, Commander of the US Naval Forces in Europe. Pershing and his officers took up suites at the **Savoy Hotel** and effectively hosted a four-day long receiving line – apart from the time when they were received themselves by the British monarch King George and Queen Mary at Buckingham Palace. Enlisted members of his party were given quarters in the barracks at the **Tower of London**. At the end of that time Pershing packed out and headed to France to pave the way for the Allied Expeditionary Force. Sim's own office grew to 1,200 (including 200 officers) at **Grosvenor Gardens,** a building between Buckingham Palace and Victoria Station.

The US Navy Headquarters Building in Grosvenor Gardens in World War I – believed to be on Armistice Day. The structure with its high entry steps still remains today. Photo courtesy US Naval History & Heritage Command via Robert Hanshew.

At Victoria Embankment next to the Thames River is the **Battle of Britain Monument.** It was unveiled by Prince Charles in front of 70 battle veterans on 18 September 2005. The names of the nine Americans who participated in the Battle can be found inscribed on one side..

At Westminster Abbey in Westminster SW1, lies the **Tomb of the Unknown Warrior.** The Unknown Warrior may be British, but in 1921 US General John J. Pershing conferred the Congressional Medal of Honor on the tomb on 17 October 1921. The medal is displayed close to the tomb.

At **St. Paul's Cathedral** is the **American Memorial Chapel**. The chapel, located behind the High Altar, is a commemoration to the 28,000 Americans who were killed on their way to – or while stationed in – the UK during WWII. Their names are recorded in the 500-page roll of honor that was presented by General Eisenhower in 1951. The roll is encased behind the high altar and Cathedral staff turn a page a day to reveal the names

In 1941 in the crypt of the Cathedral a memorial plaque was unveiled to Pilot Officer William Meade Lindsley Fiske III (pictured left, image courtesy Battle of Britain Monument) in the crypt on the southwest corner of the Cathedral's Nelson Chamber with the simple but powerful inscription: "An American citizen who died that England might live." The plaque was unveiled by Britain's Secretary of State for Air Sir Archibald Sinclair who said: "Here was a young man for whom life held much. Under no kind of compulsion he came to fight for Britain. He came and he fought and he died."

The **American Women's Hospital for Officers** was a 42-44 bed World War I installation located at **98-99 Lancaster Gate in London W2** (adjacent to the former Columbia Club site – *see standalone entry*). It comprised two buildings – one that had been loaned and the other which was rented. It was opened in 21 March 1917 by US Ambassador Walter H. Page and his wife.

The hospital was paid for by the **American Woman's War Relief Fund** (itself located at **31 Old Burlington Street, London W1** during World War I) and largely helped with the convalescence of British officers. The 13-strong staff in Lancaster Gate comprising a matron and nurses, were aided by American and British volunteers. The nurses while largely British, wore a "turned down soft collage slightly open at the neck" – a style preferred by American nurses of the time, according to the *British Journal of Nursing*. The hospital, along with a sister 250-bed hospital in the town of Paignton in Devon, was handed over to the American Red Cross in December 1917.

Rainbow Corner, The Red Cross Club was located at **23 Shaftsbury Avenue** in Piccadilly in WWII (the location was formerly a club known as Del Monico's).

Rainbow Corner was a 24-hour operation that served US service personnel from 11 November 1942 to 9 January 1946. The facility included a ballroom (dancing girls were known as the 'Rainbow Room Hostesses'). On the lower ground (basement) floor was a café known as 'Dunkers Den'.

During 1943 monthly GI usage statistics rose from 331,709 in July to 651,928 visits in November. They were catered for by around 200 paid – and up to 350 volunteer – staff.

The work done at the centre inspired a book *The Story of Rainbow Corner*. And it also featured the Churchill-inspired strapline – "Never had so few done so much for so many."

Rainbow Corner on Victory in Europe (VE) Day on 7 May 1945. Photo courtesy American Red Cross.

Rainbow Corner (like Douglas House in later years) was a 'girl magnet' of a building with women of all classes hanging around in the hopes of meeting a GI – enlisted or officer. The local streets and alleyways and door recesses close to the club and around Piccadilly Circus became a haven for quick sexual encounters (which quickly became known for obvious reasons as 'wall jobs').

A wartime report on the matter, which was only released by Britain's National Archives at Kew in 2005, labeled the more 'professional' of the women around the area as the "Piccadilly Commandos" (there were also another set known as the "Hyde Park Rangers"). The report was prepared by a Metropolitan Police Superintendent believed to have been James Arthur Cole, who had studied the issue of women congregating around areas around other US Force Clubs at **100 Piccadilly** and around the **Washington Club** at nearby **Curzon Street.** The US Military Police with their distinctive white helmets were based at **101 Piccadilly.**

If the Rainbow Club's annual birthday party bash was a notable social event then Victory in Europe (VE) Day probably put that into the shade by a mile. Regardless on both occasions in excess of 70,000 people were thought to have come through the doors in a 24-hour period.

Close to the Rainbow Club were **Frisco's International** in **Piccadilly** and **Bouilla-baisse** in **New Compton Street**. As tensions and even fights grew among white GIs over social mixing between black GIs and white British girls these two clubs were essentially given (American-driven) tacit British approval for use by black GIs.

Gen. James H. Doolittle, set up temporary quarters in **Claridge's** Hotel in Mayfair while he held meetings at Norfolk House (*see further on*) in setting up the Twelfth Air Force in September 1942. Among regular evening guests was Gen. George Patton (who apparently came armed with his famous pearl handled pistols and hunting knife).

General Eisenhower had two rooms at the **Dorchester Hotel** on Park Lane following a brief tenure at the nearby Claridge's. The Dorchester's Eisenhower Suite, as it is known, was on the floor right above the entrance. No less a personage than British Prime Minister Winston Churchill ordered that a special dividing wall be put on the balcony to give Eisenhower additional privacy. The wall remained until it was removed in about 2007 when the suite was refurbished. A faint vertical trace line indicating where it was placed can still be seen by the eagle-eyed today.

Nearby the **Grosvenor House Hotel** also on Park Lane, was doing its bit by converting its famous ballroom (known as The Great Room) into the US Officers Mess between 1943-1945. The hotel, which had been protected with some 10,000 sandbags, is said to have served some 5.5 million meals during the period.

A popular Lutz postcard showing US personnel and guests eating at the Officers Mess at the Grosvenor House. Despite attempts copyright currently unable to be determined. Original postcard from the author's collection.

The **Victory Ex – Services Club**, originally on **Bedford Row** in Holborn, listed General Eisenhower amongst its honorary patrons. It had been in a building that was used by US Forces in WWII. Today it is known as the Victory Services Club and is located on Seymour Street near Marble Arch

By 1942 American GIs were able to bunk in London at a few locations including the **American Eagle Club Dormitory** located at **16 Princes Gardens** in South Kensington. The club – a former private residence – could handle up to 200 residents a night. It's thought that the same location was later used for a US Department of Defense Dependents (DODDS) elementary school for a few years in the 1950s.

Kennington Barracks in London SE11 was, for many years, home to the USMC. Known as Building 306 by the CNAUK it occupied .11 hectares and was located on **Reedworth Street** and **Kempsford Street**. It comprised 24 x 2-bedroom 'flats' (apartments). Morale Welfare & Recreation (MWR) managed a fitness centre created out of two of the building's former car garages. The US Navy had occupied the building since 1980 but it was transferred back to the Ministry of Defence in 2004/5.

Among other locations for Marine Barracks London as the USMC was better known in its encompassing UK presence since about 1945 at least), kept living quarters at **St. Johns Wood** near **Swiss Cottage** in London NW3.

Marine Barracks London was redesignated Marine Corps Security Force Company London on 1 June 1987. US Marines also had their living quarters a period at **32 Elvaston Place** just of the Gloucester Road in London SW7. Today it is the location for the High Commissioner for Mauritius.

Today MBL is located in the London Borough of Westminster.

The Dorchester Hotel main entrance bedecked in sandbags in World War II. Eisenhower's suite was just above the lettering just seen at the top of the picture. Picture courtesy archives of the Dorchester Hotel.

Titch's Bar at the **Savoy Hotel** was effectively *the* club for US War Correspondents during WWII including Walter Cronkite and Andy Rooney and the acclaimed Edward R. Murrow who had come to London in 1937 with CBS and would go on to have two signature lines: The introductory "This is London" and the closing "Good night and good luck". So popular was the location that the correspondents are said to have caused a whisky shortage in the capital.

The Savoy had a special relationship with the Americans in wartime. The United States Entry into WWI had been celebrated in the hotel's Mirror Room which eventually became the Abraham Lincoln Room. Moreover in WWII General Eisenhower was known to have his laundry sent to the hotel. Sir Winston Churchill would even hold war cabinet meetings there.

Murrow's acclaimed live coverage of the Blitz of London came from the rooftop of the **St. Martins in the Fields** church. During the war he lived at **Flat Number 5 Weymouth House, 84-94 Hallam Street, Fitzrovia** where today there is a Blue Plaque mounted by English Heritage to his memory.

Two interconnecting buildings at **21 Cork Street** and **30 Old Burlington Street** in London's Mayfair became the combined headquarters for the 7480th Supply Squadron (Exchange Service or AFEX) following orders to consolidate operations and locate there with the Office of Special Activities in May 1953. They relocated in January 1955 to Wiesbaden, Germany.

Likewise **Norfolk House** in **St. James's Square just below Mayfair** in the SW1 postcode is worth a mention because it played a key role in the WWII mission planning. The formidable building was the WWII home for US Army, European Theatre of Operations (ETO) and Supreme Headquarters Allied Expeditionary Force (SHAEF) HQ It was used by Gen. Dwight D. Eisenhower from 24 June – 8 November 1942 for the planning of "Operation Torch" – the liberation of North Africa. And again from January 1944 – 6

June 1944 as planners prepared for Operation Overlord – the codename for the D-Day invasion of Europe. Today Norfolk House comprises refurbished offices (about 11,000 sq ft per floor over seven floors). Two bronze plaques commemorating the building's wartime role flank the main entrance. (*See picture further on.*)

During WWII several Army Air Force (AAF) units had 'bases' for the HQ of 9th Air Defense Command in **Wadham Gardens** and **Elsworth Road in St. John's Wood** in London NW3, while **Bryanston Square** in the heart of London was a Headquarters for the United States Strategic Air Forces in Europe (USSTAFE Rear) as well as Supreme Headquarters Allied Powers Europe (SHAPE) Rear and HQ Air Technical Services Command (Europe).

Keysign House, as it used to be known, was just a block to the East of the US Navy Headquarters on **Balderton Street** and was certainly in from the 1970s if not earlier for office space and dependent support for the US Navy. Another location known to many during the 1960s and 1970s was on **Sinclair Road** (and not far from the present-day Westfield London shopping centre at Shepherd's Bush). It was home for many years to a US Navy Post Exchange (PX) and also other offices including, for a period before re-locating to MoD Eastcote along with those of the Naval Criminal Investigation Service (NCIS).

High Master's House in Hammersmith was where General Eisenhower and British Field Marshall Bernard Montgomery were also said to have laid some of the plans for D-Day. The 25-bedroom house, which was invaded by Squatters in 2009, was recently up for sale.

At various locations around London and the UK were Navy designated gas (petrol) stations which were British and either had a special US pump or the price charged was set in one case at half the economy price shown on the pump. In some cases they accepted US $ or £ Sterling for the payments along with special 'Gas Ration' coupons that came in different colors for the various amounts. Personnel were given a set allowance a month. One such station was at the time **ELF** which had been located at **12-16 Rochester Row** in London SW1. For a while a second station right behind the US Navy Building and a block from Oxford Street also accepted US ration coupons.

A host of other addresses in the prestigious London W1 postcode (London's West End) became headquarters and operating sites for US Military missions. They included:

Selfridges on London's famous **Oxford Street** was where during WWII the US Army 1st Regulating Group was based. They weren't the only unit there. The store's sub-sub-basement (some 60 feet down) was home to the US Army's SIGSALY – the codename for a top-secret communications link created in 1943 around the X-System – a scrambler system set up by Bell Telephone and run by the Signal Corps to link Winston Churchill in the Cabinet War Rooms in Whitehall to President Roosevelt in Washington D.C. Both General Eisenhower and Winston Churchill were said to have made visits to the 50-strong unit known as the **805th Signal Service**. That particular area of the store was known as SWOD after the four streets (Somerset, Wigmore, Orchard and Duke) – that 'framed' that part of the building which had been constructed by American owner and acclaimed retailer Gordon Selfridge as the retailer expanded.

Selfridges was struck by a German V2 rocket bomb on 6 December 1944 which impacted operations – not only with deaths of eight American servicemen and a further 32 injured in a canteen across the street, but with damage and subsequent flooding reportedly cutting off access to the secret facility.

The author was granted permission to visit the Selfridges in late 2011 in the company of Mr. David Jarvis, the department store's divisional manager for home and food. Stepping to a hallway on the way to the store's staff locker rooms Jarvis slid back a green painted steel blast door. Down a hallway and around the corner were a series of off-shoot rooms of up to 40 x 70-feet each. Beyond them were three much smaller rooms. Each of the latter rooms still (as of 2011) had a wooden door with seven air holes drilled into the top and into the bottom.

Paul Knauf, an officer with the 805th, helped run the SIGSALY operation at Selfridges during WWII after having attended a special school run by Bell Labs in the United States. Not only did the role involve telephonic communications but soon after he was assigned there an early photo-fascimile type machine that sent pictures from London to Washington DC was installed.

In an interview with Edgar Williams of the *Philadelphia Inquirer* in 1988, Knauf (who passed away in 1999) described the vacuum tube-powered equipment used for SIG-SALY as 'immense" and said that "it filled a room". He also recounted seeing Prime Minister Churchill at the Cabinet War Rooms next to The Mall when he and a colleague went across there to do some work. His colleague even managed to get an autograph.

Now, 70-plus years on, Selfridge's Jarvis still envisions rooms filled with administration staff and weighty communications equipment. He points out an old sign reading "Gentlemen" and a stair rail plus several small rooms with wooden doors as some of only remaining items from that period. Over the years various visitors have theorized that the smaller rooms might have been some kind of interrogation or debriefing cells (flyers were known to be debriefed for several weeks if they were shot down and made it out of enemy territory but there has been no indication the author has yet discovered that any of them has recorded memories of being interrogated in the Selfridges store basement).

The rooms with the drill-holed wooden doors at the top and bottom of each door are small but, as Jarvis admits, it's only speculation.

"To think that history was being made below the department store is amazing," he said during a tour of the rooms now filled with either pump equipment, mannequins and other store shop-fitting items. "It's understood that lots of secret communications and some of the first pictures of D-Day and other key war events were fed back to Washington from down in these rooms. Gordon Selfridge, the American retailer who built Selfridges, was pleased to be able to give help to the war effort."

Over the years there has been excited talk of secret tunnels and links from Selfridges to the US Embassy at Grosvenor Square as well as to the US Navy Headquarters located between the two. But one look at the massive concrete walls that form the foundations of Selfridges would seem to belie this. (The author was asked about the same supposed secret tunnels during his visit to the former US Navy Headquarters by surveyors who had heard the same rumors but had also reportedly found nothing to confirm them.)

"We have heard these theories and while they make great stories sadly never found anything substantive that would indicated this," Selfridges' Jarvis said.

A special collection of pictures sent using the pre-cursor to the fax machine was given to Knauf at the end of the war. A number of them were for use by the press – either pool or issued by the military authorities. Knauf kept them and donated them to Drexel University in the US.

Aside from the US Embassy (which, at time of writing, is said to be the only Embassy in the world not actually owned but leased by the Americans) the US Naval Headquarters also had a significant fascia onto Grosvenor Square. Numbers **20** (the US Navy Headquarters) and **45-47 Grosvenor Square** were also homes to the ETOUSA Headquarters and SHAEF HQ respectively. The US Embassy's lease with the Duke of Westminster remains the stuff of legend. It's said that the US Government approached the Duke to purchase he Embassy Freehold. He apparently said he would be happy to as long as he could get his freeholds in the colonies back – namely the State of Virginia which had been confiscated from his family after the War of Independence!

Such is the extent of the presence on Grosvenor Square that in his special 1979 report for the US Navy entitled *London, England – the US Navy Dimension,* Jack Carter wrote: "Some measure of the extent to which the Mayfair area had become "Americanized" is indicated by the fact that between October 1945 and March 1946 the following accommodations were handed back to the British: **45 Weymouth Street**, **50 Upper Brook Street**, **25 Bruton Street** and no less than numbers **8, 14, 15, 16, 17** and **18** Grosvenor Square.

In 1945 the American presence in Grosvenor Square was fictionally captured in the film *I Live in Grosvenor Square (*which was renamed *A Yank in London* for audiences in the US)*. It starred Rex Harrison, Dame Anna Neagle and American actor Dean Jagger who played a US Sergeant in the WWII romance story.

The US Army London Traffic Office Detachment 'A', 86th Air Transport Squadron and the HQ 5th Army Airways Communications Wing were located at Station AAF-576 at **20 North Audley Street, W1.** The US Army's 7th Medical General Dispensary was located at Upper Grosvenor Street while at **83 St. George Street** the Army established the HQ for the 4th Hospital Train. And also in Mayfair the 32nd Military Police stationed themselves at one of London's most exclusive addresses – **143 Park Lane.** Picturs of GIs traveling in jeeps along Park Lane made the international press.

The Stafford Hotel (now the Kempinski Stafford London) in **St. James's, London**, was where a number of American and Canadian officers used to go when on leave. Its American Bar still proudly displays WWII and other American "special relationship" memorabilia going back to the 1930s.

The US Army's 24th Ordnance Bomb Disposal Squadron pitched up at **66 Cadogan Place** while the US Army's 16th Station Hospital was set up at **Hatherley Street.**

Other units were scattered a little further afield around London including at **Commercial Road, E1/E14** during WWII where the US Army set up one of its D-Day Medical Depots, while the US Army's 12th Port of Embarkation HQ was located at 4 **Fenchurch Street**, EC3. At **Denmark Hill, London SE5** King's College was a WWI location for the US Army's London General Hospital.

Though not a military entity and certainly not a base, one American company with UK locations and connections was certainly doing its bit for the war effort in various locations across London and the UK. The Coca-Cola Company not only supplied product to American and Canadian troops but also produced a series of advertisements showing GIs drinking their product in the company of various Allies. The UK one featured USAAF and RAF pilots and appeared in *Time* magazine on April 24, 1944 and *Life* Magazine on 1 May 1944 and the *Saturday Evening Post* on April 29, 1944.

According to Coca Cola's archive department The Coca-Cola Company Limited, a wholly owned subsidiary of The Coca-Cola Company, was registered in 1929. In 1934, the firm of R. Fry and Co. Limited of Brighton bottled for the Company and Company did the selling. The Company itself began bottling in 1935 in a small plant in **Power Road**, **Chiswick** in West London and, until 1939, other plants were installed in **Southampton**, **Manchester**, **Birmingham** and **Acton** (also to the west of London).

Sales were made from these plants, from satellite depots and by many independent concessionaires with assigned territories.

With the outbreak of WWII and with a pool of organizations called The Soft Drink Industry in operation, the Coca-Cola withdrew from what had been a slowly developing civilian market and between February 1941 until February 1948 some 70 varied companies then only bottled Coca-Cola for the Canadian and American troops. At the end of the war the product was reintroduced to the civilian market but with sugar rationing in force until the autumn of 1953 commercial sales progress was naturally restricted.

Plants at **London**, **Southampton** and **Manchester** were the key ones operated. Concentrate was supplied to mineral water firms up and down country to bottle Coca-Cola and supply the Canadian and American personnel in their areas. (Only the manufacturing facility in Bristol had Coca-Cola bottles).

With departure of Allied troops from 1946 onwards sales dropped drastically. Manchester and Southampton plants were closed. London operated one truck only, just three times per week.

Case sales of Coca-Cola during the WWII years in England grew dramatically (note the sharp fall in 1946 after the war ended):

Year	Total (Cases)
1941	128,000
1942	270,000
1943	872,000
1944	1,511,000
1945	1,224,000
1946	124,000

Coca-Cola may have played more than just a morale-boosting effort during World War II. In at least one case discovered by the author a former flight engineer who flew on board Stirling aircraft with Britain's RAF Bomber Command wrote to a British newspaper describing how he his crew would take up an empty crate of Coca-Cola bottles and drop them at night when they saw a bridge as they were flying over France.

"The Germans spent hours looking for unexploded bombs because the descending bottles made a whistling noise but disintegrated when they hit the ground," wrote LW Brock to the *Western Morning News* newspaper in 1991.

In the immediate aftermath of WWII the first of several statues was constructed in Grosvenor Square to commemorate war-related efforts. **Britain's Memorial to Franklin Roosevelt** by Sir William Reid Dick was commissioned in 1946; the money was raised from 160,000 separate donations from the British public in just six days. It was unveiled by Mrs. Eleanor Roosevelt in front of the Royal Family and other British dignitaries in 1948.

Among later potent symbols of the wartime relationship was the **Eagle Squadron Memorial** and the **Eisenhower Statue**.

The Eagle Squadron Memorial was dedicated to that squadron's American citizen volunteers. The monument, designed by Dame Elizabeth Frink and paid for by the Hearst Corporation, features an American Bald Eagle over a shaft of white Portland stone. It was unveiled on 12 May 1986.

The Eisenhower Statue in front of the current US Embassy in Grosvenor Square was actually dedicated January 23, 1989. It stands just across the road from the buildings that General Eisenhower occupied as Commander in Chief of the Allied Force (June – November 1942) and Supreme Commander, Allied Expeditionary Force (January – March 1944).

Because Grosvenor Square was the nerve center of the American Armed Forces in Great Britain it's is likely that this and another US military-related statues will remain in-situ once the United States Embassy relocates to Nine Elms, London in 2017. What also may be left behind is the distinctive Golden Eagle rooftop frontispiece to the embassy because the building has Grade II-listed building preservation status. The gilded aluminium sculpture by Theodore Roszak is sometimes known as the 'wrong way eagle' by those who know – in that the head of the sculpture faces to its left while on all US currency and emblems it faces to the right.

To the North of London AAF 575 at **RAF Hendon** (now partially the site of the present-day RAF Museum) which would become the aviation home for both a USAAF Transport and a US Navy aviation-based logistics operation during WWII.

Hendon was also where General Eaker flew into (on 20 Feb 1942) on a DC-3 from the US via Lisbon to head up the preparations for the arrival of the USAAF aircraft and men to England.

The US Military presence at Hendon was cemented when 22 August 1942 a Major F C Crowley (USAAC) was attached to Hendon with four 2nd Lieutenants. Their attachment ceased on 17 November 1942.

On 26 February 1943 the ATS service became the 2008th Transport Group (Provisional). On 15 April 1943 it was constituted as the 27th Transport Group (but more commonly referred to as the 27th Air Transport Group) and its primary function would be to ferry personnel and equipment and aircraft for onward delivery to combat units.

The 27th comprised the 519th and 520th Service Squadrons with a compliment of 120 personnel, including 108 enlisted personnel and eight officers, who provided ground support. They would receive aircraft from the Air Transport Command (ATC) which had its European headquarters near **Marble Arch** in London (and would give the unit the moniker 'The Marble Arch Line').

Later in the war the aircraft ferrying role then became the responsibility of 87th Air Transport Squadron at Warton (in Lancashire), while the 86th Air Transport Squadron became responsible for the movement of personnel and cargo with a Headquarters or shared HQ at **Cranford** (AAF-525) near Heathrow along with AAF 510 at **Heston** (also in the Heathrow area) from 7 January 1944. The operation had already expanded into Europe by 1943

By 1943 the USAAF's Air Transport Command (ATC) expanded its role from simple aircraft delivery to Hendon and other bases for onward processing to a European shuttle service with other ATC fields.

The ATC aircraft and London base relocated to **Bovingdon**, in **Hertfordshire**, on 16 October 1944 along with the 1402nd Base Unit.

Despite the USAAF withdrawal various USAAF units continued to fly into Hendon on a regular basis — in June 1945 alone 265 movements (takeoffs and landings) of US military aircraft were recorded by the base.

But the USAAF didn't have exclusive call on Hendon. The US Navy, which had used

Hendon since 1942, formalized its base relationship matters on 3 December 1946 when Utility Transport Squadron Four (VRU-4) was commissioned there. The USN continued to have a presence at Hendon in the 1950s and VRU-4 with its initial compliment of nine aircraft, was effectively the US Navy's executive officer, courier and mail delivery service across Europe and the Mediterranean. It also had a detachment and maintenance facilities at Port Lyautey in Morocco.

It would be renamed Air Transport Squadron 24 (VR24) in late 1948 by which time it had grown to about 11 aircraft and 170 officers and enlisted men. On 1 August 1950 the Headquarters transferred to Port Lyautey with detachment remaining at Hendon.

VR-24 was disestablished on 24 June 1952 and replaced by the VR-25 which also had a detachment in Naples, Italy. However VR-24 was re-established just under two years later as the operation expanded through Africa. A key workhorse of the fleet, the RD4 aircraft, was phased out by 1954 and R5D 'Skymasters' started arriving. However, Hendon was too short a field to accommodate the larger aircraft which would also have to use Bovingdon in Oxfordshire.

In February 1953 VR-25 was re-designated Fleet Aircraft Service Squadron (FAS-RON) 76 and the Naples detachment became FASRON 77. By 1 April 1955 FASRON 76 had become FASRON 200 and in September 1956 (just before Hendon closed as an airfield) it moved to **Blackbushe** (an airport on the Hampshire/Surrey borders) until June 1960 and then briefly to **West Malling** in **Kent,** until August 1960 when the last 'ready duty' flight was stood down and withdrawn.

The last American major presence at Hendon was the liaison flight of the 32nd Anti-Aircraft Artillery Brigade, present between 1956 and the airfield's closure in November 1957.

In 1948 a Spitfire aircraft with the registration G-PRXI (US registration NC74138) was loaned to the US Government by Spitfire makers Vickers Armstrong and based at Hendon for use the US Air Attaché at the US Embassy for around a year. It went on to become part of the static display at the UK's Shuttleworth Collection for a period but was then rebuilt at RAF Duxford. It crashed and was destroyed in an air display in Normandy, France, in 2001.

During their time at Hendon US military appears to have used three Bellman hangars on what was the eastern edge of the airfield. These were supplemented by a collection of Quonset huts and the site had its own entrance. The US Navy had a Commissary near the main entrance to Hendon that would also supply the US Embassy in London and a snack bar close to the flightline.

As there was no accommodation for personnel on the airfield all the officers and men were billeted in civilian homes around north-west London.

In latter years the US military acquired a housing site at RAF Hendon. The 97-unit housing area, acquired by the US military in 1987, was primarily occupied by US Navy families and was regularly patrolled by security police. However in early 1999 US European Command (EUCOM) announced that the housing area would be returned to the MOD later that year with the loss of one civilian job.

The **Defense Contract Management Agency** has continued to lease its Northern Europe office on an office park in **Loudwater, Buckinghamshire**, for many years. A staff of between 40 and 70 with just one or two military personnel review contracts, modifications and purchase and shipping orders for some of the DCMA's 334,000 active contracts (valued at about $16 billion) for everything from blankets to cars and communications, to major equipment and beyond. DCMA Norther Europe (NE) started out as a subordinate command of the Defense Management Command in 1991, at which time it was located in Uxbridge. In 1993 it moved to Loudwater and in 2000 was given its current name.

The DCMA is the Department of Defense (DoD) component that works directly with Defense suppliers to help ensure that DoD, Federal and allied government supplies and services are delivered on time, at projected cost and meet all performance requirements. Among a multitude of contract bids issued in 2012 was one issued for seven "Subaru Legacy Estate or equivalent" vehicles and two VW Touran, Ford Galaxy or equivalent for use by USAF at the flightline base at RAF Lakenheath in Suffolk, at DCMA Bristol and at BAE Systems.

Two more places that certainly should not pass without mention – even though they are not bases as such – are the two American Battle Monuments Commission War Graves locations in the UK. **Brookwood Military Cemetery**, a 4.5-acre cemetery in Surrey (part of the largest private cemetery in the UK) to the Southwest of London, is the resting place for 468 Americans who were killed in World War I. Their graves are located in the American Pavilion. The white stone Pavilion chapel is 'guarded' by four rooftop stone eagles. A brass inset eagle in the centre of the tan-stone floor on which the building columns rest is centerpiece to a room whose walls are inscribed with the names of the American servicemen from the various services killed and missing. Above the names high up are rectangular stained bass windows donated by some of the various states. Above the entrance on the outside it read: "In Sacred Sleep They Rest". Inside "With God Is Their Reward". Outside the white marble graves are arranged in four plots with an American flag at the centre.

A further 80 American servicemen from World Wars I and II are also to be found in the British Military Cemetery adjacent to the American Pavilion.

And though not as close to London, the **American Cemetery next to Maddingly Wood** in **Cambridge** contains the remains of 3,812 American servicemen and a further 5,127 names of missing servicemen. The land for the 30.5-acre cemetery was donated by the University of Cambridge.

Set back amidst the trees of the woodland-flanked road near **Nettlebed**, some 44 miles west of central London is a miniature castle-like building that is a memorial to the 343rd Corps of Engineers Headquarters & Company which was stationed in the woods from 1942. The castle monument was built by Jack Sharp of Des Moines, Iowa, who came back after the war in 1946 and constructed it at what had been the main gate to the woodland base. During their six moths there before shipping out in November 1942, the 343rd also built two sewage pump works for the local towns. The one in Nettlebed is still in operation to this day.

One other place that remains a little bit of American military presence in London – can be found in the middle of the **Thames River**. The Liberty ship *SS Richard Montgomery* had sailed to England in 1944 with 6,127 tonnes of munitions on board. Upon arrival in England it was tasked to anchor in Sheerness but dragged its anchor and ended up aground on a sandbank. During August and September part of the explosive cargo was unloaded but then the ship cracked open and later broke into two.

Because of the unexploded ordnance still on board the British Admiralty still imposes an exclusion zone around the ship, whose masts still protrude above the Thames, to this day. A survey by Britain's Maritime and Coast Guard agency was carried out in 2000 indicated the presence of some 1,400 tonnes of TNT explosive including 286 of the 2000 pound "blockbuster" bombs. Regular surveys have continued and the ship has recently featured as a potential factor in the creation of a new airport in the Thames as championed by the Mayor of London Boris Johnson. A Board of Inquiry into the accident cleared the ship's captain and put the blame on the British Harbormaster.

There continues to remain concern that the wreck is a "ticking time bomb" as one diver and munitions expert said in a 2004 article in the authoritative *New Scientist* magazine and that if the various compounds now exposed to saltwater might self-detonate creating a massive explosion and potentially a mini-tsunami wave four-feet high that could bring widespread death (perhaps up to 14,000 people in Sheerness for starters) and damage to the area. It's also been estimated that to make safe the wreck now could cost up to £30 million.

Finally, there is one additional location that, while not an American base, was certainly a US military landmark for a number of years in the 1950s.

Or rather a target!

Until about 2008 music company EMI kept a large 18-acre storage facility building in **Hayes, Middlesex** to the West of London (almost en-route between either the USAF South Ruislip base and Heathrow Airport. It was a workplace that at one point employed nearly 20,000 people.

It was investigative journalist Duncan Campbell, through his acclaimed and definitive 1984 book about the US military in the UK, *The Unsinkable Aircraft Carrier: American Military Power in Britain,* who revealed that during the 1950's Strategic Air Command (SAC) pilots flew practice atomic bombing missions against the building – flying over it at 40,000 feet twice a week as part of a three-point training mission 'run' that also took in a civic building target in Paris, France and the Leith Docks in Edinburgh, Scotland.

There has remained debate about whether crews were flying unarmed, carrying dummy concrete bombs or practicing with the real, but unarmed, thing. Some USAF SAC alert exercises reportedly did have aircraft flying 'hot' with unarmed nuclear weapons on board.

The EMI building is currently being redeveloped as a £250 million mixed-use scheme to be called "The Old Vinyl Factory".

THE PEOPLE

You can have all the history you want, but it's people and their stories that bring it all to life.

In writing this book I am reminded of the writing on a 'white board' I spotted as I walked though an empty US Marine Corps security room located of the US Navy Headquarters in London. Among a number of exhortations written in red and black dry marker was the line in bold:

COURAGE IS ONE STEP AHEAD OF FEAR

It took a lot of courage – particularly in the war years – for people from all walks of life and all parts of the US to come overseas to fight or work for their country.

Those who came either in war or peace all saw and experienced new ways of life and living. Some had amazing times, but some had terrible times – though few contributors for this book allude to those. They fell in love and married. They had heartbreak. Some saw death. Some lost friends during war and, also, during peacetime. Some got mixed up with the wrong crowd and fell into bad ways.

Trying to cover off every little historical detail and every little incident is for someone else. I expect there will be people aplenty who can come forward with additional details about additional bases, events, missions, parades, security alerts, sports, spies, incidents and moments of sadness and humor. And I hope they do. If there are those with memories to share then there are websites on which to do mentioned within this book.

Ultimately no matter how much you research or write or interview, the simple truth is that you can't do those people, places and times full justice in words or pictures. Our lives are of the moment. But it's nice to be able to try and capture just a few of the memories of those who were there. And these are their memories and their feelings from those times.

In writing this book there has been an added an unexpected bonus. Certainly, in interviewing the following people it has helped me better connect with these locations, however it has been the ability to reconnect interviewees with places and events that has been the proverbial icing on the cake. To watch someone light up as they remember something – it really is priceless. Conversely to watch – and feel—the frustration of someone who is recollecting back 70 years but can't quite remember something, is a salutary appreciation of aging and the frustrations that brings... and will bring.

Again, I can't apologize enough. Over the years as I dabbled with this book, some of my interviewees passed on before they could see the final work. It will be one of my abiding frustrations – that I never have enough time. That there was always one more story or one more aspect to try and track down.

If there is nothing else that this book has done then, in one case of one, Oliver Cork, it has helped him answer the 'whatever happened to…?' question he had about an old girlfriend. During our back-and-forth communication for the book Cork mentioned a girl he had met and gone out with a few times 'back in the day'. He wondered what had happened to her.

And then so did I.

It took me about nine calls and a few wrong numbers and apologies but, amazingly, I found a relative who got in touch with her. She called me and I gave her the 'no names, no pack drill' option of having her former beaux's details or not.

You can read the outcome in the chapter, *Transported*, further along.

I like to think that I have taken to heart what Deacon Gerald Collins wrote at the end of his chapter *Douglas House*. He referred to London as his 'Hometown" and then went on to write: "And like any hometown, it holds cherished memories where people remain young forever and are waiting for my return, if only I just look around the right corner."

It also stems from what Don Terrill, another interviewee, told me in May 2009 as we worked through his memories by phone and email.

He wrote:

"In your last email, you asked if looking back on my past experiences was fun; to tell the truth, yes and no! Sometimes, looking back, I would think if only I knew then what I know now would I have done it different? I guess everyone thinks this way; that's what life is all about. Thanks for helping me to relive my past; it was an enjoyable journey."

Let's now take that journey.

CHAPTER ONE

THE LAND

Rodney Cloke & Howard Cloke

Stonefield Estates
(Freehold landlord of the Victoria Park Estate
– South Ruislip base)
1930s – Present

(Author's Note: The 'wedge' of land between Victoria Road and the railway tracks for the central line and the main line from Paddington Station was the Headquarters for the USAF Third Air Force from May 1951 until June 1972 when it relocated to RAF Mildenhall. The land was owned by Stonefield Estates Limited, a family-run property investment company and leased to the USAF. In 2000 and then again in early 2009, Rodney Cloke, with the assistance of his son, Howard, recalled how it came to be. Rodney Cloke passed away in 2009.)

The land alongside Victoria Road was, at the time, just a field. My father, Benjamin Cloke, purchased it in the late 1930's at a price of about £500 per acre in the family company name of Stonefield Estates Ltd. The idea was to create another shopping and residential area similar to that which my Uncle George (George Cloke) had built at Kingsbury in North West London.

However World War II came along and everything stopped.

It was a triangular-shaped piece of land. We owned from Long Drive right down to Field End Road. My father put the road in – what people know as the Victoria Road – in the 1930s and the area became known as the Victoria Park Estate – the name that the USAF had before designating it the South Ruislip Air Station. Before that much of the land was used for grazing. Gypsies used it quite a bit and it was also a popular fairground site.

After WWII much of the land was leased to Bilton Estates (run by Percy Bilton) for 99 years on a building lease and they constructed many of the original industrial buildings on the site. And of course the USAF had its big site there and a five-acre parcel was leased to some road builders – I think it was a company called Meagan – for the storage of all their equipment.

The Americans wanted to carry arms and under the terms of the lease they had to get permission from the landlord. We gave it. It would have been very difficult not to give it. I recall that the USAF also constructed an atomic air raid shelter on their base.

After the Americans left one of the museum organizations from along Exhibition Road in Kensington occupied many of the USAF buildings for storage purposes until the site started to be redeveloped.

Just by way of interest, the ground opposite the old American base created legal history. In essence the case was to do with compulsory purchase – *Hillingdon Estates Co v Stonefield Estates Ltd [1952]*. The case went to the House of Lords and Stonefields won. It is cited in compulsory purchase cases around the world, so I am told, and it was often referred to in the pages of *Estates Gazette* (a leading UK property journal).

Today Stonefield Estates, now headed by my son Howard, still owns the freehold of some of the land in South Ruislip, but some parcels of land, including the former USAF base have been sold or compulsorily acquired over the years. But the last 20 years has seen much change with the arrival of retailers and other occupiers on The Victoria Park Estate and only a couple of the original buildings remain.

CHAPTER TWO

A DESIGN FOR LIFE

Paul Côté Sgt., US Army Air Corps

(3 Dec. 1942 – 23 Dec. 1945)
> 97th Service Group, Kings Cliffe

(31 Oct. 1943 – Jan. 1944)
> 8th Fighter Command – Bushey Hall, Watford

(January 1944 – January 1945)
> 8th Bomber Command – Pinetree, High Wycombe

(Jan. 1945 – July 1945)
> Supreme HQ London – U.S. Strategic Bombing Survey, Bushy Park, Teddington

(July 1945 – Dec. 1945)

(Author's Note – The name Paul Côté came to me as a name mentioned by chance in Britain. Someone thought they knew of an individual who had served at one or two head-quarters bases during WWII. That person didn't know his name but had recalled his son's name, Ed Côté, and thought that he was a noted researcher or collector of some sort in California.

Those basic clues led me on a phone call and email trail from England to the United States. Then Chris Jepsen, an Assistant Archivist in Orange County, California, came through; "I have the phone number for Ed Côté who is both a historian and collector." A phone call to the aforementioned confirmed that not only was he the son of USAAC WWII veteran Paul Côté, but that his father had served not at one but three USAAC Headquarters in England.

Ed was indeed a historian and collector; and if you could categorize collectors like ath-letes, he is of Olympic standard. He invited me to meet his family and his mom and dad. In the summer of 2009, I went to Santa Ana, California and met the two younger family generations of Côté's before Ed took me on a drive to see his father and mother.

Paul Côté was, as so many US WWII veterans are, a self-effacing man then of 90. The author spent two hours together with him and his wife, Cleo, hearing their recollections and looking at old photographs. As his story unfolds, I think you will find he was at the heart of history – but that he never really looked at it that way. For him, there was a war on and a job to do.)

Aircraft were a big part of my life from an early age. As a young man I never would have guessed that this interest in aircraft would take me to England and put me at the very heart of America's bomber and fighter efforts in WWII.

The early-age interest stemmed from the fact that my family moved from Montreal, Canada, to what was and what remains an aviation heartland – El Segundo, California – in 1924.

From that point I was hooked on design and aviation. At El Segundo High School I pursued drafting with the desire of going into engineering, which I did at Loyola University. I had three-and-a half years in at college and was also working directly for Jack Northrop of Northrop Aircraft Co. from 1940-1942 as a pattern layout man in Hawthorne, California. When war broke out, I was working on a design for Northrop's flying wing. On 7 December 1941, I was with Jack Northrop in the Northrop Loft working on the wing design when the news came over the radio of the Japanese bombing of Pearl Harbor, Hawai'i.

It was July 11, 1942 that I met my wife, Cleo Zehrer, at Northrop. I won a bracelet in a war bond drive at Northrop that was donated by actress Joan Bennett. She was gifted the bracelet by the Maharaja of Japhoor whom she had dated. I gave it to Cleo.

My three-and-a-half years at Loyola University didn't matter. The war was on and I got my draft notice and I went as G.I. Paul J. Côté 39268775 weighing 142 pounds and sporting brown hair. I was ordered to report and duly did at 6 a.m. on the 3rd of December 1942 for induction.

The time line for my transition from civilian to military man ran thus:

- Ordered to report for duty 3 Dec 1942 to become Paul J. Côté No. 39268775 U.S. Army.
- Date of entry into active service 10 Dec 1942.
- Sent to Fitzimmons General Hospital in Denver, Colorado, for medical basic-training (Job Code 521) for 3 months.
- Promoted corporal and sent to dental school (Job Code 855) at Fitzimmons General Hospital March 1943 for a duration of 8 months.
- Ordered to England October 1943.
- Sailed from New York on the HMS *Athlone Castle* 21 Oct 1943, arriving in Liverpool 31 Oct 1943.

THE DOG

The voyage from the U.S. on the British ship HMS *Athlone Castle* was a very depressing business. Many soldiers including myself were leaving the country for the first time. We could only hope that we could sometime soon be reunited with the loved ones from whom we departed. As the ship passed the Statue of Liberty she evoked mixed emotions in the soldiers. We were proud Americans ready to fight for our country yet in the minds of many, this could also be the last time we might ever see Lady Liberty again.

The *Athlone Castle* had an emergency capacity of 2,500 people. We had 7,000 on board! It was crammed and miserable. Pretty much everyone got seasick very early on during the voyage.

The most depressing incident during my time on board involved a dog. A small light-colored terrier dog, which had been brought aboard illegally, bit an officer in a mess hall area. The dog was sent to the medical department in the ship to be euthanized. My job was to hold the helpless animal down while the senior medic was to dispatch it. His method of choice was to take a large syringe and pump its heart full of air. It became a miserably tortuous affair before the job was completed. England had a strict policy of not allowing any animals into the country due to the fear of rabies.

It may have been illegal to have the dog on board for the trip to England, but that dog had been a mascot for the unit and the men took it pretty badly. They had a burial service much like you see for people; the dog's body slid off a board and into the Atlantic.

Disposing of the dog was so messy and the poignancy of that funeral added to the depression of the real job that was ahead of us.

ARRIVING IN ENGLAND

We docked in Liverpool, England. Someone came along and asked, "Can any of you drive?" I raised my hand and shortly thereafter found myself in a jeep with five other guys. We still didn't know what we were doing or where we were going; we just followed the jeep In front of us. I was driving on what, to me, was the wrong side of a country road. We could only see lights that shone through slits. Lights were limited at night because of the wartime blackout imposed across Britain. We travelled to Bovingdon – which was an Eighth Air Force Headquarters, for onward assignment. In my case, I was assigned to Kings Cliffe in Northamptonshire.

Upon arriving in Kings Cliffe I became a member of the 97th Service Group. I had trained as a dentist but I could speak French, as my family heritage was from Montreal, Canada. I helped translate for the FFI (Free French of the Interior) while waiting for a dental posting.

"Frenchies" as I knew them, would be picked up off the coast of France. They would fabricate small boats out of reeds and cover them with tar, then paddle far enough out to be picked up by an allied ship and brought back to England. In England, they would request arms, money and pepper (to confuse their scent from German dogs tracking them) and request to be dropped back into France. In France the FFI were attempting to blow up munitions dumps and create havoc with the Germans.

After some eight months of training with only a fraction of that 'in the field' it seemed that I wasn't destined to see the fruition of my dental training in action. The reason? I was summoned to the base at Bovingdon and was put on restricted orders on September 1, 1943. I was informed that I would be transferred from Headquarters and HQ Sq, 97 Service Group at Kings Cliffe to Headquarters 8th Fighter Command at Bushey Hall, Watford, England; secret codename AJAX. I was transferred with a Corporal Richard C. Boaker.

This was a worrying turn of events. When you are told to report to HQ it's usually not for your own good!

A possible reason for my reassignment was my ability to speak French. However, speculation from my fellow enlistees didn't help. They figured I was to be sent behind enemy lines. I just didn't know and that was the hard part. So you can imagine that a lot of things were running through my mind on the journey to the base.

I travelled there and it was relatively good news. Somewhere along the line someone had actually taken the time to look at my records and discovered that I was a draftsman. So, I ended up at 8th Fighter command (AJAX) at Bushey Hall where my pre-war skills were put to work.

My job at AJAX during the 11½ months I was there was to draft the sorties for the U.S. fighter planes to counter the enemy aircraft flying into England.

Our operations room at Bushey Hall consisted of two buildings. The main building, which I believe had been a school, was where the general staff was located. The single-story Quonset hut was the Headquarters' "working" area. Inside, there was a large map and the WAACS (Women's Army Air Corps Service) would use magnetic pointers to update the locations of the planes.

Bushey Hall was like a hotel compared to the Quonset huts in my previous assignment. Along the way, I was assigned as a draftsman to Lt. General William E. Kepner, who was then commanding general of the 8th Fighter Command, Bushey Hall, England. He had taken over the 8th Air Force Fighter Command in September 1943, which he led until he took over the 2nd Bombardment Division in August 1944 before becoming commanding general of the 8th Air Force at High Wycombe.

I helped plan sorties over Germany and drew flight maps. There was a first choice target and a second choice target. Once the decision was made we typed out the instructions on a black teletype.

There were about eight of us who did all the planning work for Gen. Kepner and also Lt. Gen. Francis H. Griswold (who was then Chief of Staff for the 8th AF Fighter Command). I guess you could say our small group was responsible for directing the air war effort. These were long days. We worked in shifts of twelve hours on and twelve hours off. Then, we'd go to the Mess Hall and eat. I ate anything they gave me because I never knew what I would get or when I would get it again.

We had a codebook that was salmon-colored wood held together with shoelaces, which was kept in a walk-in vault.

One other thing I remember about Bushey Hall was ice cream. We had it on base and sometimes we would share it with British children. It seemed to be their first taste of ice cream.

Of course American food and goods were very popular. I can remember being mailed tea bags and receiving them in a box mailed from home to me at my military address:

39268775
HQ. & HQ. SQUADRON
HQS 8th FIGHTER COMMAND
APO 637, C/O POSTMASTER,
NEW YORK

I took them to the home of Mr. Watling who was someone I had met. He was very old and dressed in typical British proper attire of the time. I can recall he slapped his thigh and said 'Cor Blimey' frequently. However he outdid himself when I gave him those tea bags. You would think the English had seen everything there was to be seen about tea but he and his family were genuinely amazed and amused. They'd never seen tea bags before. That day he kept slapping his thigh and saying 'cor blimey' quite a lot!

We could buy various souvenirs from the locals. During the war many cottage industries evolved to sell souvenirs to the GIs. The women made beautiful handmade cards which I would buy and send home to Cleo. I also purchased what were known as "Sweetheart Bracelets" – bracelets made from silver 'threepenny bits' as they were known.

Speaking of metal there were plenty of 'brass' on the base whom we had to salute. We certainly had to salute Generals Kepner and Griswold but when the staff encountered them we gave them a 'quickie' rather than the full straight-armed salute that was held until acknowledged. These 'quickie' salutes were permitted; otherwise, we would have been better off sewing our hands to our foreheads!

THE BUZZ BOMB

On occasion we saw dogfights overhead. The barracks were by a golf club. I would lie in the sand traps in the golf course at night next to the base and root for our side as I observed the drama overhead.

One time General Kepner was presenting decorations when I heard him say, "Look everybody!" Since we were all facing the general, we didn't see the bomb, but we heard its *phut, phut, phut.* That buzz bomb came really close to taking out a lot of AJAX personnel. I remember the blast (concussion) was such that it almost tore the clothes from my body.

While there we had buzz bombs come pretty close. One came over at night, missed the residences and crashed nearby. There was a guy in the bunk next to me who, after the bomb crashed, threw himself on the floor and rolled under my bed. He was so terrified; he just hollered, screamed, was shaking and wouldn't come out.

One of my last duties at AJAX was to help organize a confidential report that formed the heart of what came to be known as *"To the Limit of Their Endurance"*. This was to chronicle what the 8th Fighter Command did at AJAX for the first year of their presence in England.

I wasn't a party person or a big drinker. Rather than hang out in the popular haunts I would prefer to wander down some of the London streets. I would, on occasion, go into London and stop in at the American Red Cross Club near Piccadilly. I did get stuck in a theatre once during a bombing and I do mean stuck. Debris had blocked the entrances and we had to stay in there for a few hours.

Another time I was returning from London and had a major scare. I saw what looked like an incredibly tall guy but I couldn't make out the details in the darkness. My hair stood up on the back of my neck because we had been told that German saboteurs might be dropped. Thankfully the 'saboteur' turned out to be a British 'bobby' with his police helmet on.

Once, when the Germans were strafing the area, I recall shrapnel bouncing and ricocheting off the brick. One other incident comes to mind. A cook at the base figured he would 'get back' at the officers so he put soap in their soup. Some of the poor officers had pooped their pants (called 'officer pinks') so badly that they took them off and were dragging their pant legs behind them on the way back to the base from the train station.

PROMOTION

I was promoted to the rank of Sergeant on 12 Dec 1944 and was simultaneously given new orders.

This time I was transferred to the HQ of the 8th A.F. Bomber Command at High Wycombe. Code name, 'Pinetree'.

I worked at Pinetree from January 1945 to July 1945 in the bunker. The bunker was in the side of a hill. One of my jobs involved designing identification insignia for the various bomb groups, fighter groups and divisions. I made a sheet that showed all the insignia markings for the aircraft. Sometimes the Germans would acquire a downed aircraft and fix it up as the airborne equivalent of a wolf in sheep's clothing. So we had to constantly keep the fighter groups up-to-date with new aircraft markings and instructions

Some of this work has since been reproduced in the book *Battle Colors Insignia & Aircraft Markings of the Eighth Air Force in World War II*. Most people will know that when D-Day, the invasion of Europe by Allied forces, happened the planes were painted with distinctive white or black 'invasion stripes'. There were all sorts of markings that had to be updated throughout my time.

I also drafted sortie plans at High Wycombe. I coordinated routes with the aid of a meteorologist to determine where clouds were and developed flight paths that were not typical. Flight paths were not straightforward from A to B but much zigzagging was done to evade the German fighters, anti-aircraft and flack guns.

Then came D-Day on 6 June 1944. Amazingly, I was allowed to go to London on the day before the U.S. Military was to launch its greatest assault in history. They still allowed me outside despite the fact that I had at least some of the key plans for D-Day in my head!

The only time one would ever leave the bunker was to go get a bite to eat somewhere. It got pretty warm in there. Inside or outside the bunker nobody talked about how things were going; we just went about our jobs.

I also worked for General James H. Doolittle at Pinetree. Again, I worked 12 hours on and 12 hours off. One time during a raid, General Doolittle came running out of the office and I was running the other way; we collided and both the general and I hit the floor. We both got up, did the necessary apologies, gathered papers and dashed on our respective ways without further comment.

BUSHY PARK

In July 1945, I was transferred to the U.S. Strategic Bombing Survey (USSBS) team at Bushy Park in Teddington/Hampton Court to the southwest of London where I also served as a draftsman. This was under Supreme Headquarters London – we were in the same location as Supreme Headquarters Allied Expeditionary Force (SHAEF). My primary job was to go into bombed factories and facilities following their capture and make detailed drawings of damaged areas to determine destruction by certain types of bombs. At Bushy Park there were 120 draftsmen under my supervision and I checked their work before release for publication.

I got to go to mainland Europe as part of the Survey. I flew to Marseilles, France, about three weeks after the beginning of the Battle of The Bulge. Our people were interested in the capability of our bombing and what it could do to certain buildings. It was my job to assess the damage.

The trip to Marseilles was an eye-opener and a mixture of adventure, danger and disappointment for a number of reasons. I flew over and back on a C-47 wedged between various items including some 55-gallon oil and fuel drums. We landed and I got to see Marseilles and go to the famous Sulky horse races.

My main personal intentions were unfulfilled. I had hoped to see my brother, John, who had been on what they called the 'Champagne Campaign' up through Italy. I missed him by three days as he was redeployed to Okinawa in the Pacific. We didn't meet up again until after the war.

I did, however, run into a couple of "toughs" who were crooked gamblers from Texas. I also saw a black GI murdered. He had been taunting two white soldiers so one of the soldiers pulled his knife from his boot and gutted him from the bellybutton up toward his chest in less time than it takes to talk about it. There were certainly racial tensions. I was glad to be going back to London when I caught the return flight on the C-47.

Another assignment came to me as the war was drawing to a close. I was sent to Frankfurt, Germany and proceeded to Wiesbaden. When I landed in Frankfurt, the rail lines were still on fire. I took a jeep to Wiesbaden where I secured a Gestapo sub head-quarters and sent the contents back to the Strategic Bombing Survey Group. While in Wiesbaden I acquired some German bayonets; surrendered weapons had been stacked in large piles. A tank drove on top of the pile, locked a tread and ground the arms into an unusable mess.

Wiesbaden and Bayreuth didn't receive extensive Allied bombing that other German towns did. General Eisenhower, the Supreme Allied Commander and his staff picked certain areas they weren't going to level to become headquarters locations for the oc-cupying U.S. Forces.

On 8 May 1945, it was V.E. (Victory in Europe) Day and the war, at least here, was over. But not, it seemed, for me. As others went home I stayed overseas. By October it was getting pretty hard for me to be there since V.E. Day was quite a few months pre-viously. While attached to Supreme HQ London I was held over to assist in obtaining information on destruction capability of bomb sizes. For example, was it better to drop one 1,000-pound bomb or two 500-pound bombs or various combinations? This activity was actually supervised by American insurance companies.

I was frustrated and I wrote home to Cleo on 16 October 1945. "Darling, I want to get home!"

A fellow enlisted man named Baumgartner found out that his brother had been killed in the Battle of the Bulge. He became so distraught about going home to his family without his brother that he hung himself in the latrine. Looking back I think that also weighed heavily on me.

I completed eight months in dental training and 25 months as a Sgt. Draftsman in-cluding five months with the U.S. Strategic Bombing Survey. And then I was on my way home via a truck down to the port in Southampton.

On the way back to New York aboard the cruise liner *Queen Mary* the voyage wasn't much better than the one I had over on the *Athlone Castle*. I was billeted below the wa-terline at the front of the ship. This was not a good place to be either during or just after wartime when heaven knew what undiscovered mines were still bobbing about 'free range' on the oceans.

There was gambling everywhere. I particularly remember one guy who had a large amount of cash. He stuffed it down his underpants, pulled out a German Luger and waved it about menacingly making it clear that nobody was going to take the money from him.

There was something else that was equally unpleasant aboard the *Queen Mary*; the toi-lets. During wartime they cut a piece of pipe down at the waterline and hooked it up to a series of toilets. Then every so often they would open the pipes and flush everything out.

A friend of mine was violently ill on the trip back and he threw up his false teeth into one of these toilets. The worst thing was that he was due to meet his wife, whom he quickly married before he left for the war. She had never seen him without his teeth!

It took us five days and one hour from departure at Southampton until we passed the Statue of Liberty.

Once we disembarked in New York they flew me back to Long Beach, California and to Fort McArthur on 23 Dec 1945 where the formal paperwork was completed. A bus dropped me off close to 8680 Evergreen Avenue, South Gate, Los Angeles. Finally, I got my wish. My discharge was from HQ 8th Fighter Command. I received the European Theatre Ribbon (5 times), a Good Conduct Medal with Clasp, The American Theatre Ribbon, Victory Medal, The Army Occupation Medal and the Army Air Corps Badge and Rifle Shooting Badge.

The war was over and my duty completed.

In 1997 I returned to England with my son, Ed. It was an opportunity for me to revisit a part of history in which I had taken part. We tried to get into the High Wycombe base but our attempt was not successful. We walked around the entire hill (it took well over an hour), found the original main gate I had walked in and out of all those years ago and then went on and rediscovered the Red Lion pub in town.

At Bushy Park my son and I visited the grounds where the barracks, tents and command buildings were by then long gone. We also visited the nearby Hampton Court Palace. We went over there because I could recall antique guns that were hanging on the wall of the Palace in WWII. We went in to the Palace and it was the way I had remembered it during that time.

I also relocated a pub I knew well near Bushy Park. It had been the pub I went to when I was stationed there and I knew the owner well. When I went there I carried a safety pin behind my lapel that I used to pick winkles (snails) for a snack and have a drink to wash them down. When the pub was damaged by a bomb an entire rack of antique mugs had fallen down and smashed, but one had survived the and the owner gave it to me just before I left England. It was brown with a silver ceramic handle and had the image of Henry VIII and the Abbot of Reading. The inscription "Coaching Days and Coaching Ways" was marked on the bottom.

I treasured that mug so much that I brought it back to the United States – by suspending it on a shoelace and wearing it around my neck under my uniform blouse to get it home safely. My family keeps it safe to this day in a special cabinet and continue to search for others.

My son and I also made it to Bushey Hall (where some of the original base and buildings remain to this day) and that brought back many memories as well as the stories that my family can share for generations to come.

Above: **Sgt. Paul Côté with a copy of the** *Stars & Stripes* **military newspaper on 16 October 1945 at 8th AAC as he waited to be sent back home. Below left: On bike with a Sgt. at 8th Fighter Command at Bushey Hall, Watford.** Below right: Côté at drafting table for the United States Strategic Bombing Survey at Bushy Park that same year.

Above: Côté in from the C-47 he flew on board to Marseilles. Below left: tents and living quarters for USAF personnel among the trees at the High Wycombe Base. The photo processing buildings at the bottom of the base appear deceptively level (seen middle right of picture). Below right: Paul Côté (*left*) and Ben Peck (*right*) at Bushey Hall.

A family moment – historian and collector Ed Côté helps his father, Paul, into his WWII Jacket in Tustin, California, in the summer of 2009 as Paul Côté's wife, Cleo, looks on.

CHAPTER THREE

SEARCHING TIMES

Francis Xavier Atencio – Capt. USAAF

Photo Inspector, 2nd Photo Tech Squadron,
Allied Central Interpretation Unit (ACIU)
RAF Medmenham, England
1942-45

(Author's Note: Francis Xavier Atencio was 18 when he got his first job – right out of art school in 1938 at Los Angeles animation company Walt Disney Studio. He had only been there about three years and been working on animation for Fantasia (1940) and then Dumbo (1941) when he got his draft papers for WWII. For the next three-and-a-half years Atencio would embark on adventure that would take him across the US and then on to England before he would return home to rejoin Walt Disney where he had a distinguished career that would see him write the scripts for the Disneyland attractions Pirates of the Caribbean (including penning the song "Yo Ho, Yo Ho (A Pirate's Life for Me") and The Haunted Mansion in addition to numerous other Disney activities and projects. He would be named a Disney Legend in 1996 and would become known simply by his middle initial 'X'.

What follows is the story of his time at the Allied Central Interpretation Unit which grew to about 1,700 personnel and which analyzed up to 25,000 negatives and 60,000 prints a day in the last year of WWII – derived from photo reconnaissance missions undertaken by pilots of weaponless Spitfire aircraft.)

I was raised in Walsenburg, Colorado, in 1919 but my dreams lay further afield. Specifically Los Angeles where I had managed to get into art school. Some of the teachers at the school were animators and my teachers Palmer Shappe and Jean Fleury had been impressed and taken a portfolio of mine to Walt Disney. As a result, they then called me in for an interview. And at age 18 I got a job! I was so proud I was able to run three miles from the studio to my home to announce with great pride "I've got a job at Walt Disney".

Three years later I had been promoted to Assistant Animator working on *Fantasia* and *Dumbo*. But then there was a labor strike at the studio and I was out on the picket lines with the strikers even though I had no real idea why I was out there. When the strike ended they took some of us back. I was among the lucky ones who was asked back but by that time I had to say "I'm sorry, I can't. I have an appointment with my Uncle Sam!" The war had come. I knew my number was coming up. But I thought I would be back in a year to pick up my pencil.

Ha!

We did our basic training and I received my commission charged with the duties and responsibilities of Defense Intelligence.. I was sent to Harrisburg, Pennsylvania, where along with others we got our specialist training. By this time I was a 2nd Lieutenant.

Eventually as we completed our training we were told we were deploying to Europe but not what country.

In time we learned we were going into Photo Intelligence. I believe I was chosen because they thought that as a draftsman I might have an eye for small details. Moreover it transpired as we went along that we were going to England to learn from the pros – the British had become experts in photo interpretation and in using something called the Stereoscope.

The Atlantic crossing was a long trip to my mind. I managed to weather the sea voyage – it was a good few days. I can't remember much about the journey – I think we arrived in Liverpool – we arrived in daylight but didn't know where. I do remember the weather was good when we arrived. We got to some base on shore – found an officers club and the first thing I did was order a beer. That was a mistake as it was warm beer. I didn't realize at that point that the British didn't have cold beer. What a disappointment! And that was my introduction to the British.

There were about a dozen of us who made the trip over as part of the same unit. I was close with two of the other guys. One was Gordon Frederick – he was older than the rest of us and he had left a wife behind. Then there was Bert Hathaway. He was also slightly older and had a wife, too. I was the youngest. On days off they would take little trips through the British countryside. I went on a couple but it wasn't for me.

Our base, RAF Medmenham, was in the heart of the countryside about 35 miles from London. It was next to the Thames River halfway between the towns of Marlow and Henley. It was actually a base split in two by the road linking the two towns. On one side was the manor house and on the other side were our barracks.

We reported to the manor house where our work would take place. It was very impressive – especially when you got inside and were ratting around these old rooms. As winter arrived and the cold set in they would light hearth fires in some of them.

We mingled with 'the Brits'. Some of the guys remained stuffy about the whole thing but for the most part we got going with the work. As time progressed we all settled and got into the spirit of things.

We were assigned to different sections. I was assigned to an airfield section. And so it got to be my expertise to look at photographs of German airfields. We worked closely with the British – we were among them. And we were introduced to Stereoscope photography almost immediately.

The Stereoscope was essentially a three-D image taken from 2-D images that were overlapped by 60%. You could get a lot more detail from that imagery. But you also had to use a raised magnifier to look at the pictures.

Our unit had about a dozen people in it. Our work meant that we got to see airfields and airplanes out on the tarmac and bombs, etc. We would make reports about the status of the airfield based on those. The pictures from a sortie would arrive in a box and then we would take them out and go through them, interpret them and then write the reports

As the war progressed I started to see more and more pock-markings in the aerial photography. Bomb craters. The Allies were bombing more airfields. You got a sense of how the war was going through those photographs.

I had a few Eureka! moments. As the war progressed we heard rumors about a new aircraft – a fast jet aircraft. I was working on one aerial report when I noticed something unusual – burn marks on the runway. "That's the sucker," I remember thinking. Then two days after spotting those we had more photography in and I spotted a little aircraft – a Messerschmitt. "There it is!" That was a Eureka moment.

We'd be working eight-hour shifts. We'd start by getting up and going to the Mess and have our breakfast – such as it was. Usually it was the same old powdered eggs. Once a month we got a real egg. But at least we had the option of going out locally. There was a pub – the Dog and Badger – which still exists a short distance away in the town of Medmenham. There – aside from beer – you could also get a real egg on occasion!

At our barracks, which were across the main road from the Medmenham base, we had about eight guys in each. They were essentially Quonset huts. We tried to give them a 'homey' feel. A few of the guys put some 'personal items' from their girls back home or the girls they had met. One time at Christmas we put up a Christmas tree.

Of course girls were of interest to us. We went to the town of Marlow.... it was closer to the facility. We'd go to the Mess there and have a beer, associate with the Brits and when the WACs (Woman's Air Corps) came over we must have had a dozen girls that joined us. They had separate quarters but I think there was some hanky-panky.

We did overnight trips into London and stayed in quarters at the officers club near Grosvenor Square as I recall. We were well taken care of in that respect. The Brits were very good at this. I made some close friends and they'd invite us to their homes.

In my free time I'd try and get further afield – either to the 8th Air Force HQ at High Wycombe or to London. I think the main thing that sticks in my mind about London now was the bomb damage and just hoping there wouldn't be any raids while I was in the capital. If there were we'd head to the nearest shelter and sweat it out. I don't know which London Underground stations I ducked into. We could be in the shelter for 20 minutes to an hour or so before you heard the all-clear sound. Then it was back upstairs and continue on with your lives. The main thing about London to me was the bomb damage.

We were never directly bombed at RAF Medmenham that I recall, but we would get air raid warnings. Then we'd leave the mansion and go outside and sit and watch the planes come over. We see them get caught in the ground fire and be brought down. It would be like a football game – that's the only analogy I have – we'd start cheering. "They got one."

High Wycombe and the Eighth Air Force Bomber Command headquarters was a popular trip. We'd try and get there as often as we could. The reason was they had better food there. So we'd come up with any pretext we could – it was just four or five miles away at most. On a couple of occasions we had to go and make a report. Col. (later Brigadier General) Elliott Roosevelt, the son of the US President, was our Commanding Officer but we didn't see him too much. He was largely busy associating with higher echelons of command than our humble selves.

On other occasions I'd take a couple of bicycle trips with the guys and we hit a pub. I remember in wintertime it was cold and we'd go sit in anwhere warm. We had our own personal tankards for beer. The staff would put a poker in the fireplace and heat it up and then they would stick it in the beer. I still never got used to warm beer!

The social life was fun. Back home I had a girlfriend waiting for me and you could say that it was a long two years of being faithful for me. I think we all found relationships. One girl that I remember – Betty – was very nice and we'd go on picnics and we were fond of each other. But I think she had a boyfriend. We just did our best to have a social life in a wartime situation and location.

Of course things changed materially after 6 June 1944 – D-Day. Once D-Day had arrived we got batches of photography of the area across Northern France and got to see what was happening. We had ringside seats. As the war progressed we started seeing more and more damaged airfields in Germany.

We got a ration of booze a bottle of Gin and a bottle of Scotch a month. That was, as I recall, per person. At Medmenham we didn't have an Officers Mess. So we squirreled that away and when Victory in Europe (VE) Day was announced we had a real party.

I actually left Medmenham right after VE Day. By that time I had a couple of promotions and I was a Captain. Those of us who arrived early got to go home early. I left a lot of friends back there and then never saw them again.

And so it was back to the United States but not out of the military. We were scheduled to have some leave and then transfer on to the Pacific and potentially Japan. But as luck would have it the war in Japan ended, too. I had made arrangements to get to my base in Colorado where my fiancée, Mary, was. I went to her home in Denver and spent a month-long holiday there and then I wanted to go home. I reported to Fort Carson and requested to be discharged.

Mary and I got married as soon as possible in Denver. I went up to the front steps of her house and remember saying: "Are you ready to get married?" And she was ready to get married. She had a wedding dress and everything. Don't worry – we had been corresponding daily so I knew she was really ready.

Among my souvenirs that I picked up was some captured Nazi items including a helmet. One piece didn't make it back was a Luger gun. It got confiscated on the way back.

My discharge may have been as a Captain but I left the military as a Major after spending about a further year in the Reserves. By then I had certainly had enough of the military. A lot of guys had trouble getting jobs, but I had been fortunate. My job had been promised back to me at Walt Disney and they were good to their word.

So I was glad to go back to Mickey Mouse and gang. Of course it was four years later on by then. The fellows who didn't or couldn't go to war had advanced to full-time animators and I had to catch up with them. That was the hard part!

Mary passed away in 1983. I had four children with Mary: Victoria (Tori), Judy Ann, Gerald (Jerry) and Joseph. I married my wife Maureen in 1988. I believe and hope I've lived a fairly full and interesting life.

England in wartime was fascinating. I made some good British friends and we corresponded for years after. I have really nice memories of the British people and the area we were stationed in as well as Marlow and High Wycombe. I look back with fond memories.

Francis Xavier Atencio is among the staff in this picture at RAF Medmenham –photo courtesy The Medmenham Trust.

A drawing from a short autobiographical cartoon story told and drawn by Francis Xavier Atencio. The cartoon shows him 'flying' his stereoscope goggles over the English Channel to Germany to look at airfields. Right: Atencio in a relaxed moment. Drawing and photo courtesy Francis Xavier Atencio.

CHAPTER FOUR

THE 32-MONTH CONVERSATION

Juanita Wimer Folsom

Private First Class
Women's Auxiliary Army Corps (WAAC)
417th Signal Battalion, USAAF
Pinetree, High Wycombe (May 1943– July 1945)
Bad Kissengen, Germany (August 1945 – September 1945)

In 1939 I graduated from Dodge City High School in Kansas. I was 17. There was a war going on in Europe as Germany's Adolph Hitler had already marched into Poland. The US wasn't involved but we were concerned and I can remember people keeping up with events in Europe by radio and newspapers.

I was living with my family at the time in Scott City, Kansas. My parents, Charles and Virginia; my three brothers Morris, Bud and Royce; and two sisters Arlene and Dorothy. I was able to get a job with a telephone company as an operator at age 19.

It was on December 7th, 1942 that my world – and the world itself – really began to change. It seemed liked overnight the building of defense plants for manufacturing airplanes, parts, weapons, jeeps, tanks etc. sprouted up everywhere. Likewise recruiting stations popped up in every city and town. Men were enlisting when 21 years old or earlier if they had their parents' consent. Then the Draft was put into effect and troop trains were taking enlistees to different bases throughout the country for training for all kinds of military service in the US as well as overseas.

My eldest brother, Morris, enlisted in January 1942 and was in the First Marine Expeditionary Force to leave the United States. My older sister Dorothy and her husband, Reed, moved to California to work in Defense Plants. Bud, my younger brother and I also moved to California as I was able to get a transfer to the Telephone Co. in Long Beach. I moved into an apartment with three other girls who worked at the telephone company with me.

I turned 21 on 16 March 1943. It was then that one of my roommates, Lydia Weimer and I decided it would be an adventure to go into the service and see the world. We had heard on the radio that they were taking women into the service now and that some would probably be sent overseas. We decided we wanted to go into the US Marines – but the recruiting station discouraged us as they said they wouldn't be sending women Marines overseas for sometime. Then we tried the Navy and they told us the same thing. So we ended up at the Army recruiting station and joined the Women's Auxiliary Army Corps (WAACs).

And so a week after my birthday we were enlisted into the WAAC. Twelve of us enlisted girls were inducted live over the air (on radio) – there wasn't TV coverage back then.

It was a large public ceremony on a victory stand in Long Beach, California. This was the first induction of women into the service that had ever happened. It was all done in conjunction with a public fundraising event to sell Victory Bonds. The Long Beach Municipal Band played the National Anthem and other music and other musicians played and sang *The Star Spangled Banner, Indian Love Call, Some Day, Carry Me Back to Old Virginia* and *Annie Laurie.*

We all left that same evening by train for Camp Monticello, Arkansas. We had no uniforms yet and all we could take with us was: one skirt, one jacket, one pair of shoes and one change of underwear. We all had to live in these clothes a few days after we got to Arkansas waiting for the uniforms to arrive. We were issued fatigue dresses that were green and white cotton seersucker blouses. It was rainy and cold there and much different then California. We started our training anyhow still not knowing when, where or what we would be doing for the war effort.

About three days after we were in camp a farmer on the adjoining land, who was raising pigs, somehow managed to let the pigs get loose. Who did they call on to round up the pigs? The WAACs of course! It was a real sight in the mud with us in fatigue dresses chasing the pigs. It was like one of those greased-pig contests at a county fair but we got the job done. We were all a muddy mess when it was over. A week after we arrived we finally got our proper uniforms.

Our Basic Training was generally the same as the men with the orientation of military procedures, discipline, calisthenics, obstacle courses, marching in formation and lots of aptitude testing to see what we were best qualified for.

Camp Monticello was built as a POW (Prisoner of War) camp for when the US military started sending prisoners of war back to the US. We had bare and basic barracks with canvas cots. The toilets (latrines) were in a separate building with long rows of toilets with no partitions in between and a long trough of water running underneath the toilets to carry the waste away. The showers consisted of a small room in the same building with a bunch of showerheads and no partitions. We learned early that you could not be modest in the service as there is no privacy.

One very interesting thing that happened while we were at Camp Monticello was a visit by a WAAC officer – a Lieutenant Willa A. Ruditsky. She had been born in New York City and was taken to Czechoslovakia (as it was then) when she was six months old. She lived there until she was 20 years old. She had been living on a farm outside of Prague when it was occupied by the Nazis. The Gestapo hustled her off to a concentration camp.

She told us that she and 15 other women in this camp were lined up each day facing a concrete wall and questioned while their tormentors pushed their heads up against a rough wall. She suffered a broken nose and she said the wall was covered with blood. This went on for six days, seven hours a day and then they were put in small dark windowless cells that were then flooded with water so they couldn't sit down – there were no chairs or furniture in the cells anyway. They were beaten daily with rubber hoses as the Nazis we never satisfied with the answers to their questions. They had them strip to the waist so the hoses would be more effective.

One incident that she recounted happened with an old woman during a beating. She fainted and fell to the floor in the water. The Gestapo agent was wearing spiked shoes so he stepped on her face lacerating it horribly. Rudistky let her emotions get the best of her and hit the Gestapo man. Then she was beaten with a hose until she fainted.

After six days of this the American Consulate intervened and she was release in June 1939 and ended up coming back to the US that same year; arriving with 35 cents in her pocket and an English dictionary as she could only speak a few works of English. She got a job and ended up joining the WAAC when she was 23 years old and sent to Camp Monticello. Her story really made an impression on me.

After a couple of weeks at Camp Monticello and many aptitude tests they took several of us including Lydia and myself into the Commander's Office and said we qualified to go to Officers Candidate School or we had the choice of a secret assignment but they couldn't tell us what it was. We knew it had to be overseas so all us said we would take the secret assignment. We had only three weeks of basic training instead of the required six weeks and we were immediately sent to Fort Devens, Massachusetts, where we were told we were going overseas but were sworn to secrecy and not allowed to let our families know. We then found out that they needed girls for communications jobs overseas so they could replace the guys and release them for combat.

At Ford Devens we were issued gas masks, canteens, mess kits, helmets, wool-knitted caps, overcoats, rubber goulashes, long underwear, khaki-colored bras and panties (that hung almost to the knees), wool GI blankets, a GI purse. Plus we were issued impregnated clothing that was a hooded jumpsuit out of canvas type material that would cover your whole body, it was completely sealed in a plastic bag as they smelled horribly.

These were to be used in any kind of gas warfare that the Germans might throw at us. We were also issued real nice dress uniforms (they were actually dresses) that had been made for the officers but the manufacturer had made a mistake and dyed them the wrong color so they sent them to us.

We had extensive training at Fort Devens on the obstacle course and long marches in formation, plus we had several days of gas mask drills. They would put us in a small airtight bunker type of building with our gas mask and then a male GI would release gas into the building and we had to put on our gas masks fast. On the first day that we did this I was the only one that had a faulty gas mask. It leaked and I could feel my face and eyes burning so I had to exit fast.

The guy who gassed us was a Sergeant and was really enjoying his job of 'gassing the WAACS'. He was cute, too. On the last day of the gas drills he was razzing us when we came out of the bunker so some of us took out after him and chased him over a small hill and finally caught up with him. We started wrestling with him but when we threw him to the ground we broke his leg. I guess you know who got the razzing then – especially from his buddies!

After our training there we packed our duffle bags and headed for Camp Shanks, NY, which was our port of debarkation. At Camp Shanks we received a bunch of shots for Typhoid, TB, etc. At this time no woman had been sent overseas but now there were 557 of us ready to be shipped over. For some reason I recall that Canada had joined the US in the war in Europe.

So, in the summer of 1943, 650 women – mostly WAACs – boarded the *Aquitania* (a sister ship to the *Lusitania* of WWI fame). The other 93 women with us were officers, war correspondents, nurses and Red Cross personnel. There were also thousands of male servicemen aboard from the US and Canada. Many were from the Royal Canadian Air Force (RCAF).

The women were confined to one area of the ship primarily consisting of several Staterooms. Of course it was less than luxurious – for starters there were 18 of us to a room.

The canvas cots we had to use were very narrow and 'stacked' three high with hardly room to move. We had one tiny bathroom and salt water ran through the taps! We had drinking water in our canteens which we could fill once a day at a main faucet in the ship's kitchen. The only bathing we could do was 'spit baths' in salt water. You couldn't wash your hair. Some enterprising gals had brought cornmeal along so they could rub that in their hair and brush it out.

Our mess hall was actually in the ship's swimming pool with rows of tables at the bottom of the pool – no seats. We ate standing up at tables. There was no fine dining or china for us we ate using our mess kits and then washed them out, again in saltwater. We had to eat in shifts as there were so many aboard.

Our quarters were guarded day and night so the guys couldn't mingle with us. Did that stop matters of the heart? Absolutely not! We got over that by passing messages between the portholes on the outside of the ship.

It took us seven days to cross the Atlantic as we had to 'zig-zag' across to keep from being detected by German submarines or rather U-Boats as we came to know them. We had more training while aboard ship including French lessons, learning just enough to keep you from starving if you were some place where the only spoke French. The only two words I remembered – and still remember – were "biftek" and "lait" which meant "beefsteak" and "milk".

We landed in Gourock, Scotland, in July 1943. It was to be our staging area and from where troops would be dispersed to different bases around England. We had a terrific reception when we arrived in Gourock. We were the first American gals the American GIs had seen in two years. There was a newspaper headline at the time – it read: "650 WACS defy the subs" – it also became the title of a book. We were the first battalion of WACS to reach the ETO (European Theatre of Operations) and actually there were 557 of us plus 19 officers and others. A second battalion followed a couple of months later.

After two weeks in Gourock and more orientation and training a group of us who had experience in telephone communications were sent to London. The rest of the class included stenographers, cooks, drivers and communication specialists. Many of them – around 300 – were sent to Supreme Headquarters Allied Expeditionary Force (SHAEF) Headquarters at Bushy Park.

The average age of us girls was just 25 (the US Congress has passed a law creating the WAACS that required that applicants be aged between 21-45 and have no dependents and that they had to weigh over 100 pounds). Lydia, who joined with me in California, and I stayed together wherever we went. In London we were housed in a huge six-storey house. It wasn't any ordinary house. It was the former home of Lady Astor.

The home was fascinating enough though all of the rooms were full of GI bunks – there was a bathroom on every floor that was shared by 25 to 30 of us, but this was a luxury. We spent two days trying to figure out what the extra bowl was in a couple of the bathrooms were. When you turned the knob of these things a squirt of water shot up in the center of the bowl. Some guessed it was to wash your feet in… one at a time. Finally someone clued us into matters. It was a douche bowl (or bidet). It was the first time any of us had seen a douche bowl in a bathroom – we found out later that the Royalty and the rich usually had such luxuries.

We then had two weeks of training in what seemed to be the largest telephone building in the world. It was a nine-storey building where calls would come in from and be re-routed to other countries. It really was nine stories of switchboards. Among things we had to learn was the terminology they used in the UK as well as in other countries such as Belgium and France. If the line was busy, for instance, we had to say that the line was "engaged". And we had to pronounce "schedule" like "shed-ule". Instead of calling someone up, you "rang" them up. And if you were going pick someone up at 8 a.m. then you said; "I'll knock you up at 8 a.m." That got a lot of laughs amongst us!

At the time we arrived England was being bombed almost every night by the Germans so air raid sirens were a regular part of our life. All the buildings and homes had to be 'blacked out' every night. They had black out curtains over the windows and doors and there were air raid shelters every few blocks. You would have thought the shelters would have been safe havens but we were told to stay out of the shelters as so many people were killed in them during bombings. The concussion would often kill them, if a bomb dropped close and the door was left open or unlatched, the shock wave from the bomb would kill those inside.

The English were used to the bombing but it took us awhile to get used to the air raid sirens and hearing the bombs exploding or seeing the flashing in the sky at a distance. But we finally developed the attitude that everyone had – if a bomb drops on you, you won't know it anyhow. All English homes had to have a heavy table that they could crawl under during an air raid, They even inspected the houses to make sure everybody had one.

A fascinating thing about the English people to me was that a bomb could drop during the night in London and destroy a building but all the English people going by the building the next morning, going to work or whatever would stop and pick up the bricks and stack them into neat piles.

Bombs can do such weird things. I went by a place in London a couple of days after a bomb had dropped and flattened part of a four-story building. The bricks were all neatly stacked by the one wall of the building still standing, I clearly remember at the third floor level you could see a row of coat hooks and a man's coat hanging on it.

The Germans were also using 'Buzz Bombs' and then V-2 Rockets as well as Land Mines. Buzz Bombs were unmanned rockets that would come over any time, day or night. They made a 'phut-phut' sound and when the 'phut-phut' stopped, you knew it was coming down. They made a pretty good explosion but the V-2 rockets were larger also unmanned rockets that made a hissing sound and when the hissing stopped they came down and made a much larger explosion. The land mine was a huge bomb that they dropped from plane and they could flatten a whole city block.

After two weeks of training in London we were sent out to the 8th Air Force Headquarters about 30 miles away. It was a place called High Wycombe and one side of the town was the Wycombe Abbey School for Girls. It consisted of several good-sized buildings which was set in a valley in which the town was located. They used to be classrooms and offices for the exclusive girls' school. It was set in beautiful grounds with a beautiful pond and a swimming pool. A lot of the buildings were all turned into living quarters full of bunk beds. We were temporarily housed in there for about a month, The fanciest building near the pond became General Doolittle's offices, living quarters and private dining room and kitchen.

Something that happened when we first arrived and were getting settled into these rooms was that one of the girls got hold of a newspaper article from a New York newspaper that told of the WAACs going overseas and how their supply people had gone into Boston and purchased supplies for them and that among those supplies was a bunch of baby diapers but with no explanation why! We knew they had done this as they told us before we left the US that we might not be able to get sanitary pads in England – so we would have to cut up baby diapers and use them and, possibly wash them and reuse them if necessary.

The insinuation of the article was that we were all going out to have babies! It made us so mad that the newspaper article didn't explain what the diapers were to be used for, so the next morning when our captain yelled "Fall Out" none of us went out of the building in protest. We could have been court-martialed for this but when the officer saw the article she understood why. She rained hell with the brass and there was a retraction and explanation printed in the paper. (*Author's Note: One column also printed in a US chain of newspapers wrongly alleging that prophylactics had been handed out to the WAACs causing outrage among both people at home and the WAAC themselves despite assurances and denials from everybody right up to the US President*).

The room at the girls school the 12 of us were in had a fireplace and one of our girls had some popcorn. We acquired an old skillet some place, built a fire in the fireplace, put some popping corn in the skillet and handed it back to her. We told her to hold it over the fire and the corn would pop. The only thing we didn't tell her was that you needed a lid. It was hysterical watching her when the corn popped and started flying all over the room.

Next to the girls school where we were being housed was a huge tree-covered hill and on top of the hill among the trees were Nissen Huts. These huts were the corrugated metal huts with the round roof that sloped to the ground on each side. These were our permanent living quarters for the next two years. Our huts had concrete floors with two windows in each end and a potbelly stove in the centre for heat.

We had 12 bunk beds in our hut, a footlocker for each bed and coat hooks by each bed. The beds had a three section straw mattress which was slipped into a canvas-type casing. This acted as a sheet. A bolster was stuffed with straw for a pillow and two GI blankets each. We had to learn to build fires in the potbelly stove using newspapers and coke from coal most of the time. We made coffee by boiling the loose coffee grounds in a pan of water on our potbelly.

In the middle of our group of huts was one hut called the Ablution Block. This was where our latrines (toilets) and were. The Block consisted of a long row of toilets that had partitions between them and there was a small room in one end with a bunch of showerheads, a concrete floor and no partitions. That was it as we had no running water in our huts. It was hell if you had to go in the middle of the night, especially in winter. We always kept our canteens full of water as the effort to go get a drink was also considerable.

One of the huts in the middle of all the huts (both men's and women's) was the Mess Hall. We were served the typical GI chow – lots of potatoes and canned stuff from the US. We had a lot of canned spam and plenty of canned corn beef – better known as 'Bully Beef'. It was very fatty. We had no fresh milk as the cows in England were not tested for TB and we were banned from drinking any of it. There was a lot of TB in England in those days. I saw GIs crave milk so bad that they would drink canned milk if they could find any.

The fresh meat we got was mostly mutton lamb, so many of us would enter the mess hall and smell the mutton cooking and we would grab a couple of pieces of bread and some orange marmalade and go back to our huts and make that our meal. All the tables in the Mess Hall always had opened dishes of orange marmalade on the them and, since there were no screens on the windows of the hall, the marmalade usually had bees floating in it. But we pushed them aside and ate it anyhow.

We always had to take our mess kits with us into the hall, eat out of them and then wash them out on the way out of the hall. We girls pulled KP ('Kitchen Patrol) duty just like the men. A couple of us girls dated a couple of General Doolittle's cooks so we had a few goodies in our hut once-in-a-while that the others didn't get. General Doolittle's favorite salad was a piece of lettuce with sliced pineapple on and a dab of mayonnaise and grated cheddar on top. I still fix that salad today.

There were 12 girls in my hut. We called ourselves the "Smith Family" and our home the "Smith Hut". Lydia and I had managed to remain together and along with the 10 other girls through the next two years – like sisters. We came from all different parts of the United States but when you are thrown together like that you just adjust regardless of any differences.

As an example we had one girl from Philadelphia who, upon hearing how we had to keep our huts spotless (including scrubbing the floors and facing white glove inspections from the officers each week, etc.), made the remark that she wasn't going to get down on her hands and knees to scrub a concrete floor for anyone. But it wasn't two weeks before she was down there with us pulling her own weight. We had another gal who was 23 years old and had never washed her own underwear. At home she threw it in the corner of the bedroom and her mother washed it!

We did take real pride in keeping our hut immaculate. We did lots of funny stuff and had to make our own entertainment so we did a lot of short-sheeting of people's beds. We had one gal that slept so sound that we would put her hand in a pan of warm water while she was sleeping and she would wet the bed. Lydia was such a neat freak – we had to hand wash our underwear and hang in out on the clotheslines beside our hut. I used to love taking Lydia's underwear from the line, place it in really thick laundry starch and put it back on the line. Lydia is the only one of us who wrote home and asked her family to send a pair of sheets over. She had to wash them by hand but she didn't mind. She had the lower bunk and I had the upper so I used to step on her sheets and wipe my feet on them before climbing up to my bunk – that used to make her so mad!

In spite of the war and the hardships, we girls still used to have a great time together when we could. We would get packages from home with homemade items in them – we would also share them with the Smith Family. Nearly all the others had named their huts, too.

At some point after arriving in England we became a regular part of the Army. At that time you could actually get out of the service and be sent home. But nobody did. That's when we became the WACs (Woman's Army Corps).

Under this huge hill that our barracks sat on, the US military had built the 8th Air Force Headquarters – its codename was "Pinetree"). It consisted of three stories underground with an entrance into it about halfway up the hill from the girl's school. In the underground was a huge operations room. The Map Room and Weather Room were two stories high with a large map on the wall. There were several other working rooms, a small snack room and two switchboards. There was a main switchboard on the bottom level and an operations switchboard in the operations room on the second level. That is where I worked. We had a Headquarters & Headquarters VIII Bomber Command station directory – my version, which I kept, was produced on 6 August 1943. It was entitled *G.I. Joe Meets GI Jane*".

The underground bunker had armed guards at the main entrance – you had to show your identification 'dog tags' to get in. As I remember it was a large cement entrance with a large heavy door. The entry way was not brightly lit and there were guards stationed all around the area on the way to the entrance, too.

The operations room where I worked had a two-person switchboard but it was usually worked only by one of us. I pulled many double shifts (16 hours straight) because the weather in England was so cold that many of the girls would get sick with colds. I think I am the only one of the Smith Family who didn't have pneumonia. There were only a few of us who were cleared to work in the Operations Room and that is the reason I had to pull double shifts so many times.

The Operations Room was where all the 'brass' would meet including General Doolittle and General P.A. Anderson and, occasionally, General Eisenhower as he was in command of the whole ETO at that time.

They sat around the conference tables and planned all the missions; how many planes they were going to send up and where they were going to bomb. The room had one huge wall that was a map of the whole European area. It was lit up with small different colored lights indicated the various weather elements in every one of the targeted areas and also where they were going to drop bombs. They worked in conjunction with the RAF (Royal Air Force) and the RCAF (Royal Canadian Air Force).

After missions were planned, I would set up the conference calls to the different Air Force wings in England and, later, in other countries such as Belgium and France. General Doolittle was there frequently either in his office area or the Operations Room when missions were being carried out.

At times when General Doolittle came into my operations area he would lay his hat on top of the switchboard; sometimes he would stop for a minute or two asking how you were, etc. It was just small-talk type stuff. There would be 10-15 people in the Operations Room for missions. Though we didn't know any of the information you could tell when a mission had gone well or badly by the looks on their faces and by the activity.

As mentioned, our base was known as "Pinetree" and all the other bases also had code names. At the time they had what they called "scramble phones", the conversation was scrambled so the lines couldn't be tapped by the enemy. I would connect all the parties and then would say "let's scramble" and then they pushed a button on the phone that would scramble the conversations and, if you tried to listen in, it sounded like a bunch of cats fighting.

At the time USAAF were sending up the B-17s, B-24s, P-38s (nicknamed "Yippies"). Later in the war came the B-29s.

The German planes came over often looking for the 8th AF Headquarters but couldn't see anything from the air; we were well disguised. We had air raid sirens go off often when there was an enemy plane in the area. We had slit trenches all over the hill so if you were in between areas when a siren went off you dropped in the closest slit trench and waited for the 'all clear' – another siren. Our huts were all blacked out so you could see no lights from our area at night. The town of High Wycombe was always blacked out too, like all the other towns in England. We never left our hut without a flashlight; you might not find where you were going or the way back.

When we could we would get passes to go to London where they had a big service club. You could meet a lot of people there from back home. We would also get passes to other places in England that had service clubs where we could stay like Beaconsfield and a place called Maidenhead. We learned to drink warm beer as pubs didn't have ice. I recall getting invites to dances such as Harry Roy's *Lyricals at the Regal Ballroom* at the Regal Cinema in Marble Arch or to eat at Fischer's on Bond Street or the Orchard Hotel in Ruislip.

I went to all the events I could both on base and off base. Many were put on by a combination of local townspeople and the service personnel and others. These were a great diversion from daily war life, more for some then others. It was also a great way to meet the GIs from outside the camp, too.

After the Allies took over Paris some of us girls made friends with a couple of pilots who were going over on a transport flight so we 'bummed' a ride in their B-17 and went to Paris. Paris had a large service club that we could stay in. We were only supposed to stay a day and night but we were having so much fun that the pilots radioed back that they were weathered in, so we stayed an extra day! I didn't smoke but always bought cigarettes at the PX for 50 cents a carton just in case my bunkmates wanted them. I saved up four cartons and took them to Paris to sell on the black market. The first shop I stopped in I sold them to the shopkeeper for $40 which was a lot of money in those days! It gave me spending money in Paris.

All the time we were in England we couldn't tell people back home where we were or how we were living. It was always just "some place in England" and our own mail was slow in getting to us. I got one letter from my mother who told me that my older brother, who was in the US Marines, had landed Stateside and was able to walk off the ship. I hadn't received a previous letter stating that he had been shot in the lung in the battle of Tarawa in November of 1943. He was in a tank division in the Solomon Islands and was in the third wave of Marines that invaded Tarawa. He recuperated a couple of months and then went back overseas.

After I went overseas in 1943 my two younger brothers went into the service also. Bud into the Paratroops and Royce into the Merchant Marines. After I had been in England about two years I found out through my switchboard that my brother Bud's outfit was coming to England. I kept track through other switchboard operators and found out where his outfit would be landing.

My brother had only been on his base for two hours when I called him on the base phone. All he could say was: "How the hell did you find me? It is supposed to be top secret!" After that surprise we arranged to meet in London when he could get a pass. I told him how to get hold of me as I had made friends with the guy on the switchboard on his base, so he called me when his pass came through. We met at the Service Club in London and his first words to me were "You talk just like a Limey!" I guess I had picked up a little bit of an accent by that time.

Bud was stationed at a base near Swindon some 60 miles to the west of London so he arranged with an old English couple for myself and a friend named Nellie to come visit the base and spend the night with the couple. They were having a dance on his base that night and were importing some English girls from town.

My friend and I were the only two American girls so we had a ball. My brother loved to jitterbug and so we got out on the floor and started jitterbugging. They cleared the floor to watch us as a lot of them had never seen the jitterbug let alone tried it! We managed to meet up several times after that either in London or at his base until he was shipped out in August 1944, when we had the invasion of France (D-Day). Later he ended up jumping across the Rhine in the Battle of the Bulge, parachuting into what was German territory.

I had started collecting patches from different branches and divisions during my service time. I started sewing them onto my GI-issued corduroy bathrobe that I was issued nearly 70 years ago. If I saw a patch on a GI that was interesting I'd ask them for it. I even kept a small pair of sewing scissors on me to make it easier to remove! My brother Bud had taken note of my interest and had even managed to collect patches from dead German troops to add to my collection.

My other brother in the Merchant Marines spent his time in the service guarding the West Coast. There were four of us in my family (out of six children) who were in the Services at that time. My three brothers made it through the war and we were able to share stories. Sadly, Bud passed on in 1983, Royce in 2000 and Morris in 2006.

Speaking of D-Day, I was working the switchboard in the Operations Room all night and was planning on going on leave to London the next morning. I knew something big was going on because all the command brass were in the Operations Room all night and I had never been so busy on the switchboard. I remember the map was lit up over four target areas in France. I could see the maps and activity in the room but had no idea what was really going on other than it was large-scale.

When I got off at 8:00 a.m. and walked out of the underground bunker that morning the sky was filled with planes from one side to the other as far as you could see. The USAAF planes flew in formation but the RAF did not, so there were formations of our planes all over the sky and the RAF and RCAF scattered in between. It was an unbelievable sight, that was D-Day – Invasion Day. They had every available plane in the sky. Essentially that was the beginning of the end for the war in Europe. Our troops went across Europe and into Germany and the rest is history.

VE-Day (Victory in Europe) was on 8 May 1945. After VE-Day you could wear civilian clothes when we were off duty and off base We all sent home for a pair of slacks, a sweater and civilian shoes. One of the first things I can remember doing after that was getting passes to go to Torquay, a resort town on the southern coast of England. They had a big dance place there with a lot of GIs so we went to dance in our own clothes and wear our hair down, instead of the regulation two inches off the collar. It was wonderful.

Even after VE-Day the bunker was still very busy since there were movement reports and orders coming and going for the field operations.

I was fortunate enough to be able to attend a post VE-Day Service of Thanksgiving at St. Paul's Cathedral on 13 May 1945 led by the King and Queen. It was a really great gathering of all the service people, that evening they lit up all the major buildings in London.

In August 1945 several of us were sent to Bad Kissingen, Germany, which became the USAAF 9th Air Force Headquarters. We were sent there to relieve the guys who has been with the Headquarters about four years, since they had arrived in England, so they could come home. We were to train civilian girls on the switchboard so they could take over and we could home ourselves.

Our home in Bad Kissingen was in an old hotel that had been converted into GI living quarters. Germany is a beautiful country. Just a few blocks from our hotel was one of Hitler's 'baby factories'. It was large building with a big fenced-in yard and this is where they had all these small children that women had for the Fuhrer – Adolf Hitler. Hitler gave them a medal if they had a baby for the Fuhrer. In other words they would donate their baby to the German Army.

These children were beautiful kids; most of them were blonde and blue-eyed. They would come out into the play yard in formation doing the "goose-step" as they did not have enough help to de-program the kids. We would take our candy down and hand it through the fence to them. I have often wondered what happened to all those children – when the mothers gave them up they lost all identity.

When we started training the civilian girls for the switchboards I was given two girls from Belgium who were identical twins. They could not speak English and I couldn't speak their language. Thankfully I was still able to teach them enough of the telephone lingo to turn the board over to them. And, finally, after 32 months in the ETO (European Theatre of Operations), we were headed home.

In September 1945 we went by C-47 plane from Schweinfurt, Germany, to Compiegne, France, for a couple of weeks. Then from there we went to Le Have, France, where we travelled back across the English Channel on the *Marine Wolfe* to Southampton. Never had I seen rougher waters than the English Channel on that crossing.

Within the month we had boarded the *Queen Mary* with thousands of other GIs and set sail of for the good old USA. It was such a thrill sailing into New York and seeing the Statue of Liberty. As we were coming into the harbor we saw all these things flying up in the air and thought they were balloons. But we quickly found out that they were condoms that the GIs were blowing up and letting fly! The first thing they did with us when we came ashore was to herd us into a huge mess hall and fed us all a big T-Bone steak dinner. The amazing thing was that we could have all the milk we could drink, because we had been forbidden to drink the milk in England I loved that most of all.

I was then sent by train across the US to Camp Beale, CA. where I was discharged from the WAC on October 26, 1945.

It was while I was at High Wycombe and working in the bunker that I met Mike Veazey, my first husband. He was a Pilot and a Bombardier – and would perform either role depending on what was needed to fill out the flight. He came in during a mission meeting and had to wait to enter the Operations Room. As I was stationed just out side the door so we talked and he asked me out. We would sneak off base to a local pub and things progressed from there.

After the war we got married and we had a daughter, Cindy, in 1948. Mike remained in the Air Force. It was in January 1949, while on temporary assignment to England, that he hitched a ride on a B-29 to Africa. He had gone along for the ride as he had never seen Africa. But the aircraft went down somewhere near the Canary Islands based on the plane's last radio communication. Despite an extensive search no bodies or wreckage were found.

Cindy passed away in 2000.

Through the years since the war I kept in touch with most of the gals, usually at Christmas. In 1990 there were six of us from Smith Hut still alive and kicking. We did have a reunion in Atlanta, Georgia in 1990. The reunion was for the 417th Signal Battalion which we were a part of. It was a 45-year reunion and we had a great time. It was amazing how close we felt after all those years. Even a reporter covering the event said he had never covered anything like it, as he could sense the closeness we had with each other. We even had a letter from General Doolittle who has since passed away. Also a letter from the Queen of England who was a princess when we were over there.

Today, as I write this, I think there are just two of us left.

After the war I went to college and graduated from the University of Denver in Colorado. In the early 60s I started as a career in the newspaper business as an Advertising and Classified Manager. That lasted for 30 years.

I re-married twice and have a total of six children / stepchildren, 15 grandchildren and four great grandchildren.

As a hobby and for relaxation I have always loved to sew. I had grown up outside of Dodge City, Kansas and in those days, from a large family, you made a lot of your clothes and I have been doing it ever since. Even today at age 90 I still do seamstress work but only part-time! My other passion these days is playing Bridge. There are a number of us old gals that get together two or three times a week.

Every day I put on the bathrobe. It has all my own patches, along with patches from my brothers. It's my link with family and my past.

Juanita Wimer Folsom, pictured at the 8th Air Force underground bunker switchboard at High Wycombe, with her arm outstretched to the switchboard. Pictured standing behind her was her supervisor Mary Kidd. Image courtesy collection of Juanita Folsom via Bill Frey.

Playbill from one of the shows that Nita Folsom attended while stationed at 8th Air Force HQ at High Wycombe. Image courtesy collection of Juanita Folsom via Bill Frey.

CHAPTER FIVE

READY TO ROLL

Joseph Augustine Charles Kiely, Jr. 1st Lieutenant

1982nd Quartermaster Truck Company (Aviation)
Bushy Park / Hurst Park Racecourse (AF Station 586, Site 6)
8 October 1943 – 28 November 1944

There I was at Penn State University, 19 years old and having a ball when the Draft plucked me out of my Delta Upsilon fraternity house and sent me to Keesler Field, Mississippi, for basic training in the US Army.

It wasn't bad except that we had four seasons of the year every day.

Anyway, having some college I did well in the placement tests and selected Air Sea Rescue as my preferred assignment. Then one day we were assembled – all 300 of us – for "reassignment". The Captain pointed at the center of our group and said; "Break it off there. This half will be cooks and the other half (my half) will be truck drivers."

And that was that!

We boarded a troop train and two days later ended up at Camp Lee, Virginia for technical training. I applied for Officer Candidate School and was accepted. After school I became a 90-day wonder Second Lieutenant on August 27, 1943.

Then the fun began.

Several of us were sent to Olmstead Air Force Base at Middletown, Pennsylvania and reported in as directed on Memorial Day 1943. Being a holiday the Officer of the Day told us that there were no quarters on base and he instructed us to go into Harrisburg and find a place to live and the Air Force would pay the bill. Furthermore he said we were to report back in couple of days.

We did that and five of us ended up at the Penn Harris Hotel, the best in town. The suite we had was expensive and we later found out that the Air Force per diem rate was only $5.00 a day! Anyway, we lived it up for several weeks there because there was no job openings for us and each time we reported back they told us to check with them every few days. It was like being on a vacation.

Then it blew up as all good things do.

The base received a TWX (short for important teletype) that 100 officers were needed in England. (Note: We found out years later from the guy who wrote the TWX in England that they only needed 10 officers and that the typist had added an extra '0' by mistake!). They say that we won the war because the Germans ran out of carbon paper first. Ha!

We were among those rushed to the debarkation center at Camp Kilmer in New Brunswick, New Jersey and from there we boarded the MS *Sloterdyke,* a Dutch motor ship, for the Atlantic crossing.

Wow, what a voyage! At that time of the year (September) the weather in the North Atlantic was very rough and we were caught up in a hurricane.

Everyone on the ship except myself and a handful of others became seasick. I was detailed to be the Hold Officer and my post was deep in the prow of the ship where a few poker players gambled their money away because they never expected to need it again. My Relief Officer got seasick and I had to pull two eight-hour shifts.

It was like living on a sea-saw because the weather was so bad and the waves so high that our relatively little ship was like a toy in a hurricane. It rose to the top of mountain size waves and then dove to the bottom of the next wave.

Many of our bravest men were humbled by the sea. Lifeboats came loose from their moorings and banged relentlessly against the side of the ship. We were in a convoy and zig-zagged our way across 'the pond', as the Atlantic was nick-named, with destroyer escorts. The trip was otherwise uneventful and a few of us had the best food that the Dutch cooks could serve up.

When we hit Jolly Old England via the Firth of Clyde in Scotland and we docked at Grennock. Many of our soldiers dropped to their knees and kissed the ground. The Red Cross girls greeted us with coffee and doughnuts and we began to feel human again.

Our next stop was a replacement depot for redeployment in Chorley near Manchester. While there we did local convoys to move supplies. The English kids chased after our vehicles shouting "Hi Yank" and "Got any gum chum?" We obliged and they were very grateful.

It was tough to find our way around England because all the signposts had been taken down in anticipation of a highly likely German invasion. If you asked for directions the English would point and say: "It's only a ten-minute walk in that direction." On one of those convoys we were introduced to the famous 'Mixed Grill' at a Manchester hotel. It was a greasy combination of bacon, sausage, liver and kidney I think. English food wasn't the greatest but the war had a lot to do with it as everything was rationed.

My buddy, Dan Daly and I were eventually assigned to Eighth Air Force Headquarters in the Kings Canadian School in Bushy Park, Teddington, just outside of London.

After a train trip to Waterloo Station in London and then to Hampton Court Station in East Molesey, Surrey, we arrived in the dark of night in the midst of a German air raid.

We heard this 'PING-PING' sound and wondered what it was. A young lady pulled us into a doorway and told us what was going on. She said that the ping-ping sound was our anti-aircraft shells that burst in the air. The shrapnel would fall to the ground and hit the pipe railings used to queue up people for the double-decker busses that were the heart of the British transportation system then. We quickly discovered that one must be aware of what the hell is going on.

After checking in at Headquarters I was assigned to Hurst Park Racecourse (AF Station 586, Site 6) in East Molesey, Surrey, just across the Thames River from Bushey Park. The racecourse was closed at the start of the war.

I was billeted with an English family – Fred and 'Mac' McDonald – along with three other young officers. Harold 'Red' Wolfe, a Chicago Southsider; Bill Luss from Gloversville, New York & Dan Daly from Tarreytown, New York. 'Mac' wanted to know what time we wanted to be "knocked up in the morning" and Fred told us to "keep your pecker up"!

We were taken aback but found out that these were common British sayings for wake up and keep your spirits up. We all hit it off and proceeded to make the rounds of the local 'pubs'. Our favorites were The Swan in Thames Ditton and The Bell in East Molesey.

I liked the Bell Pub because that is where I met the Allison family. I liked the Swan

Pub in Thames Ditton because it was elegant and on the waterfront. It was an American and Canadian hangout. One night we saw Colonel Elliot Roosevelt arrive and spend the night. As a lowly Lieutenant I did not have any direct contact with any of the Generals.

The enlisted men favored the Cannon Hotel & Pub because it was within walking distance to Hurst Park. We certainly visited all the pubs in the nearby areas of East Molesey, Thames Ditton, Esher, Kingston and Hampton Court.

At The Bell, one of the oldest in England and a short distance from the Distillers Club in East Molesey, I would meet and play Skittles with Mr. Allison. He was the caretaker for the Distillers Club which was closed when the war started. We became close friends and I dated his daughter Stella.

They honored me by giving a party at the Distillers Club for my 21st birthday. They plied me with good scotch to the point where I was tipsy. Stella and her mother took me arm in arm & guided me back to my billet at Broadlawns, the MacDonalds residence. I had a ball and sang raunchy RAF songs to prove it and probably awakened all our neighbors.

The MacDonalds had a large home on Molember Road with a lovely yard on the Mole River. They had a chicken coop with about a dozen chickens but were not getting much egg production. Red Wolff & I decided to improve that so we fired our carbines over the coop each night and voila! The next morning we all had eggs for breakfast. Yankee ingenuity at work!

I was assigned to the 1982nd Quartermaster Truck Company (Aviation) as Leader of the First Platoon. Captain Charles F. Rowe was the C.O. and Lt. Paul Hester was the Leader of the Second Platoon. There were 99 enlisted men. We were setting up extensive trucking operations in a foreign land which was jittery and on edge for the expected enemy invasion. The British were security conscious!

With all signposts down our drivers were on their own to seek their way across the road network of the UK. The elements of fog and rain were overcome at the cost of only three thankfully minor accidents during the first year of operations. The unit had been trained for desert operations and the English weather was just the opposite.

Long hours and a great deal of hard work converted the racetrack into the terminal of operations for the 1982nd Truck Co. A motor pool was constructed and a building for maintenance refitted while grease racks and wash racks were built. I had an office / desk in what was the racecourse office.

The mission of our outfit was driving. Aside from the enlisted men and three Officers in the 1982nd we also came with 50 two-and-a-half ton GMC & Studebaker 6x6 trucks. We also had two Tractor Trailers, a Command car and a weapons carrier in addition to three Jeeps. I drove a Jeep named *Pidgeon* all over England, France & Germany.

We had excellent mechanics and had very few 'dead-lined' vehicles, mostly just those waiting for parts. The most trouble we had was due to all those missing road signs the Brits had removed. So we had to drive by the seat of our pants so to speak. I guess bombs were the hardest things to transport but it was not much of a problem.

Our own 8th Air Force and the Royal Air Force had begun large scale bombing of the Continent and every heavy bomber was needed. Because of the climactic conditions in Britain, transport planes were too often grounded causing delays in the delivery of vital parts needed to keep the Fortresses and Liberators of Bomber Command in flying condition.

It was at this time that the 1982nd was called upon to organize an Express Truck

Service between the depots and aerodromes to insure the uninterrupted delivery of the materiel needed by the ground crews to keep their planes "on operations".

You could say that we were the forerunner of UPS or Fedex! The Express Truck Service was inaugurated and grew until 1982nd drivers had become familiar with every section of England, Scotland and Wales. A Unit Commendation was received for this service from Colonel Voeller, Chief of Transportation, 8th Air Force Service Command and we were told by him that our operations became a personal interest of the Commanding General.

With the end of 1943 came the need for additional truck companies. Over 300 more enlisted men and twenty officers were attached to the 1982nd for training and subsequent activation into new units. In December the 1576th Quartermaster Battalion (Aviation) was formed and in turn, activated three new truck companies. The 2193rd stayed at Hurst Park to operate with the 1982nd and the others were sent to Brighton.

At some point I received a promotion to First Lieutenant. Late 1943 saw Private First Class Walter Balcerzak become to first man from the unit to fully cement Anglo-American relations by his marriage to an English girl. Technician Fifth Grade George Lavis followed suit shortly thereafter. Almost fifty percent of our enlisted men and Lt. Bill Luss married English girls before the unit left for France in late 1944.

Yuletide and War Orphans touched the hearts of the 1982nd and resulted in a Christmas party for about 200 orphans of the Battle of Britain. This gesture of Anglo-American relations was a great success and was highly commended by Colonel Neil Creighton who was the 8th Air Force Post Commander at Bushy Park.

From the outset of 1944 the 1982nd was called upon to assist in the gigantic task of gathering and situating mountains of supplies, ammunition and materiel for the forthcoming D-Day invasion on the Normandy beaches. Over a million miles were covered by our drivers while helping in this support our great service.

With February came the "Second Blitz", the last main attack by the Luftwaffe on Great Britain. Five small bombs were dropped within 25 yards of the barracks area and many long nights were spent in air raid shelters. After this attack there was a long lull in enemy operations over southern England until April and the appearance of the much-publicized Secret Weapon 'V-1', nicknamed the Buzz Bomb.

This indiscriminate blast bomb caused the loss of many lives and a great deal of damage in and around the London area. The 1982nd, along with the other units stationed at Hurst Park, sand-bagged every building and tent on racecourse grounds, dug slit trenches in strategic places and glued muslin over windows to counteract the terrific blast created by Hitler's newest terror weapon.

After the V-1 Buzz Bombs the Germans started sending V-2 Rockets across the channel. They were much more powerful than the V-1s but it was very fortunate that neither vehicle had a guidance system. When it did hit something it leveled things flat but most of the time they fell in open areas.

During this period our men volunteered to help evacuate injured and bombed out victims. Yet another Unit Commendation was received from local authorities for assistance rendered in emergencies. Our Battalion CO was billeted in a hotel near Hampton Court Palace and was injured by flying glass (in the butt) during an air raid. He was also injured a second time (this time different part of him) he was awarded a Purple Heart each time for wounds received due to enemy action.

I did get in to London pretty often and had a ten-day leave in Edinburgh before D-

Day. Saw all the sights but spent most of my time in the pubs and between the blackout and double British Summer Time I was pretty confused.

The 9th US Army Air Force, which had been arriving in England unit by unit, began operations in April 1944. At that time the US Strategic and Tactical Air Force Headquarters was activated under the command of Lt. General Carl "Tooey" Spaatz and included the 8th & 9th operational Air Forces.

The 1982nd was released from 8th AF Service Command and assigned to the Air Service Command, USSTAF. April saw still another combined Headquarters formed and General Dwight D. Eisenhower was name Supreme Commander of the Supreme Headquarters, Allied Expeditionary Forces. Bushey Park became the workshop of all Allied Expeditionary ground and air forces. The mission of the 1982nd was to move the Headquarters wherever when needed.

During this expansion period before the D-Day invasion the 1982nd Mess Section was feeding five times its normal capacity but again, in spite of a very difficult task, received commendations on the excellence of the Mess that was conceded to be the best in the area.

The food in our Mess Hall was typical American food with lots of Spam (canned ham) and powdered milk. We did have turkey on Thanksgiving with all the trimmings. Fresh fruit was among the missing items and the eggs were powdered, too. Our Mess Sergeant was featured in *Colliers* magazine showing him mixing dehydrated potatoes with a 1/2 horsepower electric drill rigged with paddles. He was the best cook in England. At least we thought so! The food in the Officers Mess was the same as in the enlisted mess and it was all very good.

After the invasion in June, weekly vehicle inspections were made with emphasis on both visual and mechanical perfection. The 1982nd vehicles were awarded first place among all units on two separate occasions. A silver plaque was awarded each time and a commendation received from the Commanding General.

In November the first contingent of the 1982nd, under the command of myself and S/Sgt. Cole, departed for St. Germaine en Laye, France. It was a hard time for me personally. My only girlfriend had been Stella Allison. It was getting pretty serious before I was sent to France so when I got there I had to lie to her and break it off because I was ordered to evacuate a munitions depot in the path of the Battle of the Bulge.

Fortunately for me General Patton came to the rescue and the Depot was held secure. But there was a war on and I didn't want her to be waiting for me.

The balance of the unit followed and in November 1945 with the war in Europe over the unit moved the Headquarters to Wiesbaden, Germany, where United States Air Forces Europe came into being.

As for me I left Europe in 1946 on a Liberty Ship from Le Harve, France and had a smooth voyage back to the States where I was released from active duty and returned to my home in Philadelphia, Pennsylvania. Subsequently, I landed a job as Assistant Purchasing Agent at The Franklin Institute and got married.

My wife, Margaret Alice Beverage, was from Monterey, Virginia. I met her for a short time while I was in Richmond, Virginia, in 1943 but the romance began when I was at Fort Meade, Maryland, where I was discharged and she was working in Washington, D.C .at the Headquarters of The American Red Cross.

I also joined the Air Force Reserve and five years later I was recalled to active duty for service during the Korean War. I was assigned to Chaumont Air Base, France, where I opened the Purchasing & Contracting Office. Later I was transferred to Ramstein Air Base, Germany and opened the P&C Office there.

In 1956 I was sent to Ohio State University for two years to finish my college education. I was then assigned as Air Force Officer in Charge at the IBM Corp. defense plant in Owego, New York, where they made the Bomb/Navigation System for the B-52 bomber which is still in service today. Then I attended The Air Force Command & Staff College in Montgomery, Alabama and from there to The Ballistic Missile Command in San Bernardino, California.

I was then selected to be assigned to the Inspector General at Andrews Air Force Base. Three years later I was assigned to the Pentagon and was promoted to Colonel. I served there for six years at the top echelon of Air Force Headquarters and was awarded the Legion of Merit when was I was assigned to the Defense Logistics Agency from where I retired in 1975 and moved to Annapolis, Maryland.

In 2001 I was awarded a Certificate (Diploma) at the French Embassy in Washington, D.C. for my contribution to the liberation of France.

One personally sad note for me: When I went back to England in 1990 to visit my daughter, Pat, who now lives there, I learned that the Distillers Club had burned to the ground sometime after I left England. I couldn't find out any of the details but was worried that the Allisons might have been injured in the fire.

I still don't know.

Lt. Joe Kiely in 1944. He's pictured with a moustache which, he says, "didn't last long."

In the Winners Circle of the Hurst Park – the Officers of the 1576th Quartermaster's Truck Battalion (AVN). Lt. Joseph Kiely is pictured just to the right of the sign opposite Miss Henderson, the Battalion secretary.

The gun crew picture (above) was taken at the tarmac in front of the grandstand at Hurst Park. Joe Kiely is the one on the left "pointing at the enemy".

CHAPTER SIX

WEATHER OR NOT

Anthony Pircio

2nd Lt., USAAF
18th Weather Squadron
Pinetree, High Wycombe
April – November 1944

If there was one thing World War II did it was to shorten the time I had to spend in school.

I enlisted as a private in September 1942 while I was at college at Fordham University in the Bronx, New York. As my school year progressed my classes got smaller as students joined up. And the college was also rushing people through. My graduation was moved up to January 1943 instead of June 1943.

That was great except I never made it to the ceremony. I left for my military assignment a month before I was due to graduate.

During my last few months in college I had filled out a form for cadet training in the Army Air Corps. You needed to have two years of college to get in.

I had really wanted to get into the Army Air Corps. The choices once there were pretty simple – pilots, bombardiers and navigators and meteorologists. However, when I had my physical exam it turned out my eyes were not so good. So I had to accept a ground assignment.

I had said goodbye to my family and girlfriend, Mary Smercak, the day after Christmas 1942. We were engaged but we didn't get married. I said to her: "If I can come back and we can fulfill this and get married."

I think you could say that I knew next to nothing about meteorology. Even the word was pretty unfamiliar to me. All that was to change however. And so The Bronx became Grand Rapids, Michigan and college became Meteorology School. We were stationed in a large hotel and we utilized an underground passage that connected to our classroom maybe 100 yards away.

One other difference – instead of a class of 30, 50 or 100 our class was 500 strong!

We started classes in January 1943 and finished in September 1943 with only a break of about a week to go home as I recall. We went to school every day including Sunday. Sunday also turned out to be the worst day of the school week.

On Sundays we had to study Dynamic Meteorology – which required knowledge of Calculus. But the very worst part for me was when they asked you to go up to the weather map. This giant map was up on stage. You would go up and they would 'flash' a weather synopsis and you would have to stand there and explain what was happening.

You could say I had a case of stage fright. I was a shy individual!

We got one week-long break during the entire course. We caught a train home and the thing I remember about that is that it dropped me off right in my own hometown –

Tarrytown, New York. I was the only one to get off there and it was like I had my own personal train to get me home.

In September I graduated as 2nd Lieutenant. I received my orders and I found myself heading to Warner Robbins Airfield in Georgia where I spent the next month followed by another month at Jefferson Barracks in Missouri. Then we got orders through that only stated that we would be going to a 'temperate' climate. So we requisitioned clothes for that. By this time my 500-strong graduating class had been split up and only about 100 of us headed off by rail to Fort Hamilton, in Brooklyn, New York, several days later.

And there was our surprise. We marched right onto the *Queen Mary*. I was billeted down on M-Deck (I still have the slips from my billeting). We were in a stateroom that might, in normal times, have held two or three guests. In wartime it held 18 of us. The bunks rose up along each side of the walls about four or five high. I think on that trip there were 18,000 of us going over. There were two interesting things about the trip – to me, at least. The first was that we ate our meals in the swimming pool! They had drained it, obviously and put tables in it! The second was our disembarkation point. We had all assumed we were heading towards England but we embarked at Gourock, Scotland.

My first impression of the place was not a happy one. In fact, I was distinctly angry – as I am sure were a number of others. Perhaps our view was clouded by the fact that when I got my foot-locker back I opened it to find all my liquor was gone. I had put in several bottles of whiskey. Nothing else had been taken. I'm not sure who took it but we blamed the Customs people. We were not happy campers.

On a happier note the weather was good…for Scotland. It was clear. I had naturally assumed given all my training that we'd arrive in fog! We climbed aboard a train and then something happened I won't ever forget. As we departed south people lined up down the tracks and waved us on. That was really uplifting. It wasn't just a fluke – people also waved at our train along the route.

Somewhere on the journey south we stopped at what appeared to be a newly-built concreted campsite run by the British. The cots we had to sleep on were miserable. The mattresses came in two sections.

Then we continued on south and I ended up at what I believe to be a staging area from which we received our individual assignments. I'm not sure where it was but then I spent most of those first few weeks not knowing where I was. While we were there we were served a Thanksgiving Dinner – I recall that because during the dinner we were buzzed by three B-17s. It was the first time most of us had seen that plane. I guess they were trying to introduce themselves to us! My first real weather work started at a place called Sawston. No standardized military watchtower but the home of an old and distinguished English family; a building with history and legend housed this detachment (Detachment 371). Some of us were housed in a big old stately home that had served as a vicarage. An RAF weather station had been in operation before the Americans came. The weather station was located in the attic of a building where the beams were low enough to be dangerous to anyone over six feet tall.

The weather personnel occasionally issued operational forecasts for the 66th Fighter Wing but the station's distinctive operational function was working with the Controller who made the decision on whether to fly or not.

The Officers Club was right in the middle of town and we used bicycles to get to and from it. I also got in big trouble there. I went to lean on a wall by the club little knowing it was some important wall specially made of mud – and I knocked part of it down.

On April 1944 I received orders to report to AAF 101 with the 8th Reconnaissance Wing at Pinetree – the codename for USAAF Bomber Command. I didn't know where that was either. A few days later a jeep picked up me and my equipment and took me to High Wycombe and to the Headquarters.

As we pulled up and went into the base my impression was highly favorable. There was a lovely and large open green lawn. I remember later lying on it during May – the only month that I was in England when it seemed to be sunny and warm! It was big enough that they had a travelling carnival group come with a merry-go-round and other rides.

Anyway I was driven past the main building which I quickly learned had been a girl's school. At that time I didn't realize that the building had already become a footnote in WWII GI lore. I learned about it quickly after I had settled in however.

When the 18th Weather Squadron landed in England in 1942 the first place they went was Pinetree and they were billeted in the main school buildings. The first night in the dormitories and the sound of bells echoed throughout the halls of residence. The reason? By each bed was a little button with a sign that read "Ring for Mistress"! Now for young, properly educated school girls this had one meaning, but for the GIs this was too much not to exploit. A legend thus born! I think those in charge quickly disconnected them after that.

Anyway, I was taken up the side of the hill to near the top where I was billeted in a Nissen Hut. There were about 14 of us in there – seven on each side with ubiquitous pot-belly stove in the middle.

Once I had sorted myself in I walked down the hill (probably 50 yards) to where the weather station was located in another Nissen Hut. During this period six officers and eight 'Non-Coms' (Non-Commissioned Officers) arrived to discharge the duties of forecasting for reconnaissance missions ordered by Pinetree and preparing forecasts for Headquarters, 8th Reconnaissance Wing.

Reconnaissance missions could be requested by Pinetree at any time during the day; the request designated the area to be covered, the information desired and the time when the report would be wanted.

The proposed mission was then forwarded to Operations at AAF 376 (the US military's code for its base at Watton) in Norfolk. The general plan was provided and the weather requirements and aspects were discussed with the Observer – Navigator. In the meantime the weather at 376 was also monitored and discussed and a decision would be made and a forecast for the route and the base on return.

When these were determined the mission was sent on its way.

The 8th Recon. Wing was activated as an integral organization with the task of coordinating all reconnaissance necessary to the operation of the 8th Air Force. Colonel Elliot Roosevelt assumed command and activated the Light and Heavy Weather reconnaissance stationed at Watton.

For most of our meals we ate the Officers Club located at the top of the hill. Then I went to work.

The primary duty was would be to draw the most up-to-date map. The technicians would put together the symbols on it based on the information received from the teletype from weather stations across the UK.

You would analyze where the lows and the fronts were and then you'd made forecasts. Then the information would be supplied to different airfields to either add to whatever they had (a few had their own weather units) and other places that didn't have forecasting capabilities.

The information we were providing at that time was not about targets in Europe but for the pilots, navigators and crews it was just as crucial – it was about the weather over the departure and return. In particular they were concerned about their return – they wanted to know they'd be able to see their home field runways. The weather information was provided and sent mainly to Watton and Mount Farm (in Oxfordshire).

All we would do was weather work. We worked for a large number of hours in a row so we could get a day or two off from time to time. And we never saw any of the top brass accept at parades. Once however I had a call from our top man – Colonel Paul G. Davis – he was concerned about forecasted high winds and, as I recall, wanted to know if we had more information. I believe he was being asked if they should move the planes into shelters and hangars.

I seldom had reason to go to the main building though I did get into one of the school buildings – they had an area there where you could go to play cards – and even bring a girlfriend. There were a lot of dances in High Wycombe. In the town itself and just starting up the opposite hill was the railway station. I used to walk up past there – it was where some of the girls lived and we'd walk 'em home.

I remember going to a pool hall in High Wycombe on a few occasions where they played the English version of the game, billiards with a large pool table – a game they called 'Snooker'. There I met a person who was a teacher and he and I became friends. We'd play pool and I'd bring him cigarettes as we got them but I didn't smoke. I drank the British beer but I didn't particularly like it. I also found a small bookstore where I enjoyed looking at books on Chemistry.

On occasion I also went to the movie house in High Wycombe. I saw a lot of films there but the one I remember in particular was *Going My Way* with Bing Crosby. It won seven Academy Awards in 1944. One song from that movie, *Swinging on a Star,* stuck in my head.

And on one occasion I took a girl to London and saw a show, *Arsenic and Old Lace.* We went as friends. It wasn't anything further. When it came to females I stuck to a tried and tested old military motto: Keep it in your pants! I had a fiancé.

I would write letters home – then we had V-mail. Most of them were to my fiancé and to my sister – my mother didn't' read English so my sister would read it to her. V-mail worked really well for me – one page. I never did have too much to say so that made it easier.

We seldom saw the war up close and personal, as they say, while stationed in High Wycombe. The only danger was one of those V-1s we heard once-in-a-while. The V-1 did not explode in our vicinity but one day I heard the explosion while on the phone talking to person at our air base. He said: "Did you hear that?" And I responded; "I sure did!"

I remember being on a nightshift once and I was with a group of soldiers sand we went to eat somewhere that was underground. I think it was in the bunker at High Wycombe.

One particularly memorable event at Pinetree was when Glenn Miller and his band performed before the entire personnel. I do remember this show taking place outside midway up the hill on a large flat truck. It seemed like everyone on base was there!

In December 1944, I moved to Mount Farm – with 7th Photographic Group near Oxford. We were there when D-Day happened. I remember the day well – the sky was packed with planes. As I recall on that day the entire photo group was sent on missions to photograph what was occurring during D-Day. We were kept really busy explaining the weather conditions over France.

It was at Mount Farm that we at least got some real feedback about our accuracy and usefulness. We lived in the same building as the pilots and once-in-a-while they'd complain: – *"Those clouds or that storm front you sent me to find – it was a wild goose chase."*

From about Christmas to New Year's Day in 1945 the whole group moved to Valenciennes (Airstrip A-83) in France. I went over with them and was assigned to the 27th Squadron, one of four squadrons within the 7th Photographic Group. We were billeted in a chateau that the Germans had vacated. It was amazing. We even had our own rooms! From there it was so much easier to run the weather forecasting as the aircraft had shorter distances to fly to the target areas.

The pictures that stuck in my mind came at the end of the war. The pilots had taken lower level pictures of all the damage done and it was made into a book. The images of Dresden in particular – it was hard to imagine what I was looking at had been a city.

There was one other thing that stuck in my mind. One time some of the pilots came back and they thought they would buzz the Rhine. They did buzz it but during the maneuver one of the lead planes hit a high-tension wire and crashed into the river.

Then the war was over. The squadron was made ready to go – we thought we'd be going to the Pacific. But then the news came that Japan had surrendered and the powers that be sent us back to England on the way home. As luck would have it I also caught the *Queen Mary* back home. This time rather than 18 to a room we were four or five to a suite and not the sardine situation we had coming over. When we arrived in New York Harbor we were greeted by fireboats spraying arches of water.

I returned and, influenced I guess by my training in weather, took an interest in Biochemistry. Soon after I took advantage of the GI bill for veterans and returned to Fordham University as a graduate student. In three years I earned a PhD in Biochemistry.

I returned to England with a group from the Air Force Historical Society in 1988. Our main interest was in visiting Mount Farm which was now a village called Berensfield. All that was left of the air base was a strip of runway about 50 yards long. The people of Berinsfield had a monument constructed near what was our air base in honor of the 7th Photo Recon. Group. We also visited Oxford where we had spent many hours of our free time during the war.

In 2005 I brought my son, Nicholas, on another trip with the 8th Air Force Historical Society to return to Omaha Beach and environs in France. Our home base for this trip was a hotel in Caen, France.

Sadly I never did make it back to Pinetree or High Wycombe. We didn't have time.

Oh – and one more thing. Remember Mary, my fiancé whom I had left behind in New York? Well I got back and I did marry her and as I write we've been married for 62 years.

The 325th Photographic Wing Reconnaissance at Pinetree, High Wycombe in 1944. Left to Right in first row: 1.James C. Nance 2.Donald R. Boles 3.Anthony W. Pircio 4.William H. Schemph 5.Leon A. Lowenthal 6. Paul G. Davis 7. Kenneth A. Meek 8. Charles L. Sortomne 9. Franklyn 10.Rodney A. Webb. Left to right in second row: 1. Unknown 2. David M. Sutta 3. John W. Marklin 4. William J. Pavey 5. Porter Martin 6. Osborne Richards 7. Unknown 8. Unknown 9. Andrew Couch. Photo courtesy of Anthony Pircio.

CHAPTER SEVEN

GET THE PICTURE

DWIN CRAIG

2nd Lt.
8th Reconnaissance Wing, US Army Air Corp
Daws Hill, High Wycombe
1944 – August 1945

(Author's Note: In 2001 Dwin Craig, who was then 81, was kind enough to set down his beloved trumpet and step away from his inventor's workshop table to take time out to talk about his wartime memories of his wartime assignment to Eight Air Force Bomber Command at Daws Hill in High Wycombe. Dwin came across as a man who wanted to experience life and live it and I understand that he did that to the best of his capacity until he passed away in 2007. While never having met Dwin in person but it was clear from conversations that he went about life with a twinkle in his eye. His son, also named Dwin Craig, was kind enough to fill in a couple of missing details after his death.)

My wife, Virginia and I missed the start of World War II. The fact was that we had come from Maryland to climb to the top of the Washington Monument on 6 December 1941. The climb – something like 555 feet – had been great and the strain afterwards on our muscles we never knew we had really impacted on us. We had stepped up every one of those stairs at the monument and had come down the same way. So we stayed in bed most of the day on 7th of December 1941 and it was only much later that evening that we turned on the radio and got news about the business in Pearl Harbor.

And that's why we missed hearing the Declaration of War.

Back then I was an aspiring young musician but I could see the writing on the wall. It wasn't going take too long for them to come around to my house with the letter marked "Greetings from the President…" So I dodged the draft and volunteered first. My wife worked at the Pentagon and one of the officers suggested that I should go through Officer Candidate School. They took my blood pressure about six times, I covered up the bad eye when I took the eye test and then I was in the Army Air Corps.

After that I was soon on my way to Denver, Colorado and on the start of a trip that would take me back to the East Coast and, eventually, across the Atlantic to the Eighth Air Force Headquarters at High Wycombe.

I was sent first to Lowry Army Air Corps Base in Denver, Colorado to get some training. Denver was supposed to be a great town for steaks…it was noted for it. So the first time we got leave we go to a steak house and really lived it up. That night when we returned we got the news – we were being transferred.

Two weeks later I found myself at New Haven, Connecticut, amongst the august buildings of Yale University. The military had taken over parts of the university including the Quadrangle) and was busy churning out '90-day wonders' as we Officer Candidates were known.

I recall that locals weren't too wild about us. We used to go up and down the streets singing a popular song of that time called *I've Got Sixpence. (Author's Note: There are numerous versions of this song and but words and music are credited to Elton Box and Desmond Cox.)*

And so after 90 days of training to be an officer, gentleman and a photographer, I got my first assignment down to Pinecastle Air Force base – a B-24 base near Orlando, Florida. At that time it had a mile-long runway – the longest runway in the world.

I used a K-18 camera and that required 9x9 negative film. Pinecastle was a place where they formed units to go overseas. Many of the pilots, navigators and bombardiers seemed to have already served their time in Africa and were less than keen to go overseas again.

I could have happily spent the whole war down there at Pinecastle but, in truth, I got a little bit bored. Somehow I heard that they needed photo interpreters overseas. I arranged to go to a photo interpreters school in Harrisburg, Pennsylvania, where I spent six weeks learning to tell what things looked like from the air. We learned to distinguish residences and factories and look for train lines leading into buildings as well as a little bit about the industrial/manufacturing process. During that time my wife joined me but after three months she returned to work as a secretary at the Pentagon.

And then I was on the move again – to Goldsboro, North Carolina.

It was a 'Replo Depot' (a Replacement Depot) for those going overseas. There were a helluva lot of guys there…some of who had been overseas and who were going to go back. Part or our responsibility for the month or two we were there was bivouacking. That included five-mile runs. You'd be exhausted but in the morning, instead of bugle call we had *Oh What a Beautiful Morning* being sung by Bing Crosby coming over the loudspeakers. We'd all go: "Oh Shit!" then get up and run five miles again.

Sometimes people and places make an impression on you that sticks with you for life. An incident at Goldsboro remains etched in my thoughts. There was this guy who didn't have a metal rim inside his hat. Mix him with an asshole Lieutenant who was in charge of us. This Lieutenant was checking us over and saw this guy with his floppy hat.

He told him to straighten up and asked him where he had gotten his military training." Unbeknownst to the hapless Lieutenant the guy was a veteran and staunchly replied: "Africa." That took the wind out of the Lieutenant's sails.

Sometime after that we shipped to Fort Hamilton, New York, a base that was a subway ride out of New York City. None of us knew where we were going at that point. The routine up there was to look at the bulletin board at 5 p.m. each evening. If our name was not on it to ship out, then we'd head out to town.

At the end of a month I was already halfway into next month's pay and I was hoping and praying the boat would arrive to take me out of there soon.

But the stay there wasn't wasted. I made some great friends including a young Herb Caen who went on to become a gossip columnist in San Francisco. A mutual love of jazz music sealed the friendship.

Finally my name showed up on the board. We packed our crap including a carbine rifle and headed for the docks where the *Isle de France,* the great French luxury liner that was to take us to England, soon set sail.

It was a four-day trip but it couldn't end fast enough for me. For despite the French name they had an English mess on board and the English food was awful. I found the only way I could eat it was to put mustard onto everything. I ended up putting mustard on everything but the one thing you should put it on – hot dogs!

With some 20,000 troops on board and nothing but time to kill there was only one thing to do and that was to play poker. I hooked up with Herb Caen and soon found myself at a celebrity poker game.

Amongst those there was the singer Tony Martin and several others. We got in and played with them. I lost…it was obviously the protocol thing to do!

Thankfully the ship made it across and I think we docked somewhere in Scotland. It was nighttime and I didn't much care where we docked as long as solid land lay under my feet. But then we exchanged the rock'n roll of the ocean for the bump and barge of a train heading South through Scotland and into England.

Somewhere during the night we stopped at an Army mess – where I had the best cup of coffee I'd ever had in my life. Not only because I really wanted a cup of coffee but because they had put a lot of chicory in it…a LOT of chicory…and I liked that.

I can also remember a guy looking out of the train the next morning and suddenly exclaiming; "Good grief, there's a lorry." He was trying to impress us with his knowledge of the local lingo. We saw all kinds of signs for things like 'Bovril' and for 'screwing machines'. We thought that the latter was a great idea!

Soon after the train trip South I was on my way to a school in Sudbury, near London, to learn about interpretation of airborne radar – a system being used by the USAAF. The airborne radar systems would take a 360-degree 'snap shot' of the ground below the aircraft as a radar reading known as a Planned Position Indicator, which appeared on 5" film.

It was while we were there that a Colonel came up to pick out the smartest of the bunch to work for him…and that was me! That's how I needed up at US 8th Air Force Air Corp Headquarters in the Cartography section of the Intelligence Department at Daws Hill in High Wycombe. The base was on a hill and we were housed in a building on the top of the hill where we both bunked and worked. It was a big building.

In the room where I bunked there were about 20 guys and that was about a fifth of the people up there. The USAF had a real factory operation going on there. We were all processing data and collating and sorting. It was going on all over the base which ran down the hill.

I recall that a Mrs. Menzies, a lovely lady, was retained from the Wycombe Abbey school job to take care of all housekeeping needs. She lived at officer's quarters and had large staff who took care of cleaning, feeding, etc. of the officers on base. We had Brussels Sprouts every day, cooked one of five different ways. The whole countryside stunk of cabbage!

Funnily enough for all my time there I never did see the command bunker though I heard about it often enough. We knew there was a place where they did the planning and that you had to have special kinds of clearance to get in there. And frankly 2nd Lieutenants don't do a lot of planning – except where to go after dark! We just knew it was a bunker deep down…you hear all the usual spook stuff you can think about.

One of my buddies at Daws Hill was a civilian and an electronics engineer. He was over to learn all he could about radar, which the British had invented and also to teach the British what he had learned at MIT. The British radar system – H2S – was generally referred to as "Stinky".

The trouble with being a 2nd Lieutenant at Daws Hill was that there was a lot of brass around. I saw General Doolittle on several occasions. And I know at least one other

General Officer certainly noticed me. They had an Officers Club down at the bottom of the hill. I was down there making eyes at one of the women there once when this General came over. He said, in so many words; "Lay off – she's mine."

Still I did my bit for Anglo-American relations – I had a memorable encounter with a woman in the grass around the base. I also remember drinking Gin and Bitters – I would get drunk on those – and walking up the pathway through the woods…and frequently falling over. We used to run up and down that hill to get the trains from High Wycombe to London. There was a dance hall in London where they had two bands on a revolving stage.

There were always lots of available women to dance with and lots of available women all over town. Once there was this cute little prostitute I met – I took time to talk to her – she was from the Isle of 'Somewhere' and she'd been separated from her parents.

But one of my greatest memories is the time when Glen Miller came to play at the base.

I was there standing in front of the band. Afterwards I went into the Officers Club and the piano player, Mel Powell, was playing in there and we were singing along to the ones I knew the words to. He asked if I was a trumpet player. I told him "No – I play the saxophone". It was probably the next day or so after that Glen Miller left for good on his final flight.

While being a 2nd Lieutenant was not a great thing to be at Daws Hill it was certainly the worst thing to be when you're about to rub a 'Full Bird' Colonel up the wrong way.

This Colonel I worked for had previously been a designer – designing tennis rackets. He liked to draw perspective maps of targets – how they should look to bomber crews going in from different directions. They were lovely but I decided to do some checking around and ended up talking to some of the navigators and bombardiers who had been in action.

These were some of the stressed out guys who had been reassigned to flying desks. I started asking about these targeting perspectives. Their response was: "Hell, nobody uses them." So I went in and told the Colonel that. He didn't like that and I pretty much told him to take a crap in his hat! He told me to pack my bags.

The very next day I was transferred to Elveden Hall, a huge mansion with extensive grounds at Bury St. Edmunds, located near Cambridge. Elveden Hall was the Headquarters for 3rd Air Division but had originally been home to the Maharajah Duleep Singh, then owner of the Koh-i-noor diamond) who had been raised in India and been exiled to England in 1849 before purchasing the house and its 17,000-acre estate in 1863.

It was a great place to be stationed. We were close enough to Cambridge to hop a jeep go into town and do some 'chasing'. At times, however, it could be a dangerous journey.

Some of the guys were not used to driving on the roads…and that problem became compounded when it was foggy. I can remember on one dark and foggy night hanging out of the jeep, dragging my hand on the ground and telling our driver when he had strayed from the tarmac!"

It was while I was there that I learned two pretty crucial things. One was personally important and one a little more important to the war effort (though I failed to realize the full significance of either at the time).

On a personal front I learned to drink a pint of English beer at room temperature, which was an important consideration for me. On the war effort front I got to see an awful lot of radar photography. They were using radar photographs primarily for navigation and, unfortunately they would sometimes bomb the wrong target.

While I was there I developed a technique for finding out where those bombers were based on the visual track of the aircraft. I was able to correlate the visual Planned Position Indicator (PPI) photos with timings. The PPIs took about a second to do the full 360-degree scan to create that image. Between the visual and the PPI I could tell where the bombs must have hit from the PPI. I sent out a report that went to every Squadron in the Division, not realizing the politics of it.

Even though I was probably hated by combat crews it had an effect as we noticed after awhile that the bombing got better and better because these guys knew that someone was able to check on their accuracy."

And there was a third thing I learned from my work there. As I studied the PPIs something else was becoming quite apparent to me. It was because the aircrews would turn on these radars every time they left the ground and went to Europe.

Once the images came back we could see something unusual on the south coast of England. The radar reflections of metal were growing larger and larger. Every day there would be more and more blobs. We knew D-Day was coming but nobody knew when."

I stayed at Elveden Hall until after VE-Day. After then I transferred out to another location. We hung around waiting for a ride out back home. We were all "paddle feet" (someone in the Air Corps who didn't fly) so we were the last to go home. At any rate, we waited for a ship and that took about a month or so with the requisite celebratory parties and celebratory runs into another now nameless town.

Finally in August 1945 I boarded the *Queen Mary* with 20,000 other guys – including about 86 others from the Washington DC Area. Somewhere along the line I learned to play Bridge and there was another guy and I who played as partners. On the way back we had a Bridge game going with two other older guys…we'd take turns winning using signals we'd devised.

We were playing for money. By the time we got home a total of about $1 had changed between us!

Europe was effectively a closed chapter but the war was continuing in the Pacific and that's where I immediately found myself bound for. In fact we were out at sea on our way out to the South Pacific when we found out that the Japanese had quit. Soon after that we were given a choice of staying in the military but I had enough points and it was getting silly with the officers checking everywhere for dust. I got out.

Back home I faced an inspection of an entirely different type and I had to think fast. When I got home my wife helped me to unpack my footlocker. She ran across a condom in its wrapper. I told her: "Look they made us take one with us every time we went off base. But, as you can see I never used it!"

The whole military experience taught me enough about electronics to know that I wanted to become an electronic engineer. So I took up the GI Bill and studied at George Washington University – going through a four-year programme in three years. But in the end I didn't do the Electronic Engineering degree. I got close but I ended up taking a BS in Engineering with a Physics Major. Actually it was perfect because it allowed me to become an inventor, engineer, entrepreneur and promoter.

Life since the war has seen be go from rags to riches to rags. One of my inventions related directly to my wartime job – I was able to use a television tube as a light source for printing aerial photographs. The picture tube light was 'squirted' through the negative. It wound up allowing the operator to get considerably more information out of the photograph including in the shadows. It was the automatic equivalent of 'dodging' done by photographic operators in the darkroom. Anyway it was enough to start up a company that I worked at for 10 years. Then I blew it all…

My personal life remained interesting. I managed to survive my homecoming interrogation and was married 44 years until Virginia died in 1985. We had adopted children who now have families of their own. A second wife came and went and since then I have been "happily single". I guess you could say I outlived one wife and outran another. In the meantime I picked up the trumpet and managed to get into a 17-piece band.

A few years ago I met a lady through the web who was from San Francisco. She was an Austrian survivor of the Holocaust and we wound up meeting up at various places in the US. A couple of years ago we decided to meet up in England. I had an extra day in England after she left and came back to High Wycombe. I went back up to the base.

Sadly I didn't see anything that looked familiar.

Left: **Dwin Craig in World War II.** "Here is one taken by a Hollywood photographer stationed on the base," he wrote in 2001. "My kids can't believe it was me." Right: A picture Dwin sent in 2001 he entitled 'Dwin guffaw.'

CHAPTER EIGHT

ANCHORS AWEIGH AT THE AERODROME

Joseph Tumminelli, US Navy AEAN (CAC)

Utility Transport Squadron 4 (VRU-4)
Air Transport Squadron 24 (VR-24)
Hendon Aerodrome
(1947 – 1950)

I suppose I'm probably one of the few who can still tell you about us "Yanks" at Hendon. After all I was there 34 months.

We were an odd group because we were Navy and I suppose most of the people were curious about why US sailors were far away from water and, moreover, at an RAF Aerodrome.

The squadron was commissioned on 3 December 1946 but I didn't arrive there until September '47 as I was in transit from Aviation 'A' school in Jacksonville, Florida,. to Brooklyn, NY for transit via an MSTS ship the *USS General C.C. Ballou* which docked at the German naval base at Bremerhaven for my overnight stay. Then it was a 40-mile bus ride to Bremen to be get my flight to London.

Our entrance to Hendon Aerodrome was off the Watford Way under the railroad track. The main gate was off the Colindale Underground station – the stop before Edgware.

We had one of the three hangars there along with some Quonset Huts that I assumed were put up by the crew that was already there.

The Squadron had a dozen pilots and 56 enlisted men – all aviation ratings – to service the aircraft. We flew four R4D's (C-47s) and three JRB's (Beechcraft) from a pool of US planes left in Italy after World War II. We also had a detachment in Port Lyautey in what was then known as then French Morocco. It comprised one R4D & one JRB. I worked on the electronics maintenance.

None of us lived on base. We all lived 'ashore' in rented houses or flats or in some of the hotels in London proper. Our shopping was at the US Naval Headquarters facility in Grosvenor Square for food as well as the PX items available like our ration of cigarettes and wine plus civilian clothing. Sweets were a big item that, if I can recall correctly, were rationed at that time.

We were also issued UK ration books to use on the local market. Most of us gave them to the local people we befriended. Mine went to my landlady.

We had a small dining area at Hendon in a hut that we built for coffee and eating minor foodstuffs. Since American voltage is 120V we eventually had to get a generator to take care of this particular problem. We electricians did our job even if it wasn't an aircraft.

Our working day normally started at 0800 and we were out the gate by 1600 other than the "Duty Section" who stayed overnight. All were subject to recall if needed.

Our main mission was to transport personnel, mail and cargo to where they were destined and to the Embassies being opened or reopened throughout the countries in Europe.

As the Fleet started to come back to the Mediterranean, we were instrumental in the opening the base in Rota, Spain and putting a detachment in Naples, Italy and other places. Hendon's roll became less important because larger aircraft were needed to carry out the mission. Because of this a selected crew was sent to Blackbushe Aerodrome in Hampshire so that they could service the four-engined R5D's flown in from Morocco that came from the US.

My off-base home was with the O'Malley family at 26 Finchley Lane in Hendon.

The O'Malleys were to me, at least, a typical Irish family. I recall that the father, Dan was a "dustman" and Danny, Kevin and Paul were young kids. Dan's wife, Mary, was the homemaker. She was the one to make sure I put a shilling in the meter if I lit the fire heater in my room. The boys wanted to know about America and what I did before coming to London and I, in turn, was taught that a "truck" was called a "lorry" and "petrol" was used to run it besides all the new names for parts of a car, etc. By the time I left their home and moved into a flat with an electrician friend young Dan had joined the Army and that was the last I saw of him. I did stop by at times to say "Hi" to the family as time went by.

For all of us there, London was something new with places to see especially in London proper with great shows, seeing Buckingham Palace and, of course, joining my squadron mates in the pubs around Piccadilly Circus. I know one was an underground place but the name doesn't come to mind but I do remember an eating place called the 'Chicken Inn' at the Haymarket in Piccadilly.

Two of my off-duty highlights included dancing at The Lyceum, learning and teaching different dances with the young women who also came to dance as we did. And the second was seeing Roger Bannister break the four-minute mile at Wembley Stadium in August of 1948.

I received orders to leave in July 1950 to report to Aviation Electrician 'B' school in Memphis, Tennessee.

RAF Hendon – home of the US Navy. Tumminelli and friends on a lunch break outside the NAAFI. Photos courtesy of Joseph Tumminelli.

CHAPTER NINE

THE ROAD TO HIGH WYCOMBE

Donald J. Terrill

T/Sgt (USAF Retired)
Daws Hill, High Wycombe
1952-1955

I joined the US Air Force in June 1950 when I was 17 years old. Looking back it would have been hard for me to imagine that in the next four years I would become an office clerk, then a cook, cross the Atlantic ocean, end up on a small American base in England and then get married. But that's what happened on my road to High Wycombe.

I enlisted in Detroit, Michigan, and then, along with other enlistees, traveled by train to San Antonio, Texas; a journey of three days. It was an enjoyable trip until we reached our destination. We were met by a couple of mean and tough-looking Sergeants from Lackland Air Force Base (they were called Training Instructors, or T. I. for short). They told us "we belonged to them for the next 13 weeks of training". From that moment, my life changed dramatically. I just didn't know what to expect – I felt alone!

Ironically, it was many, many years later I would eventually become one of those 'mean and tough-looking Sergeants' in charge of training new recruits for the USAF at Lackland AFB. However, that's a different story…

Because of the outbreak of the Korean War and the influx of many new recruits my basic training at Lackland was reduced from 13 weeks to just three. It sure didn't hurt my feelings! From Lackland I was then transferred to MacDill Air Force Base in Tampa, Florida, where I became an Orderly Room Clerk assigned to the Food Service Squadron; my typing job lasted just nine months when I decided to volunteer to become a cook.

I attended a six-week course at the U.S. Army Cook School on Camp Gordon, Georgia, in June 1951. Since I was attending an Army Cook School I was also taught how to set up and cook on portable Army kitchen equipment.

After graduating from Cook School I was transferred back to MacDill AFB, where eventually, a few months later, I received my orders to England.

When I received my overseas orders you'd think that since I was in the Air Force I would fly over the Atlantic. WRONG! After a week of processing at Camp Kilmer, New Jersey, I set sail on the *General M. L. Hersey*, along with around 2,000 other servicemen, from New York on the 28th of November 1951 heading for Southampton.

Due to stormy weather it took our ship 10 days to cross the Atlantic instead of usual six days. I was seasick during most of the trip across. I remember one day, while still feeling under the weather, we were ordered on deck for an emergency boat drill. It turned out that I was the only one who forgot his life jacket (lucky for me it was a practice drill).

A new world opened for me when we finally docked at Southampton on December 8th.

To cap it all we arrived at night so my first sight of England was less than impressive. It was dark and it wasn't just the fact it was nighttime – England still seemed to be running under wartime conditions with very few lights that I could see. Not only that I found out later the British were still on food rationing (six years on since WWII ended).

After completing processing (no passports needed) we were given our orders to various bases across the country; my assignment was to RAF Sculthorpe. We were put on a chartered coach and left the following morning. It rained all the way during our journey.

One thing that struck me during the trip was how few cars were on the road. However, there were many, many bicycle riders (a little different from today); it's now the other way around! Another item I found confusing were road signs stating: "HALT AT MAJOR ROAD AHEAD".

The first time it took me a few seconds to realize, it meant STOP! After that I just found the signs amusing. Narrow roads, high hedges and small picturesque villages made my first introduction to English life a memorable one.

Traveling in England those days was made more difficult because of "pea-soupers", a thick yellowish fog (this was pollution at its worst). I remember being on a bus during one of these pea-soupers; it was so thick the conductor had to walk along side shinning his flashlight (torch) to help the driver see the way. It didn't help much. We still managed to run into a brick wall! I recall learning that during the winter of 1951 more than 5,000 Londoners died due to fog-related conditions (most were the elderly).

During the next three years, I would be stationed on three bases: RAF Sculthorpe, RAF West Drayton in Middlesex near London and, finally, USAF Site High Wycombe.

RAF SCULTHROPE

My first duty base was at RAF Sculthorpe which was located in Norfolk. Upon arrival, we were issued a rifle with live ammo strapped to the stock (Korea was on the other side of the world. I don't know what they expected to happen in England). A few days later a cook, who had too much to drink, loaded his rife and fired off a few rounds. After that incident it was decided everyone had to turn their rifle back in to supply. I wonder why!

Sculthorpe was where I really gained cooking experience working in a large mess hall. We were feeding 1,500 men per meal. It was hard work but I enjoyed it! My stay at Sculthorpe lasted just around four months when, during the spring of 1952, a number of us were transferred to RAF West Drayton, outside of London.

RAF WEST DRAYTON:

RAF West Drayton didn't have a flightline. (In later years it would become a control centre for military aircraft and also civilian aircraft around Heathrow Airport). By the way, if you took a train to London it would only cost just one shilling return (British Railways gave the military discount fares).

Many facilities at West Drayton seemed old (offices, barracks, etc.), however we did have a modern kitchen. During this the time we were very short-handed, so not only did we do the cooking, but also the cleaning (dishes, pots and pans etc.). Since I had only two stripes I didn't do too much cooking.

During our off time; our favorite pastime was enjoying a few pints of "warm" beer at a nearby pub called the Cherry Tree. The landlord, Charlie, not only served drinks to the public; he would loan money to the Americans from the base and charge them one pack of cigarettes for each pound borrowed; he would then sell the cigarettes on the black market. He had a very lucrative business going until his pub was raided in the summer of 1952 by the British C.I.D (Criminal Investigation Department). He was charged, along with others, for black-marketeering. I believe the Cherry Tree Pub – or a pub – is still there (without Charlie, of course).

USAF HIGH WYCOMBE

In the late Fall of 1952 myself and about 30 others were notified of being transferred to reopen an American World War II base, located around 20 miles away. It was situated on top of a hill over looking the furniture-making town of High Wycombe. The base was used during World War II as a top-secret facility with an underground communication center used by the Allies to conduct war operations throughout Europe.

When the base we knew as USAF Site High Wycombe was reopened the underground center was also reactivated. It was nicknamed "The Hole". As it was in WWII this would again become a top-secret area of the base.

When our transfer orders came through we were shuttled by bus everyday from West Drayton to High Wycombe to work. And then back. Since the base has been closed since the end of WWII we got an additional assignment on our work menu – we had to help construct numerous Quonset huts. These would be used as barracks, toilets, dispensary, office space, mess hall, etc.

There was just one problem with that – none of us had any building experience! It was a disaster! I recall it was bitterly cold on some days and it even snowed. The snow really came in handy when we needed to level the corners of a few huts!

We were extremely slow and were prodded to work faster; but after three weeks, they finally brought in the experts – the Civil Engineers – to complete the job. We worked alongside of them and even helped erect our Mess Hall (two Quonset huts built on top of a four-foot high wall of cement blocks). After a new permanent mess hall was built in 1954, it was later used as a movie theater, commissary, supply warehouse, etc. over the years.

When the Quonset hut Mess Hall was assembled, we were allowed to start cooking inside, using several Army Field Ranges. It's one of the few occasions on which I would be caught thanking the US Army as they were the same models I used in cook school) until new kitchen equipment, such as ovens, grills, serving lines, etc. could be installed.

We also set up field ranges in a tent next to The Hole to cook meals for those working in the underground communication center. We had just one problem; there was no running water. Amazingly given its top secret designation we had to therefore descend deep into The Hole where, even more amazingly they did have running water, to do our cleaning up (trays, pans, utensils etc.)!

It was a long way down, but seemed much longer climbing back up. I remember that I was down in The Hole cleaning crockery when I heard on the radio that country singer Hank Williams had died. This was on January 1, 1953.

When our Mess Hall finally opened up for business, we discovered there was a ban on the use of fresh local produce on all bases in England such as vegetables. It was something to do with the way English farmers fertilized their fields. If my memory serves me right, this ban actually remained in effect for several more years.

All our perishable and non-perishable food supplies had to be trucked in to High Wycombe from our supply warehouse located at Denham Studios about ten miles away. I made this trip many times.

We did not have any control of what would be on a Mess Hall's daily menu; a Master Menu Planning Board met every 3 months to ensure the menu was nutritionally balanced for all bases and that supplies were available (I believe the MPB was convened at the headquarters around 15 miles down the road to London at USAF South Ruislip). Sometimes, we had to improvise if this supply chain broke down.

During World War II combat troops were issued C-Rations; these were individuals boxes containing canned food, hard candy and dried-up cigarettes; enough for one man, for one day. However millions of C-Rations were still in inventory. Therefore to help use them up we were required to have them on our lunch menu once a month. Needless to say the troops didn't like C-Ration Day.

Along with C-Rations you can add powdered eggs to our dislikes. That was because no matter what we did, they still tasted like powdered eggs. On the other hand one of the most popular breakfast dish was S.O.S. (known to us a **** on a Shingle); otherwise known as Creamed Beef on Toast. It's still popular today.

The following three stories occurred in a two-month period running up to January 1953:

I remember while we were assembling those Quonset huts, our soon-to-be fire chief decided to build a fire to make coffee. To help the fire get started he poured gasoline on the smoldering embers; the explosion knocked him to the ground. I wonder why he never became our fire chief!

Secondly, due to the underground communication facilities the Base Commander, afraid of the possibility of fire in The Hole, arranged for the civilian High Wycombe Fire Brigade to respond in such emergencies. One night, a drunken sergeant stumbling in to the barracks thought he saw smoke. Thinking where there's smoke, there's fire, he dialed the telephone operator and yelled "call the Fire Brigade, we have a fire"!

The telephone operator, thinking there must be a fire in the underground bunker, called the High Wycombe Fire Brigade on London Road instead of the base fire department. You can imagine the chaos this false alarm caused; but it sure was exciting to watch. The culprit who started everything was later discovered passed out on his bed. He missed it all!

Finally, the Christmas holidays of 1952 vividly stick in my memory, because I spent them in a local hospital. The following is what led up to that unforgettable night; it started, during one of our shuttling back and forth trips to West Drayton. I was in the Cherry Tree pub, when Bill (a GI from West Drayton) and I met a married couple ('Robert' and 'Anita') who invited us to come to tea.

We enjoyed their company and quickly became friends. When the holidays approached we were asked to join them and their family for a party on Christmas Eve. Everyone was drinking and dancing and having a good time.

At least I thought everyone was. But it turned into a nightmare.

Unbeknownst to me Bill was having an affair with Anita. Later, she related to me, that Bill told her that if she "danced with Don one more time", he would kill me. She didn't believe him. Ironically, when the Eddy Arnold song Don't Rob Another Man's Castle was played I asked Anita to dance. She did and soon after I was hit in the head from behind with a solid brass ashtray.

Bill went berserk! After he was forcibly removed from the house, Anita told me "she did the worst thing a wife could do and she was so very sorry". I was then taken to a nearby hospital where they stitched the gash in my head; the doctor then decided to keep me a couple of days for observation. By the way I still have the scar to remind me of those long-gone days.

POSTSCRIPT:

It is hard to realize that around 60 years have gone by since I first set foot in England. Of all my duty assignments, the one that stands out and remains especially near and dear to my heart was the one at Daws Hill, High Wycombe. This was where I met and married a beautiful 19-year-old High Wycombe girl Janet Green.

We met at a dance in the High Wycombe Town Hall on Friday, August 28, 1953. We made a date to go to the movies on the following day (Saturday) and arranged to meet under the clock in front of the town hall at 6 p.m.

It rained all day. Unknown to me at the time Janet had told her mother she had changed her mind and wasn't going to meet me. Her mother said: "You can't leave that poor boy standing in the rain." Somewhat reluctantly Janet went and the rest is history. We were married the following year on July 10th.

We lived with Janet's parents until I received my orders to return to the U.S. in December 1955 (fortunately not by ship). You can imagine what an emotional time it was the day we left High Wycombe. It was the natural fear of the unknown, by her and her family. To help ease their minds a bit, I promised them I would take good care of her. Since we celebrated our 55th Wedding Anniversary on 10 July 2010, I believe I kept my promise.

We left High Wycombe, by train, to Prestwick, Scotland. From there we flew to McGuire Air Force Base in New Jersey. We then caught a Greyhound bus to my hometown in Michigan. It was a long, uncomfortable two day journey. I really felt sorry for Janet; she got travel sick on the way.

When we finally arrived in Mount Clemens Janet was wholeheartedly welcomed by my family. After spending a couple of weeks with them we then headed for Tucson Arizona.

Our assignments over the years took us from Tucson, Arizona, back to England (RAF Lakenheath), Amarillo in Texas and, finally, ending up where I started 21 years earlier, at Lackland AFB in San Antonio, Texas.

After retiring from the U.S. Air Force in June of 1971 we moved back to Amarillo where we reside today. Janet, who is still a British Citizen, often travels to England to visit her family.

Above: Don Terrill, wearing his 'cook whites' standing next to a man he only knew as 'Shorty' at the base snack bar at Daws Hill High Wycombe in 1953. Below: A High Wycombe base Thanksgiving menu from 1953.

CHAPTER TEN

GEORGE & ME

Freeman Kilpatrick

SSgt, USAF
USAF Sculthorpe
(1952-55)

(Editor's Note: *In early 1951, USAF Staff Sergeant Freeman A. Kilpatrick arrived at a USAF base at Sculthorpe in Norfolk, England. The former WWII RAF base, had first been utilized by the USAF during the Berlin Airlift in 1949. Kilpatrick was part of the 3rd Air Force's 49th Air Division (Operational) and the 47th A Bombardment Wing and its B-45 nuclear-capable jet bombers which deployed that year as the Cold War took hold. But for Freeman Kilpatrick, then a 20-year old native of Atlanta, Georgia, little of this mattered – it was just a job. And it mattered even less when he and his young family were caught up in what remains Britain's worst natural disaster. He has remained incredibly modest about it – and even a little annoyed by the ensuing recognition – but what happened on the night of 31 January 1953 would result in Kilpatrick becoming one of a very select band of Americans and having to come to London.*)

I had been recruited into the Air Force in Atlanta, Georgia, where I lived. I couldn't get a job so I just did like everybody else. Basically I was young and I hardly knew nothing so I just followed the crowd. We had joined the National Guard and then along came the Korean War and we were 'Federalized'.

Somehow I ended up in communications – wire communications. It was pretty much stringing telephone lines in those days. I started out by going to Greenville, South Carolina, for six months or so where I went to school. And then came the assignment to England with the 3rd AF Communications Squadron. To me it was just an assignment. I didn't have much reaction to being sent to England itself. Again, I was young and I didn't know nothing! I knew there was an England and it was 'over there' but that was about it.

And so in the summer of 1952 I found myself heading there by a troop ship. The only thing I can remember about that voyage is throwing up for all 10 days of the journey. We eventually arrived at Southampton and from there we set out in for the assignment base USAF Sculthorpe. Not that I got to be a passenger. I was a Staff Sergeant at that point and so I had to drive one of the trucks in this convoy – taking both the vehicles and equipment up to the base. It was cold.

Remember we didn't know where we were going really. We just went where we were told. I was driving a big heavy crane – it weighed about 10 tonnes. I didn't have clue where I was but I do remember one thing: We drove along and went over a small bridge over a stream. When I looked back in the mirror a sign caught my eye – it took me a minute to decipher it backwards "Maximum Load 5 tonnes"! Too late! It quickly disappeared from view.

The only other thing I remember from that trip is that along the way we stopped off at a restaurant in some town. I'm not sure where. There were so many of us that we ate up everything they had in the whole building. It was that large a convoy.

When we arrived I was billeted on base in a Quonset Hut. We had about 12 to a hut and surprisingly they were pretty warm given the English weather. It was just another base to me – an ordinary base. Like all big bases it had the 'really important people', the 'somewhat important people' and then the 'nothing people' – us. They had B-45s flying at the time.

I was married. My wife, Sara, whom I had met in 1948 when she was 18 and when I was 19, had first seen each other at her aunt's boarding house in Atlanta. We married in January 1949 in and our first daughter, Suellen, was born in October of that same year. But when I deployed to England she had to wait in Alabama and could only come to join me with Suellen who was about two years old at that point, but only when I had located a house. You had to have a house and then your spouse and family could come over.

This was not an easy thing to do particularly around Sculthorpe. I managed to find out through some colleagues of a house that was available in a town called Hunstanton which was located about 18 miles from the base on the eastern shore of an area known as The Wash, a place that was a bit like a giant square bay. It was a stroke of luck. It was the only thing I could find. It really was hard to find a place anywhere.

As so they arrived in October 1952 and life evolved. I was working day shifts. Again I was just putting up wires to buildings and connecting phone systems. There were no computers. This was long before computers. We didn't know what a computer was!

The house we had rented was on South Beach Road. It was raised up on posts about 14 feet. We had an inside bathroom. Ours perhaps sat just slightly higher than some of the others on the beach. I think they were meant to be holiday homes but then the owners could get good money renting them out to the Americans.

It was lovely but it was so cold in there. We had a stove in the kitchen that you had to put a British 'shilling' coin into to operate it. Every time we'd start cooking something in it the shilling's worth of time would run out and we'd have to look around for another shilling.

There was only other problem – the rats. There were so many of them around. How many? Plenty. In fact, we had this one big rat that gnawed a hole in the kitchen wall and got to the bread and other foodstuffs. We named him 'Bruno'. He would come and stick his head through the hole and stare at you! He was huge.

Life settled into a routine. During the day I'd go to work. You'll appreciate just how hard housing was to come by when I say I'd have to walk up to the market square in Hunstanton and then catch a bus that would take about 30-45 minutes to get to the base. All the American military personnel who lived around there caught the bus and rode it to Sculthorpe. I could pick up food from the base Commissary. The selection was pretty good. We had a fair number of American military personnel as neighbors... which made what was to come even more devastating.

With free time I liked to ride a bicycle in the countryside. The beach was right there. We'd also visit with friends. People really did that then!! Then we slept a lot in order to keep warm since a glass of water sitting on table would freeze overnight. Never have we been so cold. We also did a lot of reading.

Sometimes in the evening we'd go up to Hustanton and watch a movie. Suellen would be looked after by a babysitter who would come in. She was a local British girl by the name of Joyce Stubbins.

At work we were in the early days of the Cold War but it wasn't something I was bothered with. It didn't matter much to me. I was only 22 years old and I didn't think much about anything except doing the job and getting paid.

And so it was on 31 January. The day started stormy and windy again for the second day in row and it continued that way. I went to work and got home. We had planned to go to a movie and Joyce had arrived to take over the babysitting duties. It was just before dark and before going I had popped out and down into the yard where I found a small body of water flowing in towards the house. I knew at once this was wrong. The water was salty and I just knew. I can't put it into words but I didn't hesitate. I ran back into the house and told Sara that something was wrong and that I was going to get around to our neighbors and warn them to get to higher ground.

Everything – and I mean *everything* – happened so fast and in a bit of a blur for me after that. I left Sara shouting at me not to go. But I went down the street and did the Paul Revere number – knocking on doors. It was all so quick. When I first spotted the water it was about an inch deep in the yard. In the course of my journey down the block the water rose to several times my height!

At some of the doors I knocked on the occupants took immediate heed and got out. Others didn't. I didn't make it far. I had been pounding on their doors and yelling "get out and get to higher ground because something bad is going to happen". Those who did get out went to the one big house nearby. It was a really sturdy block-built house that was occupied by a USAF Captain and was the biggest and strongest house in the area. Within the space of half an hour the water had gone from being an inch deep to about 30 feet deep so that just the tops of the telephone poles and roofs could be seen.

I had warned the houses I could along the block and then I wanted desperately to make it back to my own house which, though similar to many of the other beach houses, was on slightly higher ground. One of the other Air Force guys – his name was Dixon, I believe....so much time has now gone by....offered to go with back with me. I had helped his family to safety and he felt it was only right to return the favour.

We struggled to get back there. We were holding on to things and forcing our way and swimming through the surging waters. It was half pushing, half swimming and grabbing hold of a post – grabbing hold of anything to use that would push us forward.

By the time we made it back to my house it looked as though it was fixing to wash away. The water was outside the living room door. We got in and got to Sara, Suellen and Joyce. They had made a fire in the fireplace. There was no time. We got out on the balcony and climbed up on the railing and got the three women onto the slate roof. As we were doing so we heard a large whooshing sound. It was the sound of the water hitting the fireplace and extinguishing the fire. We had been lucky. Because our house sat up a little higher on the beach than others there had been a little more time... but not much more.

No sooner had we gotten onto the roof than the house disintegrated right underneath of us. We started floating around and bits of the roof started breaking off. It was a storm and it was freezing and between the rain and the waves we were all soaked through.

We managed to stay on the roof for about two or three hours. It was then that a rescue boat showed up. However the current was so strong that they were not able to control the boat and it crashed into what was left of the roof. We managed to get everybody onto the boat and get on out of there and that was that.

Once on board we found there were other people who had been rescued. What struck me was that they weren't concerned about whether they were going to live or die – at least it seemed that way to me. What they were really upset about was whether anybody had a cigarette. I was a smoker but didn't have anything. Nobody had one except one English 'bobby' (policeman) who had a pack of those ghastly British cigarettes. They lit those up and puffed on them.

Worse was to come – the rescue boat broke down and we ended up drifting around and being buffeted about by the storm. In the early hours of the next morning another boat was able to come out and find us and rescue us.

That boat got us to shore near Hunstanton and we were given glasses of whiskey. They took Sara and Suellen to a hotel in the town where they got them into bed and surrounded them with hot water bottles. Later my supervisor would come and took them home with him. They took me out to the base. Our babysitter was taken to be reunited with her family.

I had never drunk Whiskey before in my whole life and so I passed out on the ride back to the base. Once there I was taken into the base hospital. When I woke up there was no one to stop me so I left and went back to my squadron building. They had a big board up there and they were 'lining off' all the people who had been killed from the base. They already had me 'lined off'! They had to change that status.

The storm lasted the best part of the night as I recall.

(That night the gale-force storm with winds of upwards of 100 mph which had been building for two days would create a surge of water down through the North Sea – raising the sea level by an estimated 2 meters (circa 6 feet) with storm waves of another 4 meters (13 feet) on top of that – to sweep down the eastern seaboard of England. The surge arrived at Hunstanton at about 6 p.m. in the evening and moved along the coast over the course of the next three hours. It would claim the lives of 31 people along South Beach Road including 16 USAF personnel and dependent members from the Sculthorpe base. A total of 307 people died along the English coast and waters proceeded across the Channel to claim the lives of 1,795 people in Holland. Thousands of homes would be damaged or destroyed and upwards of a million people homeless or flooded out. It would become known as 'The Big Storm' and would become Britain's the worst natural disaster. SSgt. Freeman Kilpatrick would be credited by the authorities with saving lives of some 18 people as would 22-year-old Airman 3rd Class Reis Leming who, raced from the base to help. Over the course of several hours Leming, a non-swimmer, managed to make three-separate trips to isolated cottages towing an inflatable boat to rescue 27 people from rooftops. He suffered extreme hypothermia despite having donned an exposure suit in the process.)

It was probably the next day – as the reality of the disaster and our lucky escape set in – that we learned that Joyce, our babysitter, had been left an orphan. Her entire immediate family had been killed bar an older sister. We never saw her again after that night. She went to live with her sister. Someone reconnected us during the 50th anniversary of the event. Joyce Scully, as she later became, continued to keep in touch until we learned from a family member she had died not too long ago.

When I went back to where our house was – well there was really nothing left. The house and all our possessions were gone. We did get a few things back that people turned in that were scattered around. I got my insurance policy back – it had mud all over it and I kept it to this day. During the course of the next few days the landlord – a woman who lived nearby – showed up and found us and kindly told us we wouldn't have to worry about the rent!

Her Majesty the Queen, who had her one of her residences at Sandringham not too far from the Sculthorpe base, came to visit Hunstanton and see the damage the day after but I didn't meet her. But the civic wheels were moving somewhere and it was a few days later that it was announced that five people who were involved with events related to the flood (including two British policemen and a fireman) were to receive the George Medal – Britain's second highest civilian decoration (after the George Cross) for gallantry.

After that – for me, at least, was when all the really bad stuff started happening. They (the authorities) started making a big deal out of all of it. I didn't want any part of that. It had been a horrible tragedy. I didn't want any part of medals but I didn't have a choice. Suddenly I was being sent to London. We went up to London on a train. Again I missed the Queen. The Home Secretary Sir David Maxwell Fife presented us with the medals as Her Majesty was away travelling then.

Airman Leming was credited with receiving the first George Medal awarded to an American in peacetime (on 15 April 1953) and I received the second. I never met him (Leming) that I know of either then or afterwards. You have to understand things were different then, especially communications. We didn't have TV or computers and we certainly didn't have any post-disaster counseling or much help back then.

That presentation and the whole recognition thing was more traumatic to me than anything else. I just didn't like it. The base public relations guys were making me out to have done more than I did and I didn't want any part of it. A tragedy had happened. I had managed to warn some people and some of those people had taken notice. I had been fortunate to survive the event as had my family. It could have turned out very differently but it didn't. But it was a tragedy for many others including our own babysitter. I just felt like I was being used.

So I received my medal with some reluctance and some anger. And then, to cap it all, when I returned to Sculthorpe I was told they (the US Military) were going to take the medal away from me. The USAF said something to the effect: "You cannot have that medal – It's too important a medal for you to wear. We're going to keep it in Washington D.C."

They sent my medal back to Washington. I think they didn't want to see me wear it while on Active Duty. I never saw it again until, amazingly, about a month before I retired from the USAF in 1969 as a Sr. Master Sergeant. Then they returned it to me! That was a surprise.

Not only wasn't there counseling but there wasn't much other help either. We had to go out pretty quickly and find our own house – another one again. And there we found a trailer in a park opposite the base. It didn't have running water or electricity. It was hard to heat. It was pretty miserable by comparison to today but you just got on with things. That's what you didn't back then.

And life moved on. About 10 months later we faced another and even more personal crisis. Our second daughter, Tehelia, was born suffering from Hydrocephalus. After a couple of months they moved her from the base hospital up to 'The Children's Hospital in London (known these days as Great Ormond Street) where Sir Wyllie McKissock, who was the senior Consultant Neurosurgeon, and his team placed a series of shunts in her brain. The neurosurgeon used a technique that was experimental and far from effective but it did buy her time for improved medical methods to evolve. She was there for most of the first year of her life.

Since that time I have never run into a hospital or people who were as nice to her as they were there. Her head had swollen up and she moved awkwardly. As she grew she was treated very badly by strangers her whole life. We always thought that it would just be kids – but even adults reacted and treated her badly right up to her last day on Earth at age 55. She died a few years ago. We will never forget the kindness of those people at the Children's Hospital in England.

We left England in the summer of 1955. We were glad to be leaving but it felt like we were leaving a home. We love it that much despite everything that happened.

We went on as a family to other assignments including Patrick Air Force Base in Florida and then on to assignments to Morocco, back to Lowry AFB in Denver and then to Rhein Main AFB in Germany and our son, Freeman Alex Kilpatrick, was born at the USAF hospital at Wiesbaden 1965. He would go on to join the USAF himself.

When I retired at Barksdale AFB I had to fill out forms and among them was a form which asked you to list additional medals and awards. There was a small addendum of cash for any awards you won. So I wrote "George Medal" and the "Soldiers Medal". The latter medal – for "heroism not involving actual conflict with an enemy" was also awarded to me by the US Military in 1953 after what happened at Hunstanton. But again it was the George Medal that caused a problem – a junior grade officer wrote back to me in regards to that request to say "Disapproved." His reasoning – "It's not important enough!" Needless to say I didn't get the stipend for it.

So life moved on.

Sara was mad at me for a long time after flood about that night and my leaving her and Suellen alone to try and warn others. But she got over it... I think. Our daughter was three years old at the time – it was so traumatic then but fortunately she doesn't remember a thing about it.

I have never gone back to England let alone Hunstanton. I understand there is a memorial to those who died in the flooding that is located close to the town's War Memorial. I'm not sure how I would feel now about going back. Having said all that It would be nice to receive a signed photo from the Queen, but then again a few years ago I went blind. I can't see it – so it would be for my family.

I would like to think that Tehilia has made it back though. Telhia had always loved England and had wanted to go and see where she had been born. She was fascinated by it. When Tehilia died we cremated her and took her ashes to the Red River where we scattered them. The Red River goes to the Mighty Mississippi and, in turn, that eventually will mix with the waters of the Atlantic. We figure that somehow, given tide and time, she'll be getting back to England again.

And life goes on. It's fair to say that few if any of the neighbors we now have here in Louisiana know about what happened to us all those years ago or that I was awarded the George Medal. Does it matter? Anybody would have done the same thing as I did – I just happened to be the guy who got caught! Really it's not something you would bring up. It's just history.

(Note: The George Medal (GM) is the UK and Commonwealth's second level of civil decoration and is awarded for "acts of great bravery". Records from the Central Chancery at St James's Palace, which oversees the awards, indicate that, of the circa 2,050 George Medals given since 1940, three other Americans besides Freeman Kilpatrick and Reis Leming are believed to have received the award but they were awarded during wartime.

They are:

- *Ensign Milton L Sanders, US Naval Reserve. (Sept. 1943)*
- *Volunteer Neil McDowdell Gilliam, No 2 American Ambulance Unit (Sept 1944)*
- *Radioman 3rd Class Thorsten Valentine Lundquist, US Naval Reserve (Nov. 1945).*

Freeeman Kilpatrick (above) back at what remained of his house a day or two after the floods. Kilpatrick enjoyed a moment of wry humour amidst the devastation as he and his wife searched for any remaining belongings. Picture courtesy of Freeman Kilpatrick.

Above Freeman Kilpatrick joins an elite group of American servicemen – and, indeed, Americans — when he received the George Medal. Photo courtesy of Freeman Kilpatrick.

CHAPTER ELEVEN

LOCKDOWN

Earl Robert (Bob) Beeghley

7533rd Air Police Flight
7500th Air Police Squadron
Bushy Park
1952 – 1955

I enlisted in the USAF in October 1951 in Vallejo, California, with my buddy Jim McNames.

We did it on a whim.

We had taken the afternoon off from work. At that point I was three months into a job as an Apprentice Heavy Artillery Repairman at the US Army Arsenal in Benicia, California. I stopped in front of the Recruiting Office and said; "Let's join the Air Force". Jim wanted to go in the Navy but I won out and so we enlisted in the Air Force.

From there we were sent to Lackland AFB, San Antonio, Texas, via train from San Francisco for induction processing, Basic Training and further assignment after completing Basic Training.

I arrived at Lackland on October 22, 1951 to find that our quarters were tents. During processing I advised the person conducting the interview that I wanted to become an Air Policeman, (Jim & I had talked it over and decided that we wanted to be in the Air Police as we might be interested in becoming policemen in civilian life.

On January 14, 1952 I arrived at Tyndall AFB in Panama City, Florida to attend the seven-week Air Police School. We learned everything from handling all types of weapons (.45 cal Automatic pistols, Riot guns, .45 caliber sub machine guns), to the Uniform Code of Military Justice.

Upon completion of my Air Police training my orders had me going to the Far East, (read Korea during the Korean conflict), while Jim's orders had him going to Europe. We requested to see the Captain and we explained that we had enlisted together and we wished to go to the same area.

The Captain pulled out a Half Dollar and flipped it in the air and said "Call it". Jim called "Heads" and we both had orders to go to Europe.

I travelled by bus from Panama City, Florida, to my hometown of Vallejo, California, for leave prior to reporting to Camp Kilmer, New Brunswick, New Jersey. From Vallejo I traveled across the US by train to New Brunswick, reporting in on March 22, 1952.

At Camp Kilmer Jim and I were finally split up as he shipped out to England one week ahead of me. On April 22, 1952 I boarded the Troop Ship *General Alexander M. Patch* at Brooklyn, New York, destination Southampton, England. There were four others aboard the Patch besides myself who would be assigned to Bushy Park: Raymond Barker; Roland Barnes, Billie D. Bennett and James Berry; however I did not know them at that time.

Upon arrival at Southampton we were put on buses (coaches) and were driven to Shaftsbury AFB in Dorset (which incidentally later became Her Majesty's Prison Guys Marsh), arriving there on May 16, 1952. The place had, I believe, been a WWII Rest & Recreation centre for the USAAF as it was.

At formation on the morning of May 22nd a list of names was called (mine being one of them) and we were informed that our orders had been cut for us to be assigned to Bushy Park AFB. On those orders assigned to the 7533rd Air Police Flight were: Raul Montez, Columbus James, Kenneth Fausset, Raymond Barker, Roland Barnes, Billie D. Bennett, James Berry, Dwight Boughan, Clarance Brock, Paul Burch, Willie Clark, John Cleek, John Degan, Arthur Derenda, Wayne Elliott, Richard Erdman, William Tullis and Theodore Stanfield.

Also going to Bushy Park, but assigned to a different Air Police Flight (the 488th) were: Raymond Van Hook, Kenneth Beldle, John Daley, Dave Elliott, John Channell, Francis Cronin and Donald Burklow.

When we arrived at Bushy Park we were immediately ordered to prepare for inspection. The Base Commander, a Lt. Colonel whose name I don't recall, was inspecting the incoming airmen and I could hear him commenting; "You need a haircut" to several of the Airmen. When he came to me his comment was, "What the hell do you call that?" as I had not had a hair cut since leaving Tyndall AFB in early March.

He then told Capt. Freddie Vaughn, 7533rd Air Police Flight Commanding Officer, that he was to escort me to the Barber Shop personally. After the inspection I immediately went to the base barbershop – it was just a few steps away from the Main Gate – and got a crew cut.

The next morning I was asleep in my upper bunk in our 50-person barracks located in "B" Block when I heard someone call the barracks to "Attention". The next thing I heard was Captain Vaughn asked where the Airman was that he was supposed to escort to the Barber Shop.

I raised my hand. He came over, he rubbed his had over the stubble on my head and said; "We're going to get along fine".

Our duties as Base Security consisted of Main Gate Guard Duty, Base Foot Patrol, 12 Site, (a small section closer to Hampton Wick that was separate from the main part of the base and was home to the Base Stockade, Mess Hall, Barracks and a Snack Bar), Gate Guard, Commercial Gate Guard, (Day Shift hours only), Patrol Driver, Desk Sergeant, Stockade Guard and guarding a post known only as "The War Room" in "A" Block, which was a large building consisting of a main hallway with several wings going off on each side. All of the main buildings, 'A' Block, 'B' Block, 'C' Block and 'D' Block were of this type as was the building behind 'C' Block where the Central High School for military dependent children eventually was located.

We ate most meals at the Mess Hall and would go to the Snack Bar or the NCO Club for hamburgers, chili or just quick bites to eat when the Mess Hall wasn't open.

The Air Police Flight barracks was in the building behind 'C' Block until Central High School was located on Bushy Park. At that time the Air Police Barracks were moved to 12 Site so that the building could be used for student dormitories and classrooms.

Most of the Airman resented the High School students as what recreational facilities we had were taken over by them and as Air Police we had no control over their actions.

We worked three shifts: Day Shift – 8:00 a.m. to 4:00 p.m. (0800 – 1600 hrs), Swing Shift – 4:00 p.m. to Midnight, (1600 to 2400 hrs) and Graveyard Shift – Midnight to 8:00

a.m. (2400 to 0800 hrs). We rotated shifts every three days and between the Day and Swing Shifts and Swing and Graveyard Shifts we got one day off. Between Graveyard and Day Shift we got three days off.

In the colder months we wore our Class A blue uniforms with either the Blouse or the "Ike" Jacket. In the summer months we wore the khaki uniforms. On duty we wore a Sam Brown gun belt and we carried a Colt .45 Cal Automatic Pistol, (on base only) and a nightstick.

At first we kept the clip of ammo in the weapon however that changed after a "Quick Draw" incident that almost cost one of the Main Gate Guards his life. We used to practice "quick draws" with the .45 calibre Pistols and a few of us got in the habit of pushing the weapon against the front of the holster, pushing down to jack a round into the chamber and then pull the weapon.

One of the men on Foot Patrol approached the Main Gate and said "Draw". Without realizing that he had jacked a shell into the chamber while drawing the Gate Guard drew, aimed and pulled the trigger. The Foot Patrolman was struck in the wrist and the slug lodged in his chest, having missed the heart as it had deflected off of the wrist bone. After that incident the magazines of ammo were in sealed ammo pouches on our Sam Browns and were inspected at the start and end of each shift.

As Bushy Park did not have a landing strip the only airplane guard duty we got was when a VIP's plane would at Heathrow and we would then be assigned to guard the plane.

US Secretary of State John Foster Dulles' plane landed at Heathrow and I was one of the airmen assigned to an 8-hour shift to guard it. Our orders were that NO ONE, other than the flight crew was to approach the plane and the plane was not to be moved without a member of the flight crew present.

I was on guard duty and three civilian workers approached the plane and said that they had to move it so a Pan Am plane could be moved for its scheduled flight to the United States. They were informed that the plane would not be moved without a member of the flight crew being present.

About 30 minutes later a Jeep came speeding down the ramp and it was the Sgt of the Guard, S/Sgt Columbus James. He asked me what the hell I thought I was doing threatening the workers with my 'Grease Gun', (a .45 caliber sub-machine gun). I informed SSgt. James that I had not threatened anyone and that I was not armed with a Grease Gun. I had a .45 cal automatic pistol under my parka, but had not displayed it at all.

I explained to SSgt. James that the workers wanted to move the plane and I had told them it would not be moved until a crew member was present. SSgt. James stayed at Heathrow until the pilot, (a Major), arrived, but by that time the Pan Am flight to the States had had to be cancelled as no other plane was available.

Another incident was on May 1st, 1953 (May Day). The base Commander received info that a group of Communists from the nearby town of Kingston upon Thames were going to come to the base and take our flag down.

I was in town with two other guys off duty and an Air Police Jeep pulled up and we were taken back to the base and we were told to report for duty.

No sooner had we gotten back to base than we went into 'lockdown' mode.

That meant nothing and no one going in or out.

The Base Commander had a 6x6 truck backed up to the Main Gate with a .30 cal machine gun set up in the bed of the truck. I along with the other Air Police who had been off duty were assigned to walk the (base) wall. We were armed with a carbine and a bayonet, (no ammo) and we were told that no one was to come over the wall under any circumstances.

About 45 minutes into my shift I spotted a man in civilian clothing coming over the wall and just before he was to drop to the ground I stuck the point of the bayonet in one of his buttocks and told him that he had better hold on for dear life. I then called for the Sgt. of the Guard by calling out to the next sentry, who relayed it on (there were no walkie-talkies back then) and when he arrived the individual was allowed to drop to the ground and identify himself.

It turned out that he was a Captain from the 28th Weather Squadron. He had tried to enter through the Main Gate and had been turned away. The lockdown lasted for about three hours and when no-one showed up at the base we were all told to "Stand Down".

Once the 7533rd Air Police Flight was combined with the 488th Air Police Flight to become the 7500th Air Police Squadron we would work three months on base and three months in London.

When in London our office was in the enlisted club at Douglas House which was then located in Mayfair. We worked in uniform with an arm band that read "Air Police". We wore our Sam Brown Gun belts however we carried only a nightstick as the Air Force had an agreement with the English Government that we would not carry firearms off base. We also had white covers on our hats and we wore our trouser legs 'bloused'.

Our duties in London pertained to checking GIs for proper orders, assisting the London Police with any investigation that involved US Military personnel and being of assistance to GIs trying to find their way around London.

One of my first experiences in London was investigating the death of an airman and a young English lady in a hotel. The Metropolitan Police ruled the deaths as accidental. It was my first experience with dead bodies and I remember that as the wicker basket holding the airman's corpse was carried by me the lid flipped open and his arm raised straight up due to rigor mortis which shook me up for a couple minutes.

Off duty we spent most of our time at either the Swan or the Forester's Arms in Hampton Wick or at the Dolphin in Kingston on the banks of the Thames river. The drink of choice then was Pale Ale, which the pubs kept on ice for us as us Yanks were used to cold beer. (Incidentally, English beer cold tastes as bad as American beer tastes warm. My preference now when I visit England is Best Bitter).

Most of us would play darts among ourselves and with the locals and many of us joined the pub's dart teams. We would also venture into London on our days off and we also went on coach trips sponsored by the Service Club to places like Windsor Castle and Hampton Court Palace or evening boat cruises down the Thames. When we stayed on base we would go to the NCO (Non Commissioned Officers) Club to drink, play darts and on the weekends to dance as the club always had a local band come in.

Once a month the Service Club would put on a dance and young ladies would be brought in on coaches from places like Jantzen's Swimwear, (a bathing suit manufacturer).

A couple months after arriving at Bushy Park, Johnny Cleek, Clarence Brock and myself were in Kingston and we saw these two young ladies in a phone booth and we asked them what the "A" & "B" buttons were for on the phone. Of course, we knew, but...

We talked a while and then they went in one direction and we went the other...right around the block. Clarence decided to go back to the base so Johnny and I continued around and met up with the young ladies, Pamela Hazel (my date) and Diedre Ellis. We talked to them again and then asked if they would like to see a movie. We ended up seeing *King Solomon's Mines*.

After the movie they said that they would telephone us but decided not to. About a month later I was on the train from Richmond when it stopped at Teddington. As the train pulled out I saw Pam and Dee walking along the platform. I got off the train at Hampton Wick and ran all the way to the Main Gate arriving before they did. Dee had a date with another Airman who was supposed to bring a blind date for Pam, however I took his place.

At the close of the evening I asked Pam out again and she said only if I bought Johnny as Dee's date. Johnny & Dee got married a few months before Pam & I were married. Pam & I became engaged on March 20, 1954.

I had submitted the 'Permission to Marry' paperwork to the Commander, 7500th Air Base Group at South Ruislip. On the day before Pam & I were to be married the paperwork had not come back so I took the bus to Ruislip and went to the office of the 7500th Air Base Group. I was informed by a Tech. Sergeant that Lt. Col Robert E. Kaempfer, who was supposed to sign the orders, was "flying" and would not be back for the remainder of the day.

The TSgt. looked through the Lt Colonel's desk and found my request unsigned. He was nice enough to place a paper with the Deputy Commander's signature on it on the window and placed my request form on top and traced the signature of Lt. Col. Kaempfer on my paperwork so I could get married the next day.

Pam and I were married at the Registry Office in Ashford on July 3, 1954 and we went on honeymoon to Brighton.

And that was only part of my battle to get married. Of course I had to meet Pam's mother and stepfather. One problem – they didn't like Yanks!

After we got married we lived in Teddington for a couple of months and then in Surbiton for a couple more months before moving to Lingfield Avenue In Kingston.

On December 22, 1954 our daughter, Debra Lynn, was born at the USAF Hospital at South Ruislip AFB. Pam had gone into labor while I was working the Midnight shift and we were taken from Bushy Park to South Ruislip by a USAF staff car. While Pam was in labor, (bear in mind that she was the *only* one in labor at the time), a nurse walked out and asked me if my wife was the one having twins! As I clung to the ceiling by my fingernails she walked away laughing!

I returned to Bushy Park after visiting hours were over and made the trip to South Ruislip each day by Air Force Bus until Pam & Debra were discharged from the hospital.

I had only been on three other bases in the UK – near Kettering where my buddy Jim, from Vallejo, was stationed; Denham and the other was Sealand AFB in Wales when another Air Policemen and I took three prisoners to the Air Force Stockade there. We dropped off the prisoners and returned to Bushy Park the same day.

We got along great with the local Police and the London Police and we all had many friends that were on the police forces.

In November 1955 my wife, daughter and I were transported to Southampton (we were supposed to go to Shaftsbury AFB but a fire had destroyed the dependant quarters) where we boarded what was already a familiar troop ship for me, the *General Alexander M. Patch*.

On November 16th I reported in at Manhattan beach AFB in Brooklyn, New York to process out of the USAF.

I would go on to become a civilian policeman in Garden Grove, California, for 14 years and then five years with the Washoe County Sheriff's Department at Lake Tahoe, Nevada.

My enlistment with the USAF lasted four years and out of those four years, three-and-a-half years were spent in England. I enjoyed being in England and I look back at them as some of the best years of my life. I still enjoy visiting England whenever possible.

Above: The main gate at Bushy Park with Air Policemen D. Breon and J Gordon on duty. Photo courtesy of Bob Beeghley.

Above: Bushy Park as seen from the main gate early sometime in the early 1950s. Photo courtesy of Bob Beeghley. *Above Right:* The 7500 Air Police logo. Air Policemen Gonzales, Bickie & Clark.

Left: Air Policemen Gonzales, Bickie & Clark. *Right:* Two security police officers O. Dutton and J. Degan at Post 8 (known as The War Room) which was underground. "On the side of "A" Block that faced Sandy Lane was a small room that was our Post," says Beeghley. "It was enclosed and the entry door had a sliding window in it. When the duty personnel reported for duty they knocked on the door, we checked their ID to see if they were on the list and then opened the door to let them in via a large steel vault door that they opened. Once open the personnel would go down steps. We were never allowed to enter that portion, nor were we allowed to look down into it. I later learned that this was where General Eisenhower did his D-Day Planning." Photos courtesy Bob Beeghley.

Left: Formal picture of Earl 'Bob' Beeghley. *Right*: The Bushy Park base Air Police in early 1952.

USAF 'air policemen' Roland Barnes and Willie Clark clown around with actor Bob Hope at the main gate during a visit to Bushy Park. Photo courtesy of Bob Beeghley.

CHAPTER TWELVE

THE FIVE-YEAR RUN

Larry Hagman

A1C, USAF
USAF South Ruislip Air Station
1952-56

(Author's Note: Actor Larry Hagman is perhaps best known by millions around the world for his role as JR Ewing in hit TV series "Dallas" and also for his role as "Major Anthony Nelson in "I Dream of Jeanie". However far fewer people know he served for the USAF in Britain in the 1950s in a duty role that was could have been tailor made for his talents. He and his wife, Maj, whom he met while stationed in the UK, ended up living in Ojai, California. Hagman passed away in November 2012 aged 81 after having successfully reprised his role as JR in a comeback series of "Dallas". Fellow cast member and friend Linda Gray, who was by his side when he died, described Hagman as "the Pied Piper of Life". Her eulogy was apt as you will read here from a 2010 interview Hagman gave to the author in which he fondly recalled his military service in England.)

I had already been serving in the US Navy for about a year when I was drafted into the US Air Force in 1952.

You might think that would be a typical military SNAFU (*Situation Normal – All Fouled Up!* to use the polite translation of the much-used acronym) but actually it wasn't. My US Navy 'duty' was only as part of the cast of the hit stage musical *South Pacific* that was then in the middle of a two-year run (November 1951 – 1953) at the Theatre Royal, Drury Lane in London.

I had joined the cast, which included my mother (and acclaimed actress) Mary Martin who was playing the lead role of Nellie Forbush. She had invited me to join the ensemble of actors and to take the show that had been a Broadway hit, to London in 1951.

And so most evenings and occasional afternoons each week I donned my Navy garb to play Yeoman Herbert Quale and sing "There is Nothing Like a Dame" and a numerous other hits from the musical with my fellow actors.

It had been a brilliant experience and only reinforced my decision that acting was a career while opening up my world socially to life in London, which I loved.

It also marked an interesting time for me. I had not gotten on well with my stepfather. Rather than live at the Savoy Hotel, which is where my mother and he were staying, I moved out to a flat at 82 Clifton Hill in St. John's Wood with an actor friend, Theodore Flicker, who later went on to become a producer. We had been to college together in New York. There was a wonderful pub right up the street called The Clifton that I really enjoyed.

I lived the life. I was being paid about £12 a week back then at a time when the £1 was worth about $2.80! But like most actors, I hadn't given much consideration to the taxes.

So the shock of the resulting post-tax £8 a week didn't quite allow for as much high on the hog living. Still I went to places like the Buxton Club at the back of Haymarket. It was a theatrical club.

But suddenly my military service was for real. I had received my Draft call-up papers just after Theodore had received his to go into the Army. About a year earlier I had come over to England on the Cunard Line with my mother and stepfather. On board the ship was a Catholic Chaplin who was in the USAF. I knew I didn't want to go to Korea so I got in touch with him and he helped pull some strings.

Not long after I reported to the USAF Headquarters at Bushy Park in Teddington on the outskirts of London and that's where I formally enlisted. From there I was sent to Sealand, a USAF base located in Northeast Wales, where I did six weeks of basic training. Some more strings got pulled and I was assigned to Third Air Force Headquarters at USAF South Ruislip in Middlesex.

Before I was sent there I had never heard of the place. And when I arrived the base didn't make that great an impression on me. I think it was a bunch of old factory buildings – they were modern for their time I guess.

South Ruislip was a place to make regulations. There was a lot of 'brass' and you were always saluting there. There were times when you could have kept your arm in a near permanent salute if you walked around the base. It was a pain in the ass.

I was pretty much given a role in what you might term Special Services. The programme was overseen through the Deputy Chief of Staff for Personnel for the Third Air Force.

When I arrived there was a woman there who managed theatre bookings for the officers and got tickets for all those colonels and generals. I took over that role.

She also managed 'Stairway to the Stars' which was essentially a US Military talent search competition among young servicemen for singing, dancing and comedy routines. I also took on that job. We'd go out to the various bases, hold a talent show and select the winners to take part in a variety performance and then tour the UK bases with them.

I loved it. I soon managed to convince some higher-ups at USAF Headquarters in Wiesbaden, Germany, that this had a wider appeal and not long after we were touring bases in France, Germany, Italy and North Africa with about 20 people. During my time in the USAF I toured that show and about three other shows – including one I built around the actor Eddie Fisher – all over bases in Europe and North Africa during my four years there. I didn't know what I was doing much of the time but working with the talent was the fun part.

My office was located in a building on the base but it was nothing special. It was just a place to work. However, what might have made it a bit different was the number of band instruments I always seemed to accumulate in there.

I had a bunk in the large barracks on base that seemed to be in a converted factory space but I never used it. I still had my apartment in St. John's Wood and I'd take the Bakerloo line up to Harrow and then take a bus from there to the base quite often.

If you had to liken my experience to anything then I guess you could say that I was the Sgt. Bilko of the base. The thing was none of my commanders had a much idea of what I was doing. My whole aim was to keep a low profile, do what I wanted and not get into trouble. I *was* Sgt. Bilko!

I wore my uniform on base but then got out of it pretty frequently. The thing was that we had to build and keep a list of approved 'local' performers who had to meet a number of requirements set by the USAF. That meant that I *had* to go out and see a lot of shows and *had* meet a lot of local UK talent. And I translated that as I saw fit and went all sorts of places and without uniform and set my own schedule for the most part.

I had my own little kingdom and nobody knew what I did or what was going on. And I really liked that!

It was a wonderful programme. You'd do these base and regional competitions and then you'd go to the base at Bushy Park and have about 100 different acts down there and select the winners from those. It was fun.

I started out as Airman 3rd Class and during my tour I made two more ranks up to Airman 1st Class.

Of course I had an advantage. I certainly had tickets available to *South Pacific* which was one of the hottest shows in town. People had to pay for them but I could certainly see that they had the best seats.

Off duty in London it was just the best time. I had been dating Joan Collins through that period of time when I was in the service and I also dated her sister once. They were stunning girls and still are. But that was all before the alluring Maj Axelsson who had come from Sweden and worked as a clothing designer in London, won my heart.

We first met at my flat. My friend Henri Kleiman, a furrier, whom I was sharing a flat with, brought her around for tea. She had never met an American. Maj moved in about six months later. The name Maj was the Swedish word for "May" but was pronounced "my".

Eating out was an experience. The food in those days wasn't terribly good. But I was US military and had access to the Base Commissary where I could get great steaks. You could get a quart of gin for $1.25. I could also go to Germany and would bring back large chocolate éclairs and pastries which made me pretty popular. There was still post-WWII rationing in England until about 1953!

I remained a member of the Buxton Club but when I went back there in my uniform after enlisting they gave me some friendly jeering!

The shows took us around the US bases in England. I went to Burtonwood several times (it was up not far from Sealand and was a huge base located between Manchester and Liverpool. Also to bases such as Fairford, Greenham Common, Lakenheath and Sculthorpe. I went to Daws Hill in High Wycombe – that was some sort of secret base with a bunker but I never saw that.

When we toured we toured with a bus and a truck to transport us and our gear for the stage. Quite often you'd come in, do a show, get a room at the base and then leave the next day. I didn't' see much of them so all the bases looked alike to me.

At the time Burderop Park was a large USAF Hospital (7505th Hospital Group) base near Swindon in Wiltshire. Anyway we did a show there but afterwards I was left waiting around. Maj drove out in a Morgan car that she had borrowed to pick me up. It was raining like hell and she was two hours late. Well I had waited for her at the NCO club and had a few drinks. I drove but somehow drove into what I guess was one of some sort of tank training area nearby the base. The result was that we got stuck up to our hubcaps in mud.

And that's where I proposed to her and said: "Will you marry me?" A farmer came with his tractor and pulled the car out of the mud.

And we got married about a year later on 18 December 1954. But there were complications. Being in the US Military I had to get permission from my commander to marry Maj. That was quickly given. But then my request had to be forwarded to a USAF Europe Headquarters in Wiesbaden.

Well they apparently took one look at my form and read the words *Maj Axelsson* and rejected it out of hand.

But not because she was a foreign citizen.

They read her Christian name and thought that I was an enlisted man wanting to marry a major! Enlisted men marrying officers was simply not allowed. It took me about six weeks and a trip to Wiesbaden to get it all straightened out.

We had a civil wedding in Marylebone Registry Office because we were both foreigners. Then we had a full religious ceremony at the nearby Swedish Church complete with me in a top hat. I couldn't understand a word that was said – the ceremony was in Swedish. So I just said "Yes" to everything and that I think is the secret of a successful marriage!

My mother couldn't be at the wedding – she was performing in the US – but she sent an Austin Healy car as wedding gift. It looked great but only worked when the weather was perfect and the temperature was between 65-80 degrees. Anything over or under that and I could pretty much write off any travel plans – it was temperamental to say the least, but wonderful.

Eventually, a year later, I was able to introduce my wife to my mother in Paris. Mom was doing a play there. I took leave and Maj and I stayed at Le Star Hotel – I remember you could have rooms by the hour or the night!

Maj and I lived in my apartment in St. John's Wood – it was the same one I kept for all of my five years in England. She didn't have much to do with the USAF – she was not a GI wife in the normal sense. But she would come to the NCO club on occasion and also attend squadron parties with me.

As mentioned she worked in the rag trade – she was an excellent dress designer and she was making £25 a week. So we were really living like kings. We travelled and we saw all of London and did it mostly on board the Vespa motor scooter we had which I loved. It never failed me – I had bought it for about $125 at the time.

In 1958 I sold the Austin Healey it to pay the hospital bill for my firstborn, Heidi. We were getting out of the hospital and they handed me a bill which came to about $1,000 and money was pretty tight at that point so I sold it. However that love of bikes stayed with me and I still keep two (a Vectrix and a Piaggo-3MP) today.

My old roommate, Theodore, had gone to the States to serve out his time, but after I left the military it was he who gave me one of my first acting jobs in *Once Around the Block* by William Soroyan off Broadway in New York.

People ask me if I based any of my acting on my time in the USAF when I later came to play military roles in *Ensign Pulver*, *The Eagle Has Landed* and as Major Anthony Nelson in the TV Series *I Dream of Jeannie*.

I don't know. I just took the general experiences of a serviceman which, in part, was about knowing how to flim-flam everyone in a senior rank to you. And that's essentially what acting is all about, too.

There wasn't anything that annoyed me about Britain. I loved it and still do. I spent a very formative time there from 19 to 25 and I loved it. I don't know if it was because a lot of English guys had been killed in WWI and WW2 but there seemed to be a lot of gorgeous women who were all pretty horny.

When you are young and have pretty much signed your life away for four years you don't know how things are going to be. But I had, in total, five wonderful years in England – the country is like a second home to me. I never really kept any souvenirs from my time in the military but you could say that my wife is the best souvenir I could ever have!

And finally my USAF time was up. I had purchased an upright piano for £5 and when I completed my four years and got out we gave a party at my apartment. We had about 50 people come and it got raucous. Of course the police came – the 'Bobbies' were pretty nice but asked us to tone it down a bit. We invited them to "have a drink with us officer" but they said they couldn't.

In the end somehow — because it didn't involve me I'm pretty sure – the upright piano went out the window and was smashed to bits.

If anyone ever does the film of my life back then, then I'd love Leonardo DiCaprio to play me. To me he's just the best actor around right now.

Larry Hagman and his wife, Maj, on a scooter in London in the early days of their relationship. Photo courtesy of Larry Hagman/Majlar Productions.

CHAPTER THIRTEEN

THE POINT OF NO RETURN

(Prentis Lavaughan Ollis passed away on March 11th, 2006 at the age of 74. However, he was both web-savvy and historically minded and recorded his memories of his military service in England. His recollection is worth attention because of its detail and humor. that dedication can also be seen in his recording of the history of his hometown of Porterdale, Georgia.

I am indebted to Mr. Ollis' wife, Loraine and his daughter, Jan, who gave me family permission to retell his story here. It is an informative descriptions of what the command bunker at Daws Hill, High Wycombe, was like after having been largely shut up since the end of WWII. And how it was quickly developed for its Cold War role.

Prentis Ollis' fellow workmate was Charles Outram. High Wycombe provided the foundation of a life-long friendship between those several families, including that of T/Sgt. Rollie Whiteworth who arrived at what would be known as 'Lancer Control' in July 1954 and served until August 1957. That connection has continued over the years through joint holidays and regular communication. We start with his story first.)

Charles Outram

A1C, USAF
Headquarters 7th Air Division SAC
November 1953 – September 1955

I was an Airman 1st Class stationed at Carswell AFB in Texas and was working in Base Operations when I received orders to go to England on 15th of August 1953.

My wife, Nellie, was working off base at a GM distribution center. After I had 'cleared' the base, a T/Sgt stepped forward and wanted to take my place. The base officialdom agreed that I could clear back onto the base and the T/Sgt could go in my place, but said it was up to me. Where I worked and where Nellie worked they had thrown going-away parties for us already. We already had our feet out the door and so Nellie and I decided to go ahead and have the adventure.

Are we ever happy we did!

However, we still did have a slight wobble. When we left Fort Worth, Texas and were a hundred miles out we talked a little about going back but then decided we were beyond the point of no return.

In a travel plan that echoes that of my old colleague and friend, Prentis Ollis. I, too, had leave. I took travel time to Ft Dix, New Jersey, and waited for a ship to cross the ocean. This was the *USNS Upshur*, a vessel that had been commissioned in February 1953 as a troop ship.

I had departed Carswell around 1 September and didn't get to work at Lancer Control (as it was known) at the High Wycombe base until about 15 November 1953.

Part of it was due to my leave. But traveling over was a story in and of itself. When we sailed (lots of troops, 333 in compartment C1, bad weather and much sickness) it was a little longer than usual (12 Days) because we sailed past Southampton, England and went to Bremerhaven, Germany, where we got off the ship and rode a train to an old German base where we stayed about five more days.

We heard all kinds of rumors on how we were going back to England, but, in usual military fashion one day they just loaded us on a train and we went back to Bremerhaven and the *USNS Upshur* and rode it back to Southampton.

I was on orders initially to go to RAF Upper Heyford in Oxfordshire and, after checking in on base there, I was called to the Squadron Orderly Room and given TDY Orders to go to 7th Air Division Headquarters at South Ruislip for a job interview on the following Monday morning to be a controller in Lancer Control at RAF High Wycombe.

I left RAF Upper Heyford on Thursday and went to London and stayed in the Douglas House (the hotel run by the US Military) over the weekend. When I went for the interview I was interviewed by Captain Pickavance, Lt. Colonel McCourt and Colonel Swancutt. They also introduced me to General James Selser. They told me I was 'hired' and sent me back to RAF Upper Heyford to await Transfer Orders.

Turns out Captain Pickavance had been stationed at Carswell AFB working in Base Operations, the same place I had worked, before he transferred to England. He saw my name on the *USNS Upshur* passenger list and had set up my Lancer Control interview. I received my transfer orders, went to 7th Air Division at RAF South Ruislip and when I was clearing onto the base, a SSgt who was involved in the process said he was heading back to the States and he was in an apartment that was one of the best.

He took me to meet the landlady and when we agreed on the rental fee (the other person was a SSgt and I was an A1C so I wanted the apartment for a little less) I rented it and gave her a deposit. Since I was an A/1C, I had to pay Nellie's way over. The next day I went to American Express and made reservations for Nellie to come over on the French Lines ship *Liberty* sailing from December 12th to 18th and arriving at Plymouth, England.

When the time came I rode the Ship Train from Paddington Station to go and meet Nellie from the ship and we came back to London on the same train.

Our landlady's husband had passed in 1949 and she converted her half of a double house into an upstairs and downstairs; she and her three children lived upstairs and Nellie and I were in the downstairs which had a closed door to our apartment.

I started to work in November 1953 as a controller in Lancer Control at High Wycombe and immediately met Prentis Ollis.

We had shifts from 0800-1600 hours and 1600-0800 hours. We were allowed to split up the shifts and sleep in the sleeping quarters located in the bunker dependent on our duty responsibilities; if there were special exercises it was not possible. There was always an officer and two 'Non-Coms' (Non-Commissioned officers) on duty at all times; one individual could sleep from 2400 – 0400 or one from 0400 – 0800. In doing this we could go all day the next day after getting off at 0800.

Everything was there for us to do our job, communications, plotting boards, weather to do our Command Post/Lancer Control responsibilities in directing aircraft per Operations Orders.

From where I lived in Wembley, Middlesex, I travelled to the South Ruislip 7th Air Division by bus and tube system and rode back and forth to High Wycombe in an USAF car which were 1949 Chevrolet sedans and Buick station wagons in those days. The trip was about 45 minutes each way.

What a wonderful decision we made! Wonderful country, wonderful people, wonderful Air Force Family there in England including my co-workers, friends, etc.).

To top it off our landlady and her three children were the best. She was Nellie's 'English mother' while we were there and remained close throughout the years just like the friends we meet with each spring and have for some time.

Fortunately I was able to be in Europe and England many times throughout the years and could visit our landlady and her family and the beautiful country.

Nellie & Chuck Otram at the Tower of London in 1954.

CHAPTER FOURTEEN
LANCER CONTROL

PRENTIS OLLIS

S/Sgt. USAF
Headquarters 7th Air Division SAC
June 1953 – June 1956

My wife, Loraine and I had been married about one year when an opportunity arose for an assignment to England. The assignment was for Lakenheath Royal Air Force (RAF) Station in what is known as the East Anglia part of England. At this time I had not definitely decided to make the Air Force a career. In the back of my mind I was thinking of getting out of the Air Force and going to pharmacy school.

We were stationed at Rapid City Air Force Base, South Dakota, at this time. The year was 1953 and the Strategic Air Command (SAC) had just started building up their Command and Control System to manage and control their vast worldwide strategic forces. I had played a very important role in construction and developing procedures for the Command Post at Rapid City, AFB.

As a result my boss, Colonel Albert Faye, was very reluctant to release me for this assignment to England. I told him that I did not plan to remain in the Air Force for my career and planned to get out at the end of my four-year commitment. He said that he would think about it. Several days later Colonel Faye asked if I really planned to get out of the Air Force and I replied: *Yes.* He said if that was the case he would release me for the assignment, but he would not let me go to Lakenheath which he considered to be out in the "boonies".

A few days later Colonel Faye's secretary brought me a copy of a letter that Colonel Faye had written to a "Dear Bromo" at South Ruislip, England. South Ruislip was Headquarters for the 7th Air Division. It turned out that "Dear Bromo" was a Major General and the commander of the 7th Air Division, who was a classmate of Colonel Faye's at West Point. The letter to General James Selser told of what an outstanding airman I was and the good work I had done at Rapid City and that the General might find me a more important job in England than at Lakenheath.

My wife and I left Rapid City around April 1953. This provided for 30 days vacation en route to England and 10 days travel time from South Dakota to Fort Dix, New Jersey.

Loraine could not travel concurrent with me at this time. Housing was in such a short supply in England then that I had to locate and rent living quarters before the Air Force would issue travel orders to England for Loraine. Leaving Loraine in South Dakota, I proceeded on to Fort Dix, New Jersey.

I hung around Fort Dix for a couple of weeks awaiting transportation and travel orders to England. My mode of transportation would be by surface, sailing from New York on the *USS General M.B. Stewart*. This trip by boat is a story in itself.

I finally arrived at Southampton, England. On my arrival, at the first formation, my name was called. I was told that a staff car was waiting to take me to South Ruislip.

By this time it had been about two and half months since Colonel Faye had written his letter to General James Selser. The General had put out the word to find me on my arrival and deliver me to his office. So everyone was on their toes looking for me. I learned later that at his Staff Meeting the General would ask about me every few days.

When I finally arrived at 7th Air Division there was a flurry of activity. Remember at this time I was just a lowly Staff Sergeant. The Squadron Commander took me immediately to a Lt. Col. McCourt who took me to a Colonel Swancutt. Colonel Swancutt escorted me to General Selser's office. I had to explain where I had been the past two – and-a-half months.

The General wanted to know how Colonel Faye was doing. He then explained that he wanted me to work for him in establishing the 7th Air Division Command Post. Also I was to remember if there was ever any problem or that if I needed any assistance, there was nothing between us except the door and that the door would always be open for me.

After the visit with the General I was glad to find my quarters and get some rest. I got very little rest during the seven to eight days we were at sea. Also, I wanted to get busy looking for an apartment so that I could get Loraine on the way over.

I was met the next day by some of the guys that I would be working with. One was Bill Grazier who had been there for a little over a year. He took me around looking for a place to live. It was not easy finding a place. There was a real shortage of housing in England at this time. World War II had not been ended but a few years and the Brits were just trying to get their act together and a lot of the young folks were getting married.

At that time I think the average income was about three pounds per week. A pound at that time was worth $2.82 in US currency. We were given an extra allowance at that time to defray the costs of living abroad. Therefore we Americans had a little more money to spend on rent than the average Brit. This did not make for real good relations between the Americans and the Brits.

I had been in England about six or seven weeks before Loraine arrived. The first place I rented was just two rooms in a private home. Of course I would have taken anything just to get an address so that the authorities would issue travel orders and I could get Loraine on her way.

I recall this first place in that the landlord had a turtle chained to a tree in the back yard as a pet. A very good friend of mine, Walter (Willy) Church and I would go out there to the turtle and feed him beer. The turtle would get so drunk it could hardly walk.

I had located another place for us to live before Loraine arrived; it was a three-room upstairs also in a private home. I had put my name on a waiting list for a one-bedroom apartment in an apartment building in Harrow Wealdstone, which took a short time for it to become available. We enjoyed this apartment; we had some military friends who lived there and we made friends with some of the local English folks. This apartment was near Harrow-on-the-Hill, one of the schools attended by Sir Winston Churchill. We lived here for over a year before we moved to the country out near Chalfont St. Peter.

My job in the Command Post turned out to be real interesting. We were just getting involved in the Cold War and we dealt with a lot of intelligence data. The location of the Command Post turned out to be an interesting site. The name of the town was High Wycombe, England which was about 20 miles from South Ruislip. South Ruislip was a suburb of London, which was, in turn, a 20-minute ride from the center of London.

The facility at High Wycombe was underground; this place was the Command Post for the US 8th Air Force in World War II and was known as 'Pinetree'. Our modern Cold War name for it would be 'Lancer Control'. This facility contained about 40 rooms and it was three stories tall – all underground. I don't recall the number of steps down to the main operations level, but it took a little time to get down there. However, it took longer to climb out of there!

We arrived to start opening up the place and found there was about 10 feet of standing water that had to be pumped out. It took about a year to get all of the equipment installed and operating, but then we were open and operating 24 hours a day, seven days a week.

We had some of the best communication facilities available at that time. We had High Frequency Radio contact to just about any place on earth. Our telephones were remarkable for that time in history. We could pick up the phone and ask for any place in the world and be connected almost immediately.

There were huge plotting boards covered with world maps, with all types of intelligence information plotted. These plotting boards were about 20-25 feet high and 30-40 long. There were ladders attached to a rail mounted to the ceiling, where they could be moved from position to position.

I recall a very interesting map of the Soviet Union and they were building a railroad to the East across Siberia. Every few days someone would come in and mark the number of cross-ties that had been laid in the past 24 to 36 hours.

There was a Weather Central Station located next door where there were Weather Forecasters on duty 24 hours a day, seven days a week. You could get weather information for any place on Earth instantly. We had to keep posted on the weather situation for any place our planes were flying. Also we had to know the weather for bases in Germany, Spain and North Africa. The weather was always so unstable in England that we had to be prepared to divert our aircraft to these alternate bases.

Since we had this Weather Center located at High Wycombe it was a good cover for intelligence work. For example, we launched weather balloons from this location. However they were not weather balloons but spy balloons. These balloons were equipped with cameras and sailed across the Soviet Union taking pictures. When they were over the Pacific Ocean they were lowered and aircraft would pick them out of the air with tail hooks. The film would then be rushed to Lockbourne AFB (10 miles south of Columbus, Ohio) for processing. These balloons were used before the U2 aircraft came along.

England was such an interesting place to visit. We enjoyed the rich history and meeting the locals. We enjoyed London where we got our chance to see our first live stage plays. We visited such places as: White Cliffs of Dover, Stratford Upon Avon (including Shakespeare's birthplace where he did a lot of his writing, as well as Ann Hathaway's Cottage), Cambridge, Uxbridge, the Tower of London, London Bridge, Buckingham Palace, Windsor Castle and Number 10 Downing Street. It would take me forever to list all the places.

There was a pub near St. Albans, called the "Ye Ole Fighting Cocks". It had been in continuous operation since the year 795 or so. The ceiling was so low I could hardly stand upright. I had to bend over to get through the doors. Very near the Pub still stood some of the remains of the old Roman Wall. They had built a fence around the part remaining. Many of the houses within several miles either side of the wall were built of stone taken from the wall.

Note: Prentis Ollis first put his memories on-line at:
http://freepages.books.rootsweb.ancestry.com/~ollis/air-force/england.htm

CHAPTER FIFTEEN

THIRD AIR FORCE YEARS REMEMBERED

Richard Lee Wilson

Dependent/Student
USAF Bushy Park
1954 – 1957

(Author's Note: Richard Lee Wilson started writing his memories in Louisville, Kentucky, where he lives with his wife, Sue. He served almost eight years in the US Air Force, 1963-70 before resigning as a Captain. He got a day job, went to night law school and practiced law until retirement. As the son of the commanding general of the Third Air Force and High School student at Bushy Park in the 1950s, Wilson had a unique vantage point from which to observe the inner workings of the US Third Air Force from its command base at USAF South Ruislip, Middlesex.)

From 1954 to 1957 my father, then Major-General Roscoe C. Wilson, commanded the Third Air Force. His headquarters were at RAF South Ruislip. Our own quarters, on Southwick Place, about 4 blocks Northeast of Marble Arch in Central London, were technically a substation of RAF Denham. The house was a diplomatic residence.

I attended the dependent High School at Bushy Park, grades 9 through 11. The last thing a teenager wants to be is "different" from the other kids but I was very different and spent considerable effort trying to conceal from the other kids just how different my experience was.

Some caveats: I was a teenager back then. I was well aware of what was going on, but I was, as you will understand, "out of the loop" when it came to operations, missions and policy. My recollections might be wildly inaccurate. Furthermore, when an observer describes a scene he necessarily puts himself at the center of it, even though he does not deserve center stage. I have written almost entirely from memory; I have little more than a few photographs and letters to assist my recall of over 50 years. Nevertheless, I am sure that if compared with the records of the time, my recollections will prove reasonably accurate.

MY FATHER

My father was Lieutenant-General Roscoe Charles "Bim" Wilson, USAF. He spent a total of 38 years in United States military service, 32 years commissioned service and 13 years as a General Officer. He was "Mister Research and Development" for the Air Force during many of those years. My mother, Elizabeth, married him in 1929 and they were devoted companions until his death.

Father was deeply involved in the conversion of the Air Force from piston engines to jet engines, in the introduction of long-ranged ballistic missiles and interceptor missiles to the arsenal and most especially in arming the Air force with nuclear weapons.

The design patents of the P-39 Air Cobra are in his name. That airplane was built around its gun, with the engine located behind the pilot, to reposition the moment of inertia to the center of gravity, have his unmistakable engineering stamp.

During the Second World War, he was assigned to the "Manhattan Engineering District", which produced the first Atomic Bombs. Officially he was General Leslie Groves' Air Force Deputy but really he was General Henry 'Hap' Arnold's personal representative to that super-secret outfit. General Arnold commanded the Air Force. General Groves of the Army Corps of Engineers commanded the A-Bomb Project.

Groves' subordinate, Dr. J. Robert Oppenheimer, led the scientific effort and much of the engineering effort, too. Then a Colonel, my father's contribution was to redesign the B-29's bomb bays to accommodate the A-Bombs. He also designed the fins for "Fat Man" bomb, which had to fit inside the bomb bay, yet keep the weapon stable after release. Among the skeptics who thought the bomb would tumble were Oppenheimer and his fellow atomic scientists Neils Bohr, Theodore Von Carmen, George Gamow, etc, but the fins worked fine.

My father's first trip to Britain was a wartime visit to consult with Barnes Wallace on methods of very large bomb carriage and release. Dr. Wallace was the inventor of the 'Dam Busters' Bouncing Bomb and helped develop the Tallboy and Grand Slam bombs.

Three men were considered for the first atomic bomb mission. First was Brigadier General Frank A. Armstrong, the model for "General Savage" in the fictional story and film Twelve O'Clock High (but Armstrong did not have a mental breakdown, as did the fictional general).

General Arnold considered him too senior for the job.

Second was my father, then a Colonel, who had been involved with the B-29 from conception to production and who knew more about atomic bombs than any other Air Force officer. However he had never flown a combat mission. He asked General Arnold not to select him for this mission.

The third was Lt. Col. Paul Tibbets, Jr. Tibbets had served a tour with the Eighth Air Force with distinction. Most of Boeing's Engineering Test Flight section had been killed in a tragic crash of a B-29 prototype. Col. Tibbets headed the team that replaced them as the B-29 completed testing and entered combat, so he knew all about B-29's, but nothing about Atomic Bombs.

General Arnold picked Colonel Tibbets for the job; Tibbets tells this story in his book "Enola Gay".

Instead of dropping A-bombs my father went to Okinawa and rebuilt Kadena Air Base which had been damaged during the battle for that island. After Japan ceased fire, but before the surrender, he and Colonel Howard Bunker went to Nagasaki and Hiroshima to survey the damage. Their reports are in the USSBS (United States Strategic Bombing Survey).

The Japanese Commander of the Military District of Tokyo wanted to execute them as spies but they were spirited away by Shinto priests. They were wearing Kimonos and fishing from a dock on a temple island in Tokyo Bay when the battleship Missouri steamed into it for the formal surrender ceremony.

When the US Air Force became a separate service his serial number was 360. By 1950, Pop was a Major-General and was slated to take command of the Third Air Force in Britain which was then being organized as NATO's 'Nuclear Punch'. At the last minute a replacement had to be found for the Commandant of the Air War College, so instead we spent three years at Maxwell Air Force Base, Alabama.

General Francis H. Griswold went to London, took over from General Leon Johnson and built Third Air Force almost from scratch, mainly with World War II leftovers.

In 1954, with the normal three-year tour of duty completed, Pop was at last given command of a flying outfit, the Third Air Force. He had never previously commanded a Squadron, Group, Wing or Division. But by the end of our tour in 1957 the Third United States Air Force would comprise: 4,500 Officers, 38,000 Men, 1,000 Department of the Air Force civilians, 7,000 UK civilians and 35,000 dependents. Total: 85,500.

The Headquarters was located at South Ruislip. The first thing one noticed was that there were no airplanes and no runways; just a series of office buildings on either side of a main street that ran up to a main road (Victoria Road). Along the parallel back streets were warehouses, barracks, a Gym, Mess Hall and Hospital. The buildings were yellow brick, two or three stories high. "Un-prepossessing" is the adjective that comes to mind. It looked like a factory complex. The place was just kind of there.

Kind of blah.

Pop's personal airplane was parked at RAF Northholt, the Royal Air Force base just a five-minute drive from the South Ruislip base.

At first he had a C-54, a four-engined transport, but that was replaced with a two-engined Convair C-131D. Pop named it the *Elizabeth Too* after my mother and Her Majesty. Sometimes he flew and some times Major Houston, his ADC, or one of a handful of other pilots flew. The flight also had a Gooney Bird (a C-47 for transport, liaison and training work).

His personal car was a blue Buick sedan, about 1950 vintage with lots of miles on by the time he first got to step into it in 1954 but it came complete with a British driver, Mr. Charles Atkins, who kept it immaculate.

I have mentioned being different from the other kids at the high school. None of them had government quarters in London. None had a mess sergeant working in the kitchen. None of them regularly rode in a staff car. Only a very few of us, members of the Air Explorers (the aviation branch of the Scouts) ever flew aboard the *Elizabeth Too*.

The stationing of troops on foreign soil is always a difficult matter. "Town & Gown" (or airbase & civilian community) relations are sources of tension exacerbated when the troops involved are foreigners. The history of the American forces in Britain is one of remarkable amity. The British people have been largely unfailing in their generous hospitality for well over 60 years. Their tolerance has been extraordinary.

This is the granite bedrock upon which the North Atlantic Treaty Organization (NATO) Alliance rests. As much as upon weapons of war the liberties of the British and American peoples are defended by mutual good-will. Consequently British-American Relations were always a top priority for my father.

At times Third Air Force had units in the Netherlands and Libya. For a while, my father was also "MAAG-UK" (Military Assistance and Advisory Group, United Kingdom). This also carried quasi-diplomatic status. His duties were not entirely confined to Britain during the years 1954-57. Third Air Force was already "host" to Strategic Air Command's (SAC) Seventh Air Division. Their medium bomb units would rotate in and out on TDY (Temporary Duty), usually for three months at a time, as I recall.

At that time there was also a Naval headquarters at Grosvenor Square. I have the impression that they were entirely separate and that contacts between the US Navy and U.S Air Forces in Britain were minimal.

An Army anti-aircraft unit was assigned to Third Air Force for a while. My father sent them home; he saw no reason to have Anti-Aircraft Guns at American airbases when the RAF had none at theirs. Besides, AA guns do not really fit into plans for nuclear defense or retaliation.

Third Air Force was the only nuclear-armed air force integrated into the NATO command structure. The British were developing their Vulcan force and the French were then, as now, fiercely independent. USAFE (US Air Forces Europe) had air units in its Northern, Southern and Central commands. General William Tunner, architect of the Berlin Airlift, commanded USAFE. Third Air Force had medium bombers and fighter-bombers. I recall my father saying that General Tunner had given him his worst and best OER's (Officer Efficiency Reports).

I was of course never privy to specific plans but have the impression that they were to hit essentially strategic nuclear targets, as opposed to tactical nuclear targets. For example, there seemed to be considerable interest in the Murmansk/ Archangel area of the then Soviet Union, but little or no interest in the Fulda Gap in Germany – considered likely to be a key battle area if the Soviets ever invaded the West.

My father used to say that his best public relations tools were the Third Air Force Band and the Air-Sea Rescue service. I suppose that someone plucked from a potential briny grave in the Channel might well be grateful that the U.S. Air Force was there to do the plucking. The band, despite the usual problems with the (British) musician's union, would play at charitable events and put in appearances at military tattoos.

Limited by law to 36 members they were at a considerable disadvantage when it came to performing and in Europe parades, tattoos and such "showing off" occur more frequently than in the US. The Royal Dutch Grenadier Guards, for example, would parade with two hundred drums echoing thunder from the clouds. But then the Third Air Force Band would take the field, 'whanging' out *The St. Louis Blues* and bring the house down. My father tried to augment them with a volunteer Drum and Bugle Corps.

Relations between Britain and the Vatican have been, well, touchy, since the 'Eighth Henry'. The British didn't want to be rude but they did not want a rapprochement either when a Palpal Nuncio or Legate paid an official visit to Scotland (I think it was Scotland). The British didn't quite know how to handle it so the Third Air Force provided a band and an honor guard. The Legate got a proper reception but Her Majesty's Government was kept in the background, which is precisely where they wanted to be: Close but not too close. They were very grateful.

Severe storms and floods inundated portions of East Anglia,South of The Wash during the 1950s. The 81st Fighter-Bomb Wing lent its assistance to the afflicted. It was based at RAF Station Sculthorpe. That earned a lot of good will.

In 1954 the Third Air Force had a dreadful safety record. I recall father being appalled at the discovery of a color-blind mechanic; he had hooked the green hydraulic lines to the red fuel lines and a plane flamed out on the runway after burning up its hydraulic fluid. Training, personnel selection and more training underwent a thorough overhauling. The 81st Fighter Bomb Wing was a particularly mediocre outfit.

My father found Colonel Ivan McElroy to run it; figuring that if a Russian-Scot couldn't do it, nobody could. Colonel McElroy was a human dynamo whose energy invigorated the whole unit. In a year or two after Colonel McElroy took over, the 81st won the US

Air Force Gunnery Competition. No mean achievement, that.

I recall being in my father's office one time when Colonel Robert Goewey, Third Air Force Vice-Commander, came in. He had just received a thank-you letter from an airman who had been illiterate. Colonel Goewey (who later became Brigadier General) had encouraged the man to learn how to read and write. The spectrum of the training effort was that wide.

In 1954 when my father took command the Third Air Force was the smallest of the numbered Air Forces with a mediocre to poor record in combat readiness. By 1957, it was considered an elite force. That turnaround was my father's greatest achievement as a commander. It didn't happen all at once. It took painstaking effort, day after day, finding problems and fixing them. They are easier to find than to fix.

My father visited the base at Bushy Park and there found operating the 'Umpty-Umph School Squadron comma Dependent'. This being no great inspiration, my father decreed that "Until you come up with a better name, this is 'Central High School.'" And so it was.

Although I was expected to attend a British school I wound up attending Central High. A few years later I also ran into the poor harassed Major who ran the base school; only by then he was a Lieutenant Colonel and I was a Second Lieutenant passing through a course he was running at Vandenburg Air Force Base. Remembering the trouble I had caused him in London I hid from him!

At the end of WWII Barbara Hutton, the Woolworth heiress, had given Winfield House, a large mansion in Regent's Park, to the US Government. It was used as an Officers Club. It is a lovely place and huge. John Hay Whitney replaced Winthrop Aldrich as US Ambassador sometime in, I think, late '54 or early '55.

Mr. Whitney, being possessed of a considerable fortune, decided to make Winfield House into the Ambassadorial Residence. That meant we needed a new Officer's Club in London. They leased a hotel on Bayswater Road for that purpose.

My father said; "We need to give the place a name that is distinctly American but non-partizan, non-sectarian, something like 'The Columbia Club.'" Among other things the Columbia Club hosted the American Teen Aged Club of London – all Central High students. I even recall one Central High School prom being held there.

It was at the Columbia Club that I was privileged to meet Group Captain Douglas Bader, RAF (Ret.). A book about him, *"Reach for the Sky"*, was written by Paul Brickhill and the book was made into a film. Bader commanded one of the five fighter groups that fought the Battle of Britain and devised the mass formation of interceptors called "Baders", which inflicted unsustainable losses on the Luftwaffe at minimal cost. An Ace (with three or more aircraft shot down) he had a second adventurous career as a POW and escaper. He had lost both his legs in an aircraft crash, *before* the war.

There were protocol-type perks to my father's job, a few of which fell my way. I got to sit in the Royal Box at the Earl's Court Military Tattoo in London (the Royal Family were not using their tickets that day). I loved the pageantry at which the British excel.

There were other unique opportunities that fell my way. In the House of Commons there is a very small "Strangers' Gallery" from which proceedings can be witnessed. These days sessions are broadcast around the world on C-Span, but back then, you had to be there or miss it. It was hard to get tickets. However eventually my mother obtained tickets for the both of us. It was catch-as-catch can so far as the business of the day went; what was happening on the floor on the day you got to go was a matter of chance. But we had tickets and I took my seat beneath the doleful eye of a bailiff who was well acquainted with the noisy restlessness of teenaged boys.

On this day there was outrage on the floor. The day before, two Red Chinese fighter planes had intercepted and shot down a British airliner bound from Malaya to the Philippines over the South China Sea for no particular reason. It was just plain murder. And it was Question Time in the House of Commons.

The question, of course, was what, precisely, Her Majesty's Government proposed to do about this barbaric outrage. Ordinarily, HM's Minister for Air would respond, but that was *Lord* De L'Isle and Dudley so, given it was the House of Commons and not the Lords, the Government spokesman was none other than Prime Minister Sir Anthony Eden himself.

He reviewed the course of recent Sino-British relations. The Korean War was of recent, bitter memory. The Chinese were making noises about Hong Kong, which was practically defenseless. The Exchequer was near empty and the defense establishment reduced to minimums and the gist of the reply was that there was not a whole lot that Her Majesty's Government could do and that the danger of war with China, with all the cost in blood, treasure and human suffering that would entail, was quite real.

The House was tense and silent as the members chewed over this unpalatable gristle. Then there was a stir behind the Speaker's Chair and the Speaker rose. "We have just received a report from the South China Sea. Naval Aircraft of the Fleet Air Arm of the United States Sixth Fleet, while escorting units engaged in rescue operations, have intercepted and shot down two Chinese Fighter Aircraft of the same type which attacked the airliner, yesterday."

There was a gasp from the house. Then some barnacle-encrusted lump of a Tory (Conservative) back-bencher slapped the arm of his seat and said "Hear Hear!" And the British House of Commons recessed, cheering for the U.S. Navy and leaving an American boy in the Strangers' Gallery, wondering how to swallow the half-brick that had mysteriously materialized in the vicinity of his Adam's apple.

There was a great deal of diplomatic partying in London. My father hated it. His idea of a good time was a glass of wine and a good book but there was no escaping it. He and mother would frequently go out five or six nights in a week often to more than one function in an evening. Father would grumble that Third Air Force had more punch than the whole of the RAF; that the RAF had half-a-dozen five-star types to run it who didn't go to work until 9 a.m. and they could stay up all night but he had to go to work early in the morning.

One gentleman I particularly recall from the party circuit. Imagine, if you will, a tall, distinguished-looking man, tanned and slender, outstanding combat record as a fighter pilot, slightly graying at the temples, an Air Force Brigadier General and single; assigned as 'Air Attache' to the U.S. Embassy in London. Even his name was perfect: Jack Sterling. He had to fight the women off with a stick.

There was a tendency to use the same caterers at many of these affairs. My father told me of one man, a Cypriot waiter, whom he saw time and again. This man learned to see my father coming; would spear an olive with a toothpick, put it in a martini glass filled with cold water and hand it to him without a word. Quiet tricks of the diplomatic trade. One wonders how many others were sipping Ginger Ale, or cold tea and soda.

One particularly good friend was Air Chief Marshal Sir William Dickson, the commander of the RAF. He and my father would have been thrown together by official duties anyway but I am talking about genuine good friendship. He and Lady Dickson were kind enough to include me in their invitations to dinner at their townhouse and they were often at our quarters. "Dickie" Dickson had a boyhood chum named "Dicky" and they had called each other 'IE' and 'Y', respectively from the variant spellings of their nickname. Even though when I was called "Dickie" it had been spelled 'IE', when Sir William suggested that he be 'IE' to me and I 'Y' to him. 'Y' I became.

Our quarters were set up to accommodate the diplomatic reception circuit. The establishment at No. 2 Southwick Place comprised a Mess-Sergeant NCOCQ, an upstairs maid and a daily maid. Auxiliaries were brought in as occasion required.

Mother's maid was, at first, a young German woman named Ursula who had worked for Mrs. Griswold when my father's predecessor Gen. Griswold was in post. She was replaced by Wilma Claus, who had also survived Nazi Germany. Her husband had vanished on the Eastern Front. She had volunteered to take a pay cut to work in an envelope factory which got her out of Hamburg just before the devastating air raids that killed at least 50,000 people in that city. Wilma came to the US with us, went to nursing school, worked at the Home for the Incurables in Washington, retired and returned to Germany.

Molly was the daily maid. She was a Londoner technically not a "Cockney" (a person born and raised within hearing of the bells of Bow Church, London) but close to it. She cleaned up and when mother gave her instructions she would do a kind of a bob-curtsey.

Sergeant Getty was the mess-sergeant when we arrived. NCOCQ means "Non-Commissioned Officer in Charge of Quarters" so he was the Sergeant in Charge. He was a good cook and, as my parents had to entertain often, he was busy. He transferred out however and we got Sergeant Testlin.

Wilma stuck her head in my room; "The new one is all in white!" Sgt. Getty had been content with shirtsleeves and an apron. Sgt. Testlin came to work in blues, but changed to whites complete with cap before he started work. Could he cook! And bake! We got him simply because he happened to come up for reassignment at that time and if he hadn't been given a job fast he would have been whisked off to US Air Force HQ and the Bolling AFB Officers Club.

He was with us for about two months and then Pop put him in charge of Douglas House, in downtown London. Douglas House was the London enlisted men's club – a hotel for airmen and sergeants. It was a good hotel, which enlisted personnel could not otherwise afford in London. The men deserved Sgt. Testlin's talents more than we did. He had me to lunch there once and it was very nice but otherwise my orders were not to bother the enlisted men who had enough to worry about. Sgt. Testlin also had Mom and Pop there to some fancy occasion or other (Pop in Mess-dress, Mother in an evening gown) which also happened to be their 25th wedding anniversary.

Sgt. Testlin was replaced with Sergeant Wright. Sgt. Wright weighed maybe 100 pounds soaking wet and during WWII had been a BAR (Browning Automatic Rifle) man in Burma. His ammo-carrier had been wounded and stood up to go to the rear; Wright jumped up to pull him down and woke up in a hospital two months later. He had been shot nine times with three different calibers of bullet. He had scars up and down both sides. I recall he had a girlfriend whose name was Sweetie.

Auxiliary help included Lisa Arnold, a German woman friend of Wilma's who owned an antique shop and Airman Brown, who was a steward on the "Elizabeth Too". They might be called in to help serve at large parties. Duty required my parents to entertain guests frequently. They had a lot of help, but they needed all of it.

As a substation of RAF Station Denham the quarters were on the distribution list for movies shown at the base theaters. Each week we would get five movies and each week we would send them on to the next base on the distribution list. To get around British union rules I got checked out as a projectionist at the United States Information Service (USIS) office and was licensed to show them.

Television was new and TV sets were very expensive. We got our first television set about 1948 in Washington. It was a Crosley brand set, about 4 feet wide, 4 feet deep and 2 feet high; a wooden case with a ten-inch black and white screen in the middle. It weighed a ton, but we could receive ABC, CBS, NBC and DuMont Network transmissions from stations in the Washington area.

In 1950 we moved to Montgomery, Alabama. They were building a wire network to carry TV programs across the country but it was not finished. By this time we had a 15-inch set, smaller than the Crosley but still large. Montgomery had one TV station, broadcasting local programs and cinescopes. The network wires reached Montgomery shortly before we left for England in 1954 and "live from New York!" was heard on the air for the first time.

American electric power is 60 cycle 120 volts and we used a flat 2-blade plug or a triangular plug with a round ground rod and two flat blades. Special plugs accommodate 220-volt appliances, like clothes dryers and electric ranges. British power is 50 cycle, 220 volts so the US military had to use transformers to step down the current for American appliances. Radios worked. Electric razors ran at 5/6th speed but worked. Clocks only registered 50 minutes per hour. TV broadcasts had a different number of lines per inch and TVs are more "cycle dependant" than radios, so American TVs wouldn't work.

So from 1954 to 1957 we did not have a television set. That's why we had movies. It is also why we listened to the radio. BBC Radio broadcast three 'Programmes': The BBC Home Service, the BBC Light Programme and the BBC Third Programme. The Home Service was news, weather, sports, music shows from *Housewives' Choice* to the Symphony Orchestra and some drama.

The Light Program included comedy and variety shows, soap operas (I recall "Mrs. Dale's Diary"), pop music and the like; lighter fare than the Home Service. The Third Programme was experimental, avant-garde stuff, often very heavy indeed. People who have never heard well done radio drama have no idea of what they have missed and refuse to believe it can be better than TV. It can be, nevertheless.

In London I got a short-wave set. I strung antennas on the roof of the house. On the ordinary AM bands one could hear the BBC and Radio Luxembourg in English. The latter sold commercials for goods and stores in the UK, but the signal quality was very bad.

You could hear French and Dutch stations easily enough, but to hear anything in English from the Continent you had to have shortwave. BBC World Service was like a 4th BBC 'Programme'; but the Voice of America, Radio Free Europe and RIAS ("Radio In the American Sector" of Berlin) were on the air, sometimes in English. There was no Armed Forces Radio in Britain. I remember waking my parents, after hearing it on the news, to tell them that President Eisenhower had suffered a heart attack.

Stella, Lady Reading, was one of the world's more remarkable women. She headed the WVS, Women's Voluntary Services. She was made a Marchioness for her good works. Mother, as titular head of the Officer's Wives Clubs, encouraged American wives to join and/or work with the WVS, which brought her into fairly close association with Lady Reading.

Lady Reading made a needlepoint footstool with the Third Air Force Insignia on it.

She also gave us a poodle dog who lived with us for many years, totally blind for the last three or four. She named him "Barnaby", after the little boy in Crockett Johnson's cartoons, "Mr. O'Malley and Barnaby".

Mother also worked with Lady Alexander, wife of Field Marshal Alexander, but I have no recollections of her. I do recall a tree-planting ceremony with trees being planted by Field Marshal Montgomery, mother and father. Mother buttonholed the groundskeeper and gave him explicit instructions on how to water, fertilize and spray *her* tree. She missed gardening.

Prince Bernhard of the Netherlands decorated the commander of a Third Air Force unit in his country and Father telephoned the Prince's Equerry to find out how he should express his appreciation. Prince Bernhard answered the phone himself to my father's considerable confusion.

At a Fourth of July celebration at the Ambassador's Residence my father was asked to dance with Her Majesty the Queen Mother. He declined. For a man with two left feet when it came to dancing it was as great a service as any of his contributions to the preservation of Anglo-American harmony!

I mentioned Pop's staff car. The British driver was appropriately named Atkins, but Charles and not Tommy (Tommy Atkins was slang for a British solider going back as far as 1743). He was very reserved. He had been a private during World War II and had served in North Africa.

One evening he was taking my parents to a reception being given by Field Marshal Alexander, Viscount Tunis and said; "Sir, Lord Alexander is a Gentleman." From Mr. Atkins this was high praise indeed and later, at the reception, my father repeated the compliment to Lord Alexander.

Alexander excused himself, slipped away from his fancy guests and went below stairs where the servants and drivers were assembled. He called for Mr. Atkins and the former Field Marshal and former Private shared a cigarette and a few reminiscences about the desert war; then Lord Alexander rejoined his guests.

Thereby proving Mr. Atkins' assessment absolutely accurate.

For our family car we started with a Vauxhall Saloon – 'Saloon' is the British term for a 'Sedan' (it had a back seat). Then we got a Sunbeam Talbot convertible. Pop would have preferred the Sunbeam Alpine, a 2-seater sports car, but mother insisted that I had to come along sometimes and she didn't want to strap me to a fender, bless her heart. So we got the Talbot model, which had a tiny back seat. Mother also didn't like the idea of a convertible but yielded to Pop's sports car urge. It was sliver-blue, with red leather seats.

Contacts with the British aircraft industry brought Pop into association with Sir William Rootes, head of the Rootes Motor Group which included Sunbeam. Sir William was delighted to find out that Pop was already a customer. Both engineers, they got to talking about dampening vibration in four-cylinder engines and the upshot was that Sir William had his people replace the 3-speed Talbot gear-box with a 5-speed Alpine Gear box and with a clutchless overdrive gear on top of that. It was so quiet when the overdrive kicked in you would think the engine had quit, except you were zipping down the road at whatever the law allowed.

Just before we left we traded the Sunbeam in for a Ford Zodiac, one step up from the Ford Prefect. It was that car we brought back to the US and which I learned to drive. It had left-hand drive.

One evening in 1956 Sir William Dickson, Marshall of the Royal Air Force, telephoned. It was quite late and my parents were in bed. The chit-chat was rather aimless, until the second hand swept past midnight. Then Sir William informed my father that the British and French were going into Suez at dawn. They meant to prevent the canal becoming a no-man's land between the advancing Israelis and the Egyptians.

The Egyptians had lost heavily in the Sinai and had lost their Air Forces in the first hours of what was later called the Six-Day War but the bulk of their Army had never crossed the Nile, let alone the Suez Canal, and they were preparing to make a long fight of it. The British were determined to keep the canal open.

At the same time the Hungarians had risen against the Russians. They forced the Russians out of Budapest. The Russian big-wigs had a peace conference dinner with Hungarian Defense Minister Pál Maléter and Premier Imre Nagy but then arrested and later executed them. Their armor regrouped the Russians re-entered Budapest, crushed the revolt and secured the countryside.

I recall listening on short wave to heart-rending appeals for assistance from the Hungarians. One group I took to be students, broadcast in English and were begging for American "Parachute Troops". The 82nd Airborne was in North Carolina; the 101st in Kentucky; there were no others. And back then, the C-54 was still our principal heavy transport. It would have taken six months to mount an operation.

But with Britain and France entering hostilities in North Africa and with Eastern Europe ablaze with unrest the Air Forces went on alert. I recall a conversation I had with my father. He said USAFE had queried; "Have you completed your dispersal?" He said he had replied, "I have completed my concentration."

"I would have thought you would disperse", I said, "Put one airplane, armed and manned, at the end of every long runway in Britain, assuming you could provide communications and security."

"That," he said, "would have scared the British half to death." He explained that he preferred to have the planes armed, manned and ready, on their bases, where control was certain and he could get them in the air instantly. He had full confidence in the ability to detect any incoming raid, to get them off before they could be caught on the ground.

Remember, this was the mid-1950s. The Russians had "The Bomb" but they did not have much in the way of delivery systems. No Soviet Rocket Troops.

They had their copy of our B-29 and some "Bear" bombers. Their crews were not well trained, not well organized and they did not pull alerts. Their crews went home on the weekends. They slept at night. They did not appear to have the military forces in being to attempt a surprise nuclear strike and at that time they knew that if they had attempted it they would have been committing national suicide. They were full of bluff and bluster but if there was any nuclear blackmailing going on, then we were doing it to them and not the other way around.

The concentration of the British tabloid newspapers on the capture of a subaltern distracted the British public from the realization that the Egyptian Army had made a mess of British and French Army operations in Egypt. The press were more fixated on a missing British Lieutenant than the success of the Expeditionary Force. They blamed U.S. pressure for making them pull out.

They also seemed to pay no attention at all to the failure of the West to support Hungary. I was in the Air Explorers, a branch of the Boy Scouts. We set up collection cans in the Base Exchanges, Commissary and various clubs to collect money for relief of Hungarian Refugees. I still think ours was the only action. The rest was just talk.

But Nasser closed the Canal and the Jordanians blew up the pipeline and that meant oil shortages. Most famines are not caused by lack of food; they are caused by lack of transport. That was the situation in 1956. There was plenty of oil but there was a shortage of tankers to haul it around Africa. The British asked the U.S. for "petrol". They meant crude petroleum.

We offered all the gasoline, diesel fuel, heating and lubricating oil they needed. The difference between crude and refined petroleum products, in tanker tonnage required, is about 40 to 1. They did not want to cause unemployment at their refineries, so they declined our offer. We then declined to ship them crude oil in the quantities they requested.

With really extraordinary efficiency the British instituted gas rationing. From decision to ration to delivery of ration books into consumers hands was a matter of three days. The books had obviously been printed in advance and the type of rationing scheme to employ had been decided in advance, as part of a contingency plan. Every vehicle had a basic allotment of 200 miles per month.

To recoup the revenue to the British Treasury lost by diminished sales, the tax was increased, boosting the price of a gallon from about forty cents to a little over a dollar. As far as I could tell it worked pretty well. There was an inflationary impact, of course; everything slowed down but nothing stopped.

In the press and in public statements by politicians the British hid behind their ambiguous word "petrol", which blurs the distinction between petroleum and gasoline, to make it appear that the U.S. was unwilling to supply them. It was grossly unfair and had to be suffered in silence.

One politician seriously proposed requiring the U.S. forces in Britain to supply their own POL (Petroleum Oil and Lubricants) instead of drawing on British stocks. My father was all for it; he would have had plenty of gas and its cost would have been a tenth of what we were paying them for it, but the idea went nowhere.

There was a cartoonist who loved to poke fun at the Americans in Britain. His specialty was drawing portly sergeants, whose stripes started at their cuffs and went up to their shoulders. I recall one of his from this time; it showed a Texan and wife and a tank truck with a big bow tied around it, tagged for shipment to their son in the Air Force in Britain, saying; "That ought to take care of the dear boy's holiday driving."

Giles' cartoons disappeared for a while and the paper explained that he had been taken ill. My father sent some flowers to the hospital, "From his Friends and Victims in the Third US Air Force" and got a note back expressing genuine surprise that the Americans thought his cartoons were funny. Of course we appreciated his humor; what were we, Prussians?

Another Giles' cartoon showed a bunch of Americans on a subway train pulling into Wimbledon, with one of his patented sergeants asking a startled matron "Is this the way to the Little Mo Show?" Maureen Connoly – 'Little Mo' – dominated women's tennis in the mid-Fifties and scandalized the tennis world by wearing frills on the panties beneath her tennis skirt.

One time, the senior diplomats who normally dealt with such matters being occupied elsewhere, my father represented the U.S. in negotiations involving some sort of appropriation to the British for defense. The British were suggesting that funds for project X be spent instead on project Y. My father's position was that if project X was not to be undertaken, Congress having appropriated money for project X, then the money should be returned to the US Treasury. He quoted the British reaction with amusement: "Oh! Her Majesty's Treasury would take a very dim view of that!"

One subject was very sticky; in negotiations with the British at this time as I recall. The U.S. was providing subsidies for the British Armed Forces. We were concerned that the funds and/or material provided be employed in a manner supportive of U.S. foreign policy and security objectives. Yet, for historical reasons, the United States cannot employ, or seem to employ, foreign mercenaries. How things were done was of as much importance as what was done and appearances were as important as reality.

There was a chronic shortage of quarters for personnel and dependents. One particularly involved series of negotiations produced 'Tobacco Housing'. Surplus American tobacco was swapped to the British for 'Frozen American Dollars'; the Dollars were then spent to erect housing. I recall my father's saying; "I suppose the first street will be named Tobacco Road." But I don't think it was. I do recall that the quarters were in part carpeted with wool, which, under British law, had to be British wool.

Unfortunately, the American Wool Act requires the U.S. Air Force to buy American wool exclusively. A Government Accounting Office investigation recommended that "the officer responsible be held accountable" which gave my father considerable indigestion, as might well be imagined. Fortunately no action was taken on the matter and I got to go to College!

My father was a killer of a Gin Rummy player. He was really very good at that game. He had a protocol aide, a WAF Lieutenant named Romanovich, andon long flights and they would play Gin Rummy. These games were played for money, usually a penny a point, but they kept a running tab and somehow, never got around to settling up.

I remember that one time we were visited by Mrs. Briggs, wife of General "Buster" Briggs, old and dear friends of my parents. Somehow or other, she got to take a trip on the plane assigned for my father's use (normally civilians don't fly on military aircraft; she must have had flight orders of some kind or other) and as soon as they took off, my father and Lt. Romanovich began playing Gin Rummy. Quickly Lt. Romanovich was losing.

As the Lieutenant lost – and remembering what it was like to try to live on a Lieutenant's pay – Mrs. Briggs grew alarmed. "Don't worry it'll all go on the tab," Lt. Romanovich assured her.

"Well how much is the tab?" Mrs. Briggs wanted to know. So they did some figuring and it came to a little over $20,000 – several year's pay for a lieutenant! Mrs. Briggs was appalled and when they landed she pressed a $20 bill into the Lieutenant's hand and wouldn't take it back!

I got to fly on that plane occasionally when they were on diplomatic missions. It's the old story of the travelling Indians: If there were just braves, it was a war party; but if squaws were along, it was a peaceful hunting party. So bringing dependents along on diplomatic trips was the modern equivalent of travelling "with squaws". The Air Explorers also got to use it occasionally without my father along. I recall flights to Rome and Scotland in it.

On December 17, 1960, some years after we had left Britain and while I was a student at the University of Maryland at College Park, I believe it was that very airplane that crashed. It had been transporting 7 crew and 13 military dependent student passengers from the University of Maryland's Munich branch to return them to their families in the U.K. for the Christmas holidays and it crashed into a trolley car full of holiday shoppers in the streets of Munich, killing everyone aboard and 32 more the ground. It was a horrible tragedy.

(The plane, which was by then the transport for Maj. Gen. Ernest Moore, the 3ʳᵈ Air Force Commander of the time who was in the nearby town of Garmish. As soon as he heard he travelled to the crash site. The plane reportedly clipped the 120 foot high St. Paul's Church in Munich and crashed into Martin Greiff Strasser, a street near the church. A street car with passengers was covered in burning fuel and other local people on the ground were killed in the accident which resulted in a death toll of about 60 according to the Stars & Stripes newspaper of the time (18 December 1960). A plaque to the 13 students was placed in the Library of London Central High school and, when it closed, was despatched to the American Overseas Schools Historical Society in Wichita, Kansas).

The name Napoleon Jackson Green comes to mind. Though he bore the names of three Generals, he was a 21-year old airman who went berserk one day on 24 August 1955 at the air base at Manston near Ramsgate in Kent. Green was reportedly to be questioned that day about an alleged theft and a separate incident. He left a note reading "Today I die", armed himself, shot and killed a colleague, then shot an RAF lad who was leaning a bicycle against a building. He shot others as he headed out of the Main Gate including a couple who worked for American Express. He commandeered a car and had himself driven down to the beach.

The Air Police and unarmed local police were in pursuit. He holed up in some rocks on the beach. Greene was later found dead, probably, but not necessarily, by his own hand.

Three people had been killed and seven others injured. This was not the kind of incident that endeared the U.S. Air Force to the British public. "Pack 'em up and ship 'em out" is a perfectly reasonable response to that kind of behavior.

Greene was an extreme, of course, but there is always drunkenness and rude behavior and lewdness; always something happening to upset the natives. British sentries, at that time, generally were unarmed. If an armed guard were deemed necessary, it was my understanding that they would post two men; one with a rifle and the other with a bullet. In extremis they were to confer and decide whether or not to use the bullet.

Ergo a petty thief detected trying to enter a British military compound when called upon to halt, would run. If the poor sod tried that on an American sentry there would be gunfire and bad press.

The British feeling was that you ought not shoot a man for stealing. The American feeling was that he should have known better than to run.

"Yoicks!" and "Talley Ho!" It's the Grand Old County Hunt. No sport like a blood sport. Tradition and all that. Slip yourself into your hunting pinks and a stirrup-cup inside and it's after the fox! But the dogs follow the fox under the fence around the ammo dump and there, between wily Reynaud and the Master of the Hounds, is Airman Givadam from Detroit. The Airman has a carbine and the Grand Old County Hunt is over at least for this day.

Such things have been known to cause letters to be written to *The Times*.

A more serious economic problem was caused by the noise of jet aircraft operations. If someone books a holiday cabin by the sea and spends his nights repeatedly awakened by afterburners cutting in he is likely to take his next holiday elsewhere. The local economy, dependent upon the tourist trade, suffers and the tempers of neighbors who are not tourists are more than likely to fray, too.

One effort to improve community relations was the brainchild of the RAF Liaison

Officer. It was simply to appeal for understanding. If people are given the facts, are told what is going on and why and if the efforts to ameliorate the noise and other problems are explained to them, they just might understand and forgive. At the very least their complaints will be informed complaints, based on fact instead of erroneous supposition. What the Liaison Officer suggested was that they hire some "Absolute Smashers" as public relations operatives.

When complaints were received, the complainer would be visited by young lady who, incidentally, was a raving beauty and she would present a briefing on the mission and operations of the US Air Force in Britain with particular attention to the gravamen of his complaint and calm him or her down. They hired three Absolute Smashers. It was virtually impossible to argue with these charmers; difficult, indeed, to be anything but charmed by them. It worked beautifully.

Except one time. One of the Smashers called on my father in his office and after some polite chat began to talk about their work and how well it was going. It then dawned on my father that the Liaison Officer had sent her around to make a pitch to increase the funding for his operations! It was, as my father thought, a nice try, but as he suggested to the LO; next time come yourself and be direct.

There was an Air Force Hospital at a place of whose name I am uncertain. It may have been Burderop Park. The ground had been commandeered during the war and the hospital treated the wounded being hauled out of Eighth Air Force Bombers returning from Germany. The war had been over for a decade and the owner wanted the land back, but the hospital was still there. The hospital was, if anything, busier than ever. I do not recall exactly what happened, but I believe the owner eventually forced the relocation of that hospital.

London was still full of "bombed lots" at that time. I recall there being one on Bayswater Road, two blocks from Marble Arch. One of the most visible scars of war was the Church of St. Clement Danes, on an island in the Strand. (The church was made famous by the children's rhyme "Oranges and Lemons, say the Bells of St. Clements" A few gaunt arches pointed skyward above a heap of burned stones, as busy traffic divided on either side to pass around it.

It was decided to rebuild St. Clement Danes as a memorial to the RAF to 'The Few' to whom so many owe so much and to the many more who perished in later operations. The people responsible approached my father and asked if the US Air Force would contribute as a memorial to the American Airmen who gave their lives in the WWII. He agreed to try to raise the funds for the organ and then he found out what the organ cost: £75,000 pounds ($113,500 in 1953). Given 50 years of inflation that's about $978,000 in 2013 money.

Third Air Force then numbered about 30,000 men. Persuading them to donate an that much money was a tall order. It took an awfully long time. There were fund-raisers and fund-raisers and more fund-raisers. But they did it. They also endowed two scholarships to King's College Cambridge, to train organists to play it. (*Author's note: Among reported fundraising efforts was a 1956 football game between the London Rockets team from USAF South Ruislip who took on the team from Wiesbaden for the USAFE Championships – an event covered by the BBC.*)

The Seventh Air Division of the Strategic Air Command (SAC) was a 'tenant' on Third Air Force bases. I recall my father's being very sorry for the 7th AD commander. SAC was terribly over-centralized. The man had no discretion to act. He had to do as he was told and exactly as he was told, no more and no less.

The SAC commander was then General Curtis E. LeMay. During WWII General Le-May had invented the 'Combat Box' formation and pattern bombing tactics used by the Eighth Air Force. In fighting the Japanese it was his decision to lower the bombing altitude, to switch from day to night bombing, from formations to individual attacks and from high explosives to incendiaries. The weapons and tactics are obsolete but the decision-making process is one that every air officer should study, carefully.

I first met General LeMay in London. He came with Mrs. LeMay and their daughter, Janie, partly to inspect the Seventh Air Division but mainly to attend the London premiere of movie titled *Strategic Air Command*.

The day before the premiere, the movie producer, Mr. Sy Bartlett, hosted the LeMays at Pinewood Studios. Mrs. LeMay and Janie had their hair and make-up done by real movie make-up artists. Janie was about a year older than I but a lot more sophisticated. My parents were hosting a big party for all concerned: RAF and movie people, US Air Force and visiting VIPs. Bartlett telephoned. He was in his office and Miss de Havilland had unexpectedly dropped by. Would it be all right to bring her to the party?

Miss Olivia de Havilland? Why sure, we can always water the soup! By all means, fetch her along! For those younger readers who can't immediately place her she played Melanie Wilkes in *Gone with the Wind* and numerous other roles. Kin to Sir Geoffrey de Havilland of De Havilland Aviation (maker of the famed De Havilland Mosquito) a 'Brit' born in Tokyo and naturalized American citizen, she fit right in. She was Joan Fontaine's older sister. Both were big-time movie stars.

As the party got rolling I was deputed to take Janie to the movies to get us both out of the way. We went to the Odeon, a movie palace at Marble Arch. Janie fell asleep in her chair. The party was over by the time we got home. Janie told me she slept on her stomach for a week to preserve her hair-do.

General LeMay was almost pathologically silent. He was so silent that when he spoke it was almost startling. The 'Great Stone Face' was inexpressive because his face was paralyzed on one side by a nerve disease. General William H. Blanchard, who was one of his deputies at the same time my father was, had the same affliction.

LeMay habitually wore flying boots because he had gout. Half the Air Staff, including my father, had gout, too. He was slightly hard of hearing. "Liberty Ears" the old-time flyers called them; deafness from listening too long to the whine of the Liberty engines used in post-WWI Aircraft.

The movie *Twelve O'Clock High* is about a War II B-17 Bomb Group Commander who finds out what limit of endurance might be expected of bomber crews by pushing himself to it and a bit beyond. To my mind the central character is about 50% Frank Armstrong, 25% Curt LeMay and 25% John Gearhart (Commander of the USAF's 95th Bomb Group during WWII who became chief of the National Security Council's Military Assistance Advisory Group to the United Kingdom in the 1953).

The book was by Bernie Lay, who had been there as a public information officer and the movie was produced by the aforementioned Bartlett. It later became a television series.

Now the same team had produced *Strategic Air Command* starring Jimmy Stewart, former Eighth Air Force bomber pilot and an Air Force Reserve officer. This had to be a friendly piece of propaganda and it was.

In the movie there is a character based on LeMay who had a wonderful line: "Don't bother me with your petty problems; I'm interested in results!" Managers take note: That's not how to inspire subordinates to greater effort. So General LeMay and my father went off to one of the SAC bases (possibly the USAF base at Lakenheath). My father asked General LeMay what to do with some surplus construction equipment and General LeMay pulled this line on him leaving him sputtering.

The next day father's Secretary came to him to interpret his own handwriting on some communication. "Is this word 'impressive' or 'impression'?" she asked and he pulled the line on her. She burst into tears and it cost him a bouquet of flowers to persuade her he was kidding.

General LeMay's face may have been inexpressive but not when he had a cigar in it, which was most of the time. Mrs. LeMay was an exact opposite: voluble, effervescent. They were devoted to each other.

I was sent with them as escort on a shopping trip. I was what – 15, maybe 16 – years old. I knew my way around Regent Street and Bond Street, Savile Row and the Burlington Arcade. They went to a recommended tailor in Bond Street. They selected cloth and the General was measured for a suit.

Later, while I was still in college, I had gone to the Pentagon for some reason (probably some errand for Pop) when General LeMay saw me and said; "Come along."

The Pentagon is filled with art. He was going to see an art vendor who had a picture for sale. We got to the corridor where the picture was hanging and it seemed an ordinary landscape to me. General LeMay was silent, as was his wont.

During the USAF's Regensburg-Schweinfurt Raid in World War II part of the raiding force had flown on to North Africa, instead of returning to England in order to spread the defenses. LeMay had commanded that group. The vendor explained that this particular painting had been sent from 8th AF Headquarters to decorate the new 15th AF Headquarters in North Africa and it had been holed by flak en route. There was a little torn flap in the picture, about the size of a nickle.

"It could be repaired ..." the vendor said, but I interrupted. "Sir, I believe that the General would rather have the hole than the painting." LeMay's cigar oscillated vertically, indicating agreement.

After the WWII LeMay organized the Strategic Air Command (SAC) as the nation's nuclear punch. At the time, all was calm, all was bright... and all was unready. As SAC Commander General LeMay went to one of his bases, maybe in Florida, one Sunday afternoon and was met by the Wing Commander. LeMay said; "Execute your War Plan."

The wing's war plan was to move their B-29's to England and to conduct atomic bomb strikes against the USSR from there.

The result was chaos.

There were frantic phone calls in all directions to unbelieving people with people trying to decide what gear to pack while others packed it and yet others trying to find transport planes and which mechanics go and which stay. And what about the Mess Hall and are we going to take doctors along? How many bombers are fit to fly and do they have aircrews and where the hell are the targets and it is going to take the whole ordnance squadron to take care of the bombs and maybe the Air Police should provide security? And did anyone think about how the wing is going to communicate with SAC Headquarters?

A year later SAC HQ could call up one of its wings and say; "Execute your War Plan" or "Exercise your War Plan" at any hour of the day or night. Then the right people would be notified in the right order by passing a single code word and everybody would have a list of precisely what to do and when to do it. And they would do so while Inspectors-General stood around with stop-watches, taking notes.

For the premier of *Strategic Air Command*, my Air Explorer Post was mobilized to serve as ushers. In the lobby, I heard a familiar voice ask; "Why Dickie, what sort of uniform is that?" I was able to introduce the Air Chief Marshal of the RAF, Sir William Dickson, to my high-school buddies who were mightily impressed.

The Seventh Air Division was very helpful to Third Air Force in one unexpected way. SAC didn't play football. USAFE did. By swapping people with equivalent skills to 7AD for some of their larger, faster men the Third Air Force was able to reinforce its football teams. Somehow or other the better players got transferred to bases in the London Area and the "London Area Rockets" won the Third Air Force championship.

The Continental champs were the Wiesbaden Flyers and they came to London for a championship game. I think it was played at White City Stadium. The British turned out en masse to see these fellows dressed up like spacemen play something they called 'football'. But didn't look at all like the football they knew. The Stadium was packed but was mainly silent; they did not understand the game at all.

There were, however, sizeable numbers of Americans from the Continent, to cheer on the Flyers and plenty more from Britain including the entire student body of Central High School to cheer for the Rockets. London upset Wiesbaden 38 to 0. I think that the Airman Mitchell who starred for the Rockets may have been Bobby Mitchell who became a star running back for the Washington Redskins.

Quite a large number of American officers and airmen met and married British girls. Others met, courted and jilted British girls or were jilted by them. Some of those feminine attentions were diverted to American interests from British boys who were jealous. "Over-paid, over-sexed and over here!" was their complaint and not much to be done about it. I don't know the numbers but I have the impression that it was on the order of several thousand marriages, annually.

There is this aspect to it: Youthful affairs of the heart gone awry, however painful at the time, pass away. The jilted boy, British or American, and the jilted girl eventually mature and settle down usually to life with someone else. The marriages endure and produce things like grandchildren which tend to bind British and Americans together in family kindred of affection reinforced by blood. In the long run it is a good thing. In the short run it hurts.

One matter of considerable concern for the US Military were, ahem, *undesirable women*, who wanted to marry a GI simply as a means of getting into the U.S. where they would likely dump their husbands.

Interracial marriages were another problem. This was the mid-fifties, remember. Segregation was the norm in the U.S. The Civil Rights movement did not really begin to achieve results until the early Sixties and, even now, it is difficult. The love-struck British girl who wanted to marry an African American and to live with him in the States usually had no conception of the kind of trouble she was heading into. No idea at all.

The stresses and strains of an interracial marriage in that day and age were such as could be expected to shatter the strongest relationships. Add to that the adjustments to be made by a British girl to a new country, a new society, thousands of miles from home and family and you have a prescription for heartbreak. Compounding the problem were the "*undesirables*", who found in black GI's particularly vulnerable targets for exploitation.

Command authorities investigated requests for permission to marry as best they could, trying to sift out the undesirables. They used chaplains to try to dissuade the parties from inadvisable marriages. But there are limits to the ability of commanders to say "No." Fundamentally, if an American wants to marry, whether he is in the service or not, he has a right to do so.

A number of the 'undesirables' probably slipped through. A larger number of inadvisable marriages were solemnized. One hopes that the majority of the marriages produced stable, loving and lifelong relationships. Ultimately one wishes the newlyweds, in all sincerity, the very best of everything.

Other servicemen arrived in Britain already married and childless. Some of these, unable to have children, found a short-cut to adoption. It worked like this: Wife would announce that she was expecting. After awhile, she would go to Dublin, ostensibly to visit a laying-in hospital. Her husband would visit and they would pick up an infant Irish orphan or foundling and return to Britain, usually by way of Northern Ireland and register the child as their own.

This was too good a thing to keep quiet for long. The practice became known and the British authorities stepped in. The dangers are manifest. Infants could be kidnapped and sold to these people, or pimps could sell the offspring, whether the natural mother approved or not. The charges were serious: Kidnapping, smuggling aliens and fraud. The potential for heartbreak was profound. Some of the children had been with their adoptive parents for as long as two years before discovery.

My father was one of many who sighed with relief when a Learned Judge decided "Her Majesty's Government cannot condone, but it can relent..." What had been done was not undone. But the Black Market in babies came to a sudden and absolute end.

There is a surprising tendency for Americans abroad to cling together, to socialize only with other Americans abroad. In Germany or Japan, this seems to be caused by a language barrier, but I think it is much more than that. There is no language barrier in Britain. Yet the contacts between the American servicemen and their dependents with the British civil community were, I venture to say, 90 percent business. They were words exchanged with a bus conductor or ticket vendor, perhaps a conversation with a landlord. I do not understand it. I cannot explain it. I merely observe it and say that it is so. Americans abroad. Americans aloof. I also thought it was stupid.

I have also heard that times have changed and that this cliquish tendency has eased, considerably. If so count that change as an improvement.

I thoroughly enjoyed my years in Britain. I developed a love of history, which I retain to this day. Some things were dreadful, but many were wonderful. Most of all I think I learned to appreciate and to value the diversity of the British people, diversity surprising in so small an island. But to a degree not enough Americans and not enough Britons realize, their history is our history; their literature is our literature, their Common Law is our Common Law, their language is our language and their liberties are our liberties.

Heading Home:
We left London in 1957. My mother and I returned aboard the ocean liner *United States*, which was an experience in itself. One dressed for dinner aboard ship: Tuxedo, every night. Wilma was aboard, too, and so was our Ford Zodiac) One of Pop's favorite subordinates was also aboard: Colonel Ivan W. McElroy, who was returning to the US after commanding the 81st Fighter-Bomb Wing at RAF Station Sculthorpe.

Once we were at sea his wing staged a fly-over of the ship, forming a great big "M" as a final salute. We were assigned seats in the dining room and I remember the others at our table were quietly impressed when mother received a bon-voyage telegram from the wife of the US Ambassador to London. The crossing took five days.

My father retired from the Air Force in 1961, four years after leaving London. He ran a research company in Boston for a few years, then retired for good to Kentucky. He died in 1986, aged 81. His remains are buried in Zachary Taylor National Cemetery in Louisville. Mother died in 2007 at age 100 and is buried beside him.

CHAPTER SIXTEEN

FRESH WOODS AND PASTURES NEW

William 'Bill' Cooper

Student, London Central High School
Bushy Park, Teddington
1955 – 1957

It has gone back to being fields and groves now and rightly so but when I knew Bushy Park AFB it was a microcosm of your typical air base just without the runways.

I first set eyes on it in September 1955 as the coach – 'bus' to us Yanks – drove through the gates set in the curving red brick wall. For a while I would only know it as Central High School (CHS), a one-storey rehabbed WWII building.

A few months later we – that is Bushy Park and I – began a more intimate acquaintance; one that I still treasure. Here's how that happened:

Our family had traveled to England that Fall from Alexandria, Virginia, because my Dad, Lt Col William C. Cooper, had been assigned as Base Commander at Denham Studios in Buckinghamshire.

That's right, 'studios'.

The place belonged to J. Arthur Rank and had operated as a film studio until 1952. A host of movies including: *Goodbye Mr. Chips*, *Brief Encounter* and, appropriately enough, *A Yank At Oxford* had been made there before and during WWII.

By 1955, however, the place was used for warehousing whatever it was the USAF wanted to warehouse there. The big sound stages were still there. And still there in all its Art Deco glory was the office that supposedly had belonged to Sir Alexander Korda, who'd built the studios. It became Dad's office. I loved it. So did he.

But that didn't last long. Dad moved to a smaller, more ordinary office but a bigger, more wonderful job. In early 1956 he became Base Commander at Bushy Park in Teddington. He remained there until Fall of 1958 when our family returned to the States. Both Dad and my Mom, Margaret (or "Peg" as he and most of our friends called her), reveled in it. To me it became a mixed blessing; mostly wonderful.

We – Mom and Dad and my younger brothers John and Dennis – lived at 'Willows', a house on the Thames in Datchet, near Windsor. But my weekdays and some Saturdays and Sundays were spent at Bushy Park in Teddington where I was in the last two years of high school. It was in a place unlike any other. They are my memories, so they aren't perfect, but I'll try to paint it as it was.

The Bushy Park base grounds were divided into two parts surrounded by the much larger Bushy Park itself which was, as it is today, largely trees and open grassland. On the smaller part, over toward Kingston-upon-Thames, were the quarters and a Mess Hall and club facilities for enlisted troops.

On the larger of the two parts were most of the buildings for the five resident organizations at that time. The London Central High School was not the only function, despite what we students might think. Besides the school, the base housed, at that time, the 28th Weather Squadron (MATS), the 9th Air Rescue Group (MATS), a dental facility, which, I believe, served Air Force needs for much of the London area and the USAF band for the entire United Kingdom. Supporting these were the usual Chapel, BX, Enlisted Mess, NCO Club, Officers Club (with the world's best bar – I'm talking atmosphere here, I didn't imbibe at the time!), Movie Theater and stage, tailor shop, ball fields, parade ground and so forth.

There was also a British organization on base; The Kinsman Bureau, formed during WWII to help US service personnel find accommodation in England, had an office not far from Dad's. Mrs. Patricia Staines, the bureau's on-site rep, continued the group's work of welcoming us Yanks and helping make our stay there an easier one. I 'Googled' the Kinsman Bureau many different ways in recent years. Nothing. Not a sausage, as they used to say back then. Things pass.

Here then, in miniature, was every air base I'd ever known, hiding quietly behind that red brick wall next to the village of Teddington and across the park from Hampton Court. And LCHS wasn't the only school. There was an elementary school as well.

That's the way the Base was set up, but, of course, the important part was the people. So many there were and so wonderful.

There were the teachers. To name them all and their varied characters would take up too much space, but they had wonderfully uncommon traits that they shared in common. They treated us like young adults, which we were… sometimes. They were all interested in teaching us as much as possible and – mostly – making it fun. Their courses were the equivalent of most college courses. They became our friends.

The administrative staff at Central High, including the USAF folks who made sure the school had what it needed (and yes, even the principals we loved to make fun of), were right up there with the teachers.

The other base organizations seemed to be staffed with characters from the perfect movie about the Air Force, a sort of latter days *Twelve O'Clock High*. It was the Cold War, but these were the folks who'd won WWII. They were still celebrating even as they proceeded through the current bit of trouble.

I'd drive to Bushy with my Dad almost every morning of the school week. Those were wonderful trips. We talked about everything under the sun. Sometimes these were serious discussions but most of the time we'd find ourselves laughing. A particular chance for laughter came the day when, halfway there, Dad wondered aloud what he'd done with his briefcase. Then he slowed the car very carefully. The briefcase had enjoyed a brisk ride on the front bumper of our Chevy wagon. Dad loved it. He told the story on himself often thereafter.

Each morning we'd pass by Runnymede of Magna Carta fame and Cooper's Hill with its air memorial and then the backside of Heathrow Airport, finally coming into Teddington and through the main gate in the brick wall which remains today, sans gate – about the time the first coaches with students were arriving. In the afternoon I'd ride back on the coach most of the time.

That was for us town students. There were dorms on base near the school for other students from all over England; one for the boys, one for the girls. Most of the dorm students would commute on weekends to their home bases, though some were from so far north that they'd only go home during longer holidays.

During senior year my riding home on the coach changed and I often went home with Dad at the end of his day. My steady girl, Alice Moore, was a dormitory student whose father was stationed at Lakenheath near Cambridge. By staying late I got to spend more time with her. It also meant more time with Dad. I treasure both memories.

The coaches for the town students lined up next to the school in the afternoons. And there was Lieutenant Virginia Sweet (yes, that was her name) making sure all was in order. I mention her because she had a presence about her. Then, too, there were the silver wings over her left pocket. What were those wings? Women in the military or most anywhere else didn't wear wings.

Dad answered my question. She had been one of the civilian ferry pilots, the WASPs (Women's Airforce Service Pilots) who had delivered all types of planes from the factory line to the airfield. He seemed to take pride in having her there. Lt. Sweet rose through the ranks and left the USAF as a Lieutenant Colonel. She passed away in 2009 and has posthumously been awarded a Congressional Gold Medal of Honor along with 1,000 of her fellow WASPs.

There were other surprises waiting in the far future. Dad's 'exec', a young Lieutenant Don Gruenwald, a quiet type who put up with the admittedly mild shenanigans of the commander's son, became the president of a university. Thanks for that James Dean movie poster, Don. There's more to that story but I'm not telling it here!

The coaches, those marvelous English vehicles that carried many of us to school and back home, were nothing like our US buses. They were at once less powerful and more artistically interesting than our buses at home. Lovely Art Nouveau sort of designs driven by small engines with whiny gears, which nonetheless seemed to do the job.

There was a bus – since it was a USAF vehicle and much more prosaic than the English coaches, I'll call it a bus – from and to the South Ruislip base at least once a day. To me South Ruislip didn't have much more than Bushy Park, except a Commissary, some two-storey buildings and some generals. Still, I remember riding on it now and then; probably a quicker way to get to some girl I knew.

We had parades at Bushy. Lots of them. Why not? We had the Air Force Band right there. The 751st U.S. Air Force Band – better known as the Third Air Force Band – the only American military band in the United Kingdom. Any excuse would do for Chief Warrant Officer Louis Kriebel's goodly musicians to march around playing rousing tunes. Armed Forces Day, held each May, was always better because of them. I think the local Brit populace thought every little base had a big band.

That brings to mind some of the special dedications and occasions. One such was the day the elementary school's 2nd Grade classroom was filled with lots of USAF and RAF brass. They came in to dedicate a plaque over a fireplace (in a classroom!). It had been Dwight D. Eisenhower's office during WWII. For the dedication Ike sent an autographed photo especially addressed to the members of the 7533rd Air Base Squadron, Dad's unit at Bushy. Then there was the visit by Miss U.S.A. and the entertainment by the Skidmore College Choir. I've still got the Skidmore mug and pennant.

Even in summer, and particularly in that magnificent, cool English summer of 1956 the Bushy Prk base called to me. I would travel with Dad or take a circuitous train route and wind up at Bushy. Often that summer my close friend Barry Smith, LCHS Class of '56, and I would play game after game of ping-pong in the 'O-Club' (Officers' Club). We got very good at the game, or so we thought. Barry's Dad was Top Sergeant at the base and yet another good man who was willing to put up with my antics.

Another close friend, Karl Phaler, Class of '57, lived nearby in Teddington. Lots of chess games there. Karl and our good friend and classmate John Soulé and I had been thrown together early in our junior year when the school saw fit to send the three of us to Cotton College, a private secondary school in Staffordshire, for discussion and debate with some of their Sixth Form. There is no truth to the rumor that the high school staff picked the school's three biggest troublemakers for this assignment!

Well perhaps we weren't troublemakers but there were pranks. It was John Soulé who came up with the idea for the most elaborate. Paul G. Francis, who taught American History, had mentioned that he would talk about the effect of barbed wire on the settling of the West. John had suggested that he could bring in a bit of barbed wire as an example. That was Friday.

Monday afternoon, just after lunch, a rather large group of students and teachers gathered at the doorway to Mr. Francis's history classroom to watch as he entered. The room had been wrapped, pillars to windows, with barbed wire and a rather loud tape of specially recorded anti-aircraft activity was playing. The good and long-suffering Mr. Francis, who had already had the delight of joining us on the Cotton College adventure, simply said: "You've got ten minutes to straighten that up."

There were other lesser entertainments, some spontaneous. At least one is worthy of note: My funeral. It seems that Soulé and I and Bill Douglas (another classmate and a great friend to this day) who had helped with the barbed wire affair, were headed to class – again Mr. Francis's class – after lunch. We spotted a few lillies in a pitcher in the school office. Two minutes later Douglas and Soulé carried me into the classroom, my eyes closed and the lilies clasped in my hand and then laid me to rest on Mr. Francis's desk.

The imperturbable Paul G. Francis said he didn't do funerals and that they should bury me somewhere in Bushy Park. We took this as approval for playing hooky. Now Bill Douglas swears this all happened. I have no recollection of it. Perhaps an effect of my temporary demise.

After that summer of 1956 the world and England and Bushy had two crises to deal with. Autumn brought the Hungarian Revolution which was quickly put down by the USSR. The year's end also brought petrol rationing which lasted till May of 1957. The major contretemps involving the United Kingdom and Egypt had resulted in radically reduced petrol supplies. Bushy did its part in the aftermath of the Hungarian Revolution by helping refugees who were housed in the local area. I did my part in the Suez oil crisis by not driving our Chevy. As Base Commander Dad had access to as much petrol as he needed. He decided to use no more than was needed to get to work and back and perform other official duties. London's wonderful train and tube systems came to my rescue.

Some of the other things I remember are a result of the base being there but aren't about the base itself. There were all those official occasions the surrounding towns felt necessary to share with us. Heaven knows how many 'mayor makings' – that's what they called them – I went to. Lots of pomp and circumstance and hors d'oeuvres. Mom and Dad even got to meet the Queen at a ceremony in nearby Twickenham. The local dignitaries became our acquaintances. Two became good friends of our family. Alan and Ann (Nan) Brock were the Mayor and Mayoress of Richmond. I remember many wonderful times with them. And it was Nan, a good Scotswoman, who taught me that the word 'Scotch' was to be used only for the drink; the people were 'Scots'.

Kingston-upon-Thames was right across the river from Teddington, not a long walk from the base. A favorite place for after school walks for the dorm students and those of us who were fortunate enough to have a late ride home. The first thing one noticed when crossing the bridge into Kingston was the odor of the tannery. I don't recall that I ever knew exactly where the tannery was but one could tell it was there, somewhere. Once you got past that smell, or used to it, you could walk up river a short way to a great coffee house. There we learned terms important to our later lives, words like 'espresso' and 'cappuccino'.

All the Base buildings, with the exception of the slightly taller movie theater, were single-storey. That worked better when they were covered with camouflage back in the war. It also preserved a bit of the park-like character, at least from outside the wall. Some of the buildings still had blast walls in front of the entrances. Bombs, you know.

There was also what appeared to be a bomb shelter out behind the movie theater. It was a great place to make out. Which reminds me, for some reason the girls' dorm had a barbed wire topped fence around it. This did not stop at least one guy – no, not me – from sneaking in to visit his honey. I kept telling Dad to have the fence torn down. He finally did…after I graduated and left for college!

About half of the classrooms in the LCHS building had double doors opening to the outside. Often enough they'd be open during class in the early Autumn and late Spring; the best kind of air conditioning and the only kind in most of 1950s England. I remember one particularly fine day in 1956 when we all trooped out of Miss Hynes' classroom, sat on the small grade school carousels and, swaying back and forth, chanted Vachel Lindsay's 'The Congo'.

One other thing about the school building is memorable for its sheer silliness: The yellow line down the middle of the main hall and the arrows. At the beginning of senior year, Autumn 1956, the Principal, who shall (deservedly) remain nameless here, decreed that the way to cut congestion in the hall was to remove the lockers, paint a yellow stripe down the middle of the hall and paint arrows on either side to show us where to walk. This really happened. You can imagine how much attention we paid to it. Several of us did point out that the arrows had the traffic running on the right-hand side and that we were in England.

If I could return to one place on the base at Bushy Park it would be to the O-Club bar. A small room it was with a red and white striped canopy over the bar itself and walls of dark blue with white stars. I'd sit there eating crisps – the little packets each with its twist of salt so you could salt your own – and drinking Coke (yeah, just Coke) and talking to whoever had just gotten off duty, while I waited for Dad.

There were the requisite one-armed bandits for entertainment of the inebriates. Winners of jackpots (these babies were rigged to pay a lot more times than those in Vegas) would write their names and the date on the stars. I didn't win a jackpot but I couldn't resist writing "W W Cooper Born 1955" on one of them. That's right 1955. England meant that much.

That England – and the Bushy Park base – are no more. Bushy has gone, like Milton's singer in *Lycidas*, to "fresh woods and pastures new." I walked in those pastures with some other former students back in 2000. No remnant of the base remained. But there were memories. Later, back home, remembering those early days and that later visit, I wrote the following poem, which seemed to me to sum up so much of what Bushy and the other home bases were to those of us who knew them:

A PEPPERMINT STRIPED FLAG

I stood in my father's old office and saw that all
was grass and not merely our flesh it seems,
for the buildings and walks and streets had fled as well,
leaving only the winding, high brick wall,
and it so changed that I could not even tell
whether its lost main gate was the stuff of dreams.
Still, in the everlasting park, the trees
stood quiet guard over worthy memories
of splendid girls and boys, good women and men,
in a world where the lights had just come on again.
And here, through the main gate's fog washed piers,
loomed coaches full of youth that had yet to be
lost in the careless days and the careful years
that lurked, just biding their time, across the sea.
But for now we had class and the long-awaited bell,
and holding hands and necking in the old
bomb shelter behind the theatre and the untold
ache of young fingers unclasping in farewell.
Back then, with contrails trellising the sky
in the sharp spring twilight over London Town,
I stood at what passed for attention, across from my
father's still office, lost in the last few bars
of retreat, as the fading blue uniforms drew down
and caught and gently folded and bore away
that banner, brimming with stars,
at the end of our day.

Lt. Col. William Cooper (Left) and Mrs. William Cooper (Right) at a military wives function at Bushy Park.

The 25 Nov 1956 dedication of the plaque in General (later President) Eisenhower's old office, which had become a grade school classroom at Bushy Park. *L to R*: Lt. Col. William C. Cooper, Commander Bushy Park; Maj. Gen. Walsh, Commander 7th Air Division; Air Marshall Lord Tedder, Commander-in-Chief RAF; and Col Zimmerman, Chaplain 3rd Air Force. *Below*: The plaque up close.

The USAF Band performing at Bushy Park – Lt. Col. Cooper considered them to be one of the great PR assets of the USAF. Photos courtesy William "Bill" Cooper.

Mr. William Mercer, Director of Civilian Personnel at Bushy; Miss Charlotte Sheffield, Miss USA and Lt. Col. William C. Cooper, Commander Bushy Park. Photos courtesty of William 'Bill' Cooper.

CHAPTER SEVENTEEN

SNATCHES

Martha G. Kelly

Teacher, DODDS
Eastcote Elementary School
Bushey Park, Bushy Hall
London Central High School
(1956 – 1986)

A handful of stories that were either memorable or amusing remain in my memories from my time as a teacher in the Department of Defense Schools system at schools around London.

EASTCOTE ELEMENTARY SCHOOL – A SPORTING CHANCE

"In this school the girls will sing songs and the boys will kick balls."

Those cringing words were spoken by the principal at the first faculty meeting of the new junior high to be established on the Eastcote base around 1956.

They proved the impetus to my avowed aim to conjure play areas for all the 7th and 8th graders.

We sweet-talked the Supply Sergeant into ordering two basketball stanchions, two tetherball poles and the markings for Four-Square games areas in the front parking area. We convinced him they were as vital as blackboards and pencil sharpeners for the children's education.

"But that's the bus area," he protested.

"When are the buses there?"

"Well, in the morning and after school, of course."

"Fine. Playground off limits then – but during the school day, it's The Playground."

Wonderful – perfect for Grades 4-6 and quickly usurped by their teachers.

We were moved to the back walkway. We were given two wings of the building that had once purportedly been created to be a hospital. Our only playgrounds were small areas between the building wings. By the back walkway was a hill above boarded bomb shelters.

The grassy knoll was to serve as Grades 7-8 playground, study corner and general 'hang out' area. A water tower loomed on the hill and atop was about a 10-foot flat area. We erected a high jump pit in this space that the boys monopolized.

I still recall watching one student, Alice White, wait until after the bell rang and everyone had scurried down to class, take a two-step run-up and clear the height none of the boys could manage.

Heigh ho.

Physical Education classes were held in the Multipurpose Room – endless exercises and forward rolls, dodge ball and 'crusts and crumbs'.

Driving in the base by the Lime Grove entrance one day I noticed an entrance gate leading – where? I stopped and followed some dog-walkers into an absolutely huge, lovely and luscious green park. It was Warrendar Park.

We soon secured permission to walk the youngsters up to this oasis where they could run and play their fill. It was the treat when the classes were good. 'We'll go up to the park this afternoon.'

HOORAY!

Then we discovered a nearer park area – out the front of the Base across from the Field End Road entrance. The worry was the speeding autos threatening from both directions. Our solution was to line the students up along the curb with a teacher at either end. At the whistle's toot, they would all walk swiftly across the street. Everyone safely across. Great!

Along a rustic path – and there was another marvelous play area – park, wooded area, corners to explore and through that clearing – the River Pinn! Too exciting and adventurous – they loved it!

Then the answer to the junior high PE teacher's prayer was the Highgrove Swimming Pool. It was erected next door to the Base. Arrangements and permissions organized we took our classes there and taught the youngsters to swim – splendid! More than one young lady who began fearful of putting her face in water ended by diving in and swimming a length....well, a width.

Eastcote Junior High – cramped classrooms, non-existent play area, equipment shortages and a hallway that gently sloped, throwing the youngsters into a natural bounce as they descended to the Cafeteria.

"NO RUNNING IN THE HALL!"

"They're not running, Mr. Principal, they're trying to keep their balance..."

What makes a school? It is the enthusiastic children, their helpful parents and the dedicated teachers who make a school – not elaborate physical *accoutrements*. Every youngster from those years that I am still in touch with has excelled in their chosen field.

What would Health and Safety say to our solutions to the junior high play/exercise/physical education solutions? I shudder to think.

DEATH AND LIFE AT EASTCOTE

At the Eastcote Junior High Parent Teacher Association (PTA) meetings in the mid 1960s Sergeant Hazus, the USAF 3rd Air Force Mortician and I always whispered about making babies; we were both trying to. We only saw each other in the hospital corridors at the South Ruislip Base or at the monthly PTA meetings.

The sergeant and his wife were looking for a little sister for their son, Richard, who was in my class. My husband and I were hoping for a sibling for our two lads. While the majority of parents were discussing reading problems and Science tests, Richard's father and I were comparing monthly cycles and ovulation periods.

Sergeant Hazus was suddenly thrust into the limelight when he became the local hero; the wife of a USAF serviceman ended up on his slab in the Base Mortuary atop the Eastcote base hill. Perhaps a heart attack? Maybe a stroke?

The canny mortician, however, noticed a bluish tinge to the deceased's fingernails; he called in the authorities. They, too, were suspicious and after an investigation, it was discovered that what tears the serviceman might have were distinctly crocodilian. He had apparently begun an affair with a British girl and had secreted the arsenic in his wife's food. Alas, his wife eventually had taken ill and died. Sgt. Hazus was hailed as the astute Hero of the Day, alerting the officials to what initially proved to be a mysterious death.

A three-way success story: After a subsequent trial, the airman was sentenced for the murder by poison of his wife. When the Sgt. Hazus and his wife were finally planning a trip to Germany for the purpose of adoption, Mrs. Hazus discovered she was pregnant. And my third son, Jesse, was born the following summer.

CONVENIENCE SHOPPING AT SOUTH RUISLIP

In the 1960s every Friday late afternoon the South Ruislip Base Commissary and Post Exchange were jammed with customers; military and civilian personnel from all the surrounding Air Force and Navy bases in the London area.

A civilian employee from the US Navy Headquarters in London who shared a London mews house with fellow worker recounted this story: Every Friday after work at Navy London they would whip out to South Ruislip Base to shop at the Commissary and the PX. His flatmate was always in and out of the Commissary so quickly that he entrusted the food shopping to him.

Still the speed of his flatmate's task completion amazed him and he finally asked his housemate the obvious question: "How come you're finished in the Commissary when I'm just starting in the PX?"

"Well, come along and watch," his roommate suggested nonchalantly. "What kind of meat do you want? No, I'll get it – you just wait here at the checkout."

Off he speeds towards the back of the Commissary, sans cart, where he promptly hand gathered up a week's supply of steaks, mince and chops on his way. In a trice, and again to the amazement of his now astonished flatmate, he was back with a full cart. They checked out, paid up, stopped at the base Gas Station and were quickly back on the road to London.

"So now will you explain how you shop so fast?"

"I don't," came the reply. "I let others do it for me. The main thing is the steaks or the type of meat. As long as I have selected what type meat we want", the rogue explained, "I go to the back of the Commissary and find a near-full shopping trolley that's unattended. There are just two stipulations: That there's no purse nor baby in the trolley. I throw our meat on top of everything else, grab it and head for the shortest checkout line.

"We have our choice of main course; I've discovered that 85% of items in the random basket are what I'd buy anyway – potatoes, washing up liquid, Kleenex, coffee – and the other 15% are interesting surprises."

Cue loud guffaws, a hearty dinner of steak, French fries and green beans, then off to Piccadilly and a late show; heady bachelorhood in 60's swinging London!

Martha Gail Kelly standing at the back of the school (later known as Block 1) to the left (mostly off picture) at Eastcote sometime in 1956 or 1957. Behind her up on the grassy knoll was the area off limits to the students – the 3rd Air Force medical facilities including the Morgue (Block 4). Below: Her grand-daughter, Caitlin, who accompanied the author on a site visit just after the base was closed down in 2006, seen standing in almost the same location.

CHAPTER EIGHTEEN

TOUCHDOWN

Archie Tatum

Technical Sergeant, USAF
1956 – Communications, High Wycombe
    Air Station
1957 – 1969th Comms. Sqdn., USAF South
Ruislip
1963 – 1969th Comms , USAF South Ruislip
1969 – 1969th Comms. Sqdn., USAF South
Ruislip
1973 – Retired from USAF
1975 – Communications, US Navy
Headquarters, London
1975 – US Navy Special Services,
NAVACTUK
    RAF West Ruislip, DOE Eastcote,
    RAF Daws Hill and US Navy London

It was a delivery to my local railroad yard to pick up supplies that set me on a course for a career in the military and a life in England.

My hometown of Denison, Texas, didn't have a lot of employment at the time I was coming out of high school and my mother and father had separated, so I went to stay with my father and step-mother in Fort Worth, Texas. I needed a job, so when the fruit and vegetable company of H.T. Pritty was hiring. I jumped at the opportunity. Could I drive? Sure. Did I have a driving license? Uh…sure.

But I didn't.

It wasn't long after I got the job they sent me down to the railway yard to pick up some items in the company truck from one of the train carriages. But I made the mistake of backing up and parking a little too close to the train. When I was loading the items the train started moving. And so did my truck. It got pretty scraped up on the side panels.

Well I was a bit scared and, at that point, quite a bit more fed up with life in general. I drove the truck back to the company warehouse, I left the keys hanging inside the truck and left it. I then headed home and told my dad. He only asked me one question: "What are you going to do?" It was at that moment I told him: "I'm going to join the Air Force."

The short story is that I did just that. It was to be the beginning of life changing journey that would take me on three tours of duty at USAF South Ruislip as part of a 21-year career with the USAF before retiring to try out civilian life and then going to work for another 22 years for the US Navy's Morale Welfare & Recreation office in London.

It would also see me meet a girl, marry her, play football, try a pizza business, manage a band and experience what I believe were some of the best times to be based with the US military in London.

The first leg of the journey was relatively short: On 14 July 1952 I travelled 'down the road' to Lackland Air Force Base in Texas to undergo Basic Training. After that I headed to Francis E. Warren AFB in Wyoming to attend a Communications Centre specialist school. By the end of the year I had graduated and, upon completing the course, I was assigned to the 922nd Air Communications and Warning base at Goose Bay, Labrador.

I was in the USAF so it was logical that I travelled to it by troop ship! It was my first time on a ship and it was packed with Army, Air Force and, of course, US Navy personnel. A great mix. We were headed for Goose Bay in Labrador, then England and then Germany.

Our second day on board and I was having the GI standard fare – SOS (the abbreviation for S*** on Shingle or Same Old Stuff (aka 'chipped beef')) when the ship made a really hard turn. Our trays slid down several places past us with some landing on the floor. Suddenly I wasn't having SOS but what looked like peas and carrots.

It was colorful and interesting but there was only one problem – that had been the previous night's dinner. With some stellar timing one of my tablemates had just been sick on his tray and it had then slid down in front of me.

What a treat!

On board we bunked in hammocks. I had the bottom one of the trio of stacked bunks. It was the least distance to fall out – and believe me plenty fell out on the trip. After three days we made it to Goose Bay – somehow I don't think the Navy were in any hurry to get us there. We disembarked by shuttle and, once ashore took a truck on a 10-mile journey to the top of the 'hill' where the 922nd was based.

Goose Bay was a one-year remote tour of duty posting and after that I was sent to the 78th Operation Squadron at Hamilton AFB, California.

It was while I was stationed there that I stared hearing stories about some of the USAF bases in Europe. Man it sounded great. I thought; "I'll have some of that!" And so when I came to England in 1956 for the first time it was by plane to RAF Mildenhall. We piled off the plane and to various buses. Some were stopping at Mildenhall. Some personnel were going to Lakenheath. It was the first time I heard I was going to a place called "High Wycombe".

It was a drive up through London that seemed to drag on forever but I remember thinking how green the land was. We eventually arrived at the base gate, went through an ID check and were processed in and to our dorm rooms. I remember they had a pot-bellied stove in the middle with about 14 of us sleeping in the same room.

The next morning we walked from the barracks to our cafeteria – I think it had been an officer's quarters sometime before I arrived. And there I was asked a question I hadn't heard since I started in the US Air Force: "What are you going to have?" It was great to have a choice. I had waffles (they were actually pancakes) and milk. And to top it off you could go back for seconds!

It was only after we arrived on station that we learned we had been assigned to work in 'The Hole' and that it was an underground command bunker. We had a two-striper show us the way to our job, which was a walk further down a hill and along a long thin road. I was wondering where in the hell we were going. It was about 150 yards before we went through a pass gate and into a bunker area and through a large, thick steel door. The switchboard, where I would be working, was just inside the bunker entrance.

I had learned how to work a teletype machine and switchboard at Goose Bay and the system was much the same as it was in the bunker. I had two ways of answering the calls that came into the Switchboard: "High Wycombe 2601" was one and "Sea Bell" was the call-sign that I would use for answering military calls. Sometimes we were busy and sometimes we were quiet.

The switchboard comprised a series of lines. Each one had an incoming call button and a light would flash for an incoming call. We'd take it, plug it in and ring two times with a key. When the caller hung up the lights would come back up and we'd disconnect. I was working a day shift that ran from about 4 p.m. to Midnight. There were two other shifts and there were four of us on each shift.

When there was an alert or crisis the entire switchboard would light up. We also had a buzzing sound for added priority when it was the Commander's line. That would take priority over everything.

Life initially revolved around work, eating and sleeping. After work we'd go have dinner. I remember it was fish on Fridays and liver on Thursdays. It was the military way of doing things! Of course that didn't last long. We were eager to get out but money at our pay grade then was tight – back then I was on the princely sum of $98 a month.

There was a pub – there are just a few in the UK! – that we went to in High Wycombe. We didn't have money though – the US military back then paid us in script and we weren't allowed to use it in town. However the bartender at the Cherry Tree pub in High Wycombe ran a script tab – we'd give him the script but come pay day we'd get the script back and pay in pounds for it. Compounding the problem was the fact I only got paid once a month in those days.

Getting to town was interesting. The base sat at the top of the hill above town. The normal way would have been take a long walk to the front gate and a longer walk out to the main road and down to town and reverse the process coming back to base. However I was quickly indoctrinated into the secret base shortcut. Some enterprising predecessors had cut away a section of fence that ran along the lower end of the base above the town it would regularly be used by pushing it out and then pushing it back into position. Thankfully we never got caught using it!

And there was London. Initially I didn't go because I didn't really know how to get down there and I always seemed to spend all my money before I could go.

Everybody said it was better to go with someone who knew how to get around there. But when I went I made up for lost time! The train would take about 40 minutes to Marylebone Station and from there a world of possibilities lay before us.

The other thing that struck me back then was the weather. The blistering heat of Texas was a joy by comparison. I had never seen fog before. You had a to develop a sense of direction – especially on the walk across the base to work. You didn't hurry because it was more than likely you'd trip on a curbstone or fall into some ditch.

It was easy to get into a boring routine but I didn't want that. And I had arrived at a fortuitous time. They had put some information up about recruiting football players for the London Rockets – the name given to the sporting teams based at the Headquarters at USAF South Ruislip.

"I'll have some of that," I thought. Subsequently I called and then showed up for try-outs with about 30-40 others. The Base Commander at that time loved sports and wanted sports on the base. I think if he couldn't have had sports teams there he would have shut the base down back in 1957!

The London Rockets wanted me but there was a problem. I had to be at South Ruislip to train. Solution: They could take me on TDY (Temporary Duty) if my Commander approved it. And so in mid-1957 I relocated and I changed jobs. Football training was at 2:30 – 5:00 p.m. each day but I needed a job that would be flexible with that. I could not just up and leave the switchboard – and so I was moved into working at the Commissary so I could leave early with little impact.

I had played football in high school a little but I was a big guy and weighted 190 then and they made me a tackle. We went up and down that field training on the 'sleds'. It wasn't long before I got my nickname "Goose" after the Harlem Globetrotters player Reece Tatum who had that nickname.

That first season we got third place playing against other US Third Air Force bases including Wethersfield, Chicksands, Upper Heyford and Lakenheath.

After the football season ended I moved back to the 1969th Communications Squadron – back on the switchboard but this time at the Headquarters at South Ruislip.

The difference here was they had a bigger board (with 8-9 of us) and there were civilian women on the day shift along with a civilian supervisor. But I wasn't at the switchboard for long. They moved the military guys to the Teletype department sending and receiving messages. I could type 40-60 words a minute and within two weeks they had made me a shift supervisor.

There was a definite cachet to being on the London Rockets. I could pretty much get what I wanted no problem. For instance we got to cut to the front of the line in the Dinging Hall. But I couldn't get into trouble. The Base Commander wanted his football players to be 'Al '.

The Communications Building was right next to the NCO club. I could get in to get a beer but I couldn't afford to have a mixed drink. I wasn't allowed because I was a football player. If you did drink something stronger than beer then someone would probably report you and it would be a case of extra laps!

The second football season kicked off in real celebrity style. We were told we were to play a game at no less than Britain's famous Wembley Stadium. It would be against the mighty USAF Rhein-Main base team from Germany. Moreover we were to have celebrities present.

Come the day and come the celebrity – it was actress Jayne Mansfield. They had made me Captain of the defensive team that day and we lost the coin toss. So Ms. Mansfield would kick off and I had to hold the ball. The first time she tried to kick it she missed. I had to whisper to her: "Just take your time and look at the ball before you try to kick it." She did. The ball didn't go very far to be honest but she was in heels! Two players had to link arms and form a human chair to carry her out on to the field.

I had a good season that second year. In a game against the USAF base Chicksands also in England we pulled out a last minute upset win 22-18. We were watched by about 2,000 people. By then I weighted 225 pounds and I managed to run eight yards for a touchdown. I also was named to the All-UK team as a tackle.

We won the UK Footfall crown but didn't win the USAF Europe crown. Then I was working in the 28th Weather Squadron. Part of the reason for my success was my adjacent line-guard Roger Miller. He weighed 250 pounds. He'd take out the first two guys and I'd go to work on the rest.

We also played at the famous White City Stadium which had been an Olympic Stadium and eventually became part of the BBC Television's main site in the mid-1980s. I also tried out for the London Rockets Basketball team but those guys were too good, so I played basketball for the Squadron team.

Because South Ruislip was a USAF headquarters we got to see some of the stars including boxers Joe Frazier and Muhammad Ali and the then young British singing sensation Lulu.

So it was also going pretty well. And so was my personal life. In the early days there had been a few girlfriends – but they were really interested in commitment. In reality it had not been long – about three months – after I arrived that I did get down to London. Myself and some of the other guys had walked from Marylebone Station to Hedda's Club in Berwick Street. The owner, Hedda, was a said to be German woman who reputedly lived in style at London's chic Chelsea Embankment.

Beryl was one of the girls waitressing there. It was her second job as she was working as a receptionist for a medical practice on Harley Street during the day. We met and spoke and I think I'm safe to say we hit it off right way. I told her I was going to come back. And I made it my business to stop there as soon as I could get back to London.

We would meet at the Lyon's Corner House in Piccadilly where they served tea, coffee and biscuits. It was a cheap place to eat and hang out. It cost just six or eight British pence for a cup of tea and you got a biscuit with it.

I needed an angle and I had it! Football season was starting and since I had made the team I invited her out to USAF South Ruislip. "There's a football game on. Come out and watch," I told her. "Mention my name and they'll let you in at the Front Gate."

I met her at the base snack bar. It was there I told her that I wasn't going to watch with her because I was playing in the game. We got together after that and not long after I met her mom and dad.

Her granddad initially tried speaking Swahili to me. He had been in the British Army and for some reason thought I was from South Africa! I had to say "Excuse me but I don't understand a word you're saying." That's when Beryl told him I was a Yank. I was invited over for dinner with her parents and I came armed… with a bottle of Scotch. And I made friends for life.

Love's true path isn't always smooth. And that was certainly the case for us. There was the small question of race. Being a mixed-race couple back then was not the norm. In fact so much not the norm that it would cause us trouble for years. I even had a run-in with my Squadron Commander about in 1962 when I was stationed at Goose, Bay Labrador for the second time (I had been there in 1952) when he told me I was making a grave mistake.

I used to go to the NCO club at the 922nd AC&W Sqdn. The black guys were okay about it but I'd catch flack from just about everybody else including the white guys and the Hispanics. It would come in the form of cat-calls and more intimidating actions. Eventually I got in fight with two white guys. It was a 'come outside and say that' fight. I took them both on and it was a beating…for both of them.

The Security Police came and I got locked up. That's when the Squadron Commander and I 'talked'. He felt I should know better. My response? "With due respect, Sir, you can never know what's its like for me – for us." I think he got it, though, as he only gave me an oral reprimand.

I don't know how much flack Beryl got – she was very good at keeping it from me. I don't know if I was the first around there in a mixed-race relationship but they had "never "had any problems" until I came along!

1962 proved a year of highs – Beryl and I were married at a registry office in London. And lows. In order to stay in the UK I had to take a year-long isolated duty station so that I could have my pick – or at least a better shot at my pick. I had been assigned again to the base at Goose Bay, Labrador. And it was a long and lonely assignment. The misery was compounded by the death of my mother. The Red Cross got me a 15-day emergency leave but when I went up to the flight pad there was nothing doing – we were socked in with bad weather. Moreover I couldn't call anybody to tell them. And so I missed it my mother's funeral. By the time I could leave it was over and so I didn't go back at all. It was one of those things.

By the time I came back to USAF South Ruislip in 1963 – things had changed in regards to the equipment – for instance they had a Plan 51 Communications System and we were on the Autovon telephone system which had started to be introduced in 1958.

And then I got a posting to Vietnam in 1968.

I was at Tan Son Nhut Air Base for three months and then they wanted people to go to Korat in Thailand. Like a fool I volunteered. I was one of three who made that mistake. We went right to the border of Thailand and Vietnam. We were in a small communications detachment.

Around six months after arriving there we got overrun and had to get out of there like mad. I had to burn the classified documents and other stuff and destroy the safe. We could see the flack coming over the runway thanks to the tracer bullets. US Special forces and police came through and pushed them back. Then our guys gave us clearance to return: "You can go back now – just watch out for mines!"

We had to carefully pick our way back to our office that was right next to the fenceline. It was a tough time though and not just because of that. We had guys who would go up on flights to do 'Comms' work and not come back.

We had not long been married but the racism issue never went away. One incident stands in my mine. In 1968 before I went to Vietnam we decided to take a holiday trip from Scott AFB in Illinois to California. We weren't far down the road when we were pulled up by the cops. To say they didn't take kindly to a mixed-race couple was an understatement. Beryl was driving at the time and were pulled over for 'speeding'. The policeman said we had to follow him to court. We were charged a $120 fine that had to be paid there and then or spend a weekend in jail. We paid but it wiped out our trip.

One other thing stands in my mind from that time. In May 1962 I had to fly back to Labrador for my assignment. In route we landed in Keflavik, Iceland, to refuel as many aircraft did back then. However the pilot reported a problem with the aircraft and that we would have to camp out over night.

Little did I know that would cause a major diplomatic problem. The US military didn't want me to leave the aircraft. Apparently they had an agreement with Iceland that no blacks would be stationed there! They didn't have a place on base for me and, in the end, I was taken, by escort, to a hotel and kept there overnight before being escorted back to the aircraft and hustled out of there the next day.

I returned to South Ruislip in 1969. Communications had changed by then – the switchboard had moved out and communication was largely through computers. They didn't have any communications role in my pay grade so they sent me to 2180th Detachment at RAF Uxbridge. It was a 'Comms' post then manned just during the day – I would go in clear up messages and distribute them. Eventually I got back to South Ruislip where I was now faced with floor-to-ceiling high banks of computers. We'd open up the doors and take the tapes out and change them. I just recall lots of magnetic tapes and taking messages and putting them in the Translation Room.

But the South Ruislip base was going to close. We had people in Communications who knew about that before the bigwigs did. So you'd hear talk: "They're going to shut this place down." I wasn't worried but I didn't want another isolated duty site. Matters became clearer when Col. Hart (the Base Commander) was reassigned and Col Miller took over. They started closing portions of South Ruislip down little by little. I think the 7500th stayed there the longest.

It was in 1972 as South Ruislip was closing that I became involved in the music scene. A few guys from the base were trying to get a group together. I heard about them and went to see them practice – they met up off-base at the house where one of them lived. Marvin Harper, Ulus Strayhorn, Roger Hatcher, Ken Marsh and Rick Osborne were the guys. It was a group born out of the US military. One of them was an Airman 3rd Class, another a Tech Sgt., another an Airman 1st Class, another an Airman 2nd Class and the fifth was a US Navy Seaman 3rd Class.

I thought they sounded pretty good and set about getting them some gigs – including a record deal. And of course a band name: Boss Effect. It came about because I said to them – "I'll be the boss, you be the effect!"

Johnny Goodison (Johnny B. Great) became a singer with a group called Brotherhood of Man in 1969. He was appearing at RAF Chicksands. I approached him and told him I had a group looking for a record deal. "If they're any good, I'll give a deal to them," he said. He then heard them and said: "You've got four weeks to get them in shape to record a record."

And we did. It was a song called *Brother Hold Your Head Up High*. We recorded it up in studios in Finchley, London. It went to Number 17 in the British charts and we started getting gigs. But they still didn't have equipment and we had go out and buy it and we also needed some backing singers. We were getting £300 a gig but split between five guys and any back-up singers and it wasn't much. Then Strayhorn went back to the States and Ken Marsh left to get married. Still it was a tremendous 18-month run with them.

In 1973 I got out of the military and didn't have a job. I bought into a pizza operation called Cass's Pizza. I decided the Air Force bases needed pizza. I took a pizza oven around and made fresh pizza at the NCO clubs but it was a lot of work with 12-hour shifts and travel and, pardon the pun, it didn't pan out! I needed a job and found some work coaching at Brunel University in Uxbridge where I had been playing some pick-up games. I ended up coaching the women's basketball team there for a time.

In 1975 I joined the US Navy working in Communications on the third floor at the US Navy HQ at North Audley Street in London. This was a different type of communications – now I was doing reproduction using the press machines on six days shifts. It was like an early-day copy shop.

A chance encounter when I returned the US Navy bus I had checked out to take people a sports event saw me join the US Navy as an Athletic Director. That would lead to 22 years of work with the US Navy's Morale Recreational and Welfare (MWR) working at the bases at West Ruislip, Eastcote, High Wycombe and at the US Navy Headquarters at North Audley Street in London.

During the time with MWR I wore three different hats: As an Athletic Director, as a coach for NAVACTUK's Men's & Women's teams and as the NAVEUR Sports Director for all US Navy bases throughout the UK & Europe. That meant bases from Edzell in

Scotland to Keflavik in Iceland, Brawdy in Wales, St. Mawgan in Cornwall and the bases around London including the US Navy Headquarters and the bases Eastcote, High Wycombe and West Ruislip.

Somewhere in between all that I managed to create the softball field at RAF West Ruislip and it made me proud when one of the MWR directors wanted to name it after me – even though it never happened. We also sorted out a racketball court and got it refurbished. We were based at West Ruislip but we'd work all over the place.

I retired in 1997 and Beryl and I stayed in England not far from the former West Ruislip base which is now mostly covered in housing. One way or another, England and these bases around London had become an integral part of my life.

Archie Tatum in 2012.

Archie & Beryl Tatum in the 1960s.

BOSS EFFECT

Boss Effect as they were. The band, managed by Archie Tatum, went on to have a Top 20 hit in the UK.

CHAPTER NINETEEN

PATHOLOGY

Iris Margaret White (nee Madden)

Secretary to Chief of Pathology
7520th Hospital
DOE Eastcote (USAF)
(1956 – 1965)

I ended up at the base at Eastcote because I wanted a local job. I had resigned from the British Ministry of Labour and National Service which had its offices in London. I lived in South Ruislip and found out about the job at Eastcote. I went to the local Labour Exchange where they told me the there was a job working for the American military. It sounded interesting and so I went along.

The entrance to the base was from Lime Grove with the Cafeteria for the GIs conveniently by the main gate. They served the most delicious hamburgers there.

Our building was, as I recall, the third one of from the gate. It was an extremely interesting place to work with the first half being the USAF and further down the hallway offices for the British Government's Board of Trade. My job was to type up Pathology reports on all specimens sent from the main USAF South Ruislip Hospital and all autopsies that were performed at Eastcote of deceased members of the USAF and their dependents.

In our building there was also a Dental Clinic in one wing and Dental Lab in another on the opposite side and a Psychiatric Wing on the right and our Pathology Lab and Mortuary on one wing and then the different labs on the opposite wing. The hallway down the middle of the building led to the offices of the British Government's Board of Trade.

You wouldn't think working in a Pathology Lab could be uplifting but in many ways it was really a happy time for me. I loved my job and the people I worked for at the lab. I really got to know the people and their families and hearing about their lives in the USA and watching them getting to know the UK.

And little did I know that it would become an eight-year job that would bring me into happy and sad times and would also introduce me to my future husband when he was assigned to the laboratory.

In the office, as far as I can remember, we had the old white walls and the metal desks. I do remember we all had pale green doors.

One of my fellow workers was Mr. Sutton, who was British and who was a former chauffeur who used to drive the wealthy around in his younger days. He worked in our Sterilization Unit, sterilizing the dishes and other equipment used in the lab chemistries. Another was a Dr. Dybowski who was I believe Polish and was a Bacteriologist. He was a very good friend of ours and we're delighted when we met up again when we were both in Texas years later.

Dr. Speicher was one of the Pathologists, the other was Dr. Frost. They were the last pathologists before I left Eastcote. They were among many kind colleagues.

There were also several men during my time who had the toughest jobs to my thinking – the morticians: Among them were Mr. Hazus and Mr. Raglen and they were really wonderful at their jobs, sad though it may be. They were not military but civil servants. One of them dealt with a murder — I also knew about that as I was the one who did all the typing of the autopsy at that time, the surname was Marymount.

I believe the bodies of deceased persons came in ambulances but as I could not see from a window I honestly do not remember, they were just brought straight to the back entrance of the block. We normally just had one regular ambulance driver going to and from the hospital at USAF South Ruislip – no more than about two miles away.

One of my most personally distressing memories was the day I came back from vacation in Spain to find we had the bodies from an aircraft that had crashed in the North Atlantic. It was one of the grimmest memories I remember. We had so many bodies in the Mortuary. It was really a sorrowful sight.

While the aircraft disaster was one of the most difficult memories it was in a way equaled by having to type up autopsies on crib death babies. That was really hard.

I did see some of the bodies, but it was not my favorite thing to have to go into the mortuary, especially when it was a child or infant. I hated the noise of the bone saw. Sometimes it echoed into a room which I first had which was next to the Mortuary. Some days we had lots of specimens from the hospital and not a lot of autopsies at times which was good. I know there has been some talk of bodies of soldiers from Vietnam coming through our Mortuary but, as far as I recall, bodies of the soldiers from the wars never came through the mortuary there.

I then moved around to Pathology; no bone saw noise – just the smell of formaldehyde which I got used to eventually and, of course, different specimens being set over and brains in various jars around, which was extremely interesting. Dr. Speicher was great and I often watched him dissect a specimen and show me cancer at different times.

One special memory was my 21st birthday party that they gave for me and it was at the lab with plenty of cookies and a cake. It was a total surprise.

A1C Edward D White Jr. was assigned to the Pathology Lab at Eastcote when I was holidaying in Spain. I returned and did not know who the new guy who was working on Bacteriology so asked my friends in Pathology to find out about him. He then asked me to attend a Thanksgiving dinner at the main hospital in South Ruislip and it developed into a romance and an engagement ring. We were married at St. Paul's Anglican church in Torrington Road, Ruislip Manor.

The staff from the Pathology Lab shared in our wedding day. We had spent many great hours at the local bowling alley, which for us was at North Harrow, where I believe a Safeway supermarket took its place in later years.

Edward had arrived in England in 1962 and found that bowling was quite new and that no 300 game had happened yet. In Harrow a cinema had been converted into a bowling arena and he joined a local league. He was a great bowler. His average was about 180 and in time he was asked to coach the Harrow Bowling Team which competed in a travelling league). At the time he was an E-4 in rank and ended up coaching a British team that was headed by someone known as "Bunny". The team was invited to a big event at the West Ruislip base bowling lanes.

There was security in force, but it was 'Bunny' who vouched for my husband and got him in to take on teams of US officers and senior civilians. The British team had prepared quite well and, in fact, demolished the USAF teams. It turned out that Bunny was the local Lord Mayor and he stayed friends throughout our tour of duty.

We made many good friends there who also shared in our wedding day. When we came out of the church it was to find an honor guard holding up bowling pins over our heads. It was a picture that appeared in the local paper.

Our favorite pub was The Plough in Ruislip. We had lots of British friends and we used to gather there. For one of my Edward's birthdays, he was presented with his own tankard from behind the bar from the pub owner at that time. That was a wonderful surprise.

There were the day-to-day things that happened in the lab that, looking back, were humorous in hindsight. For instance in the lab we used to cook chili in a big pot over a Bunsen burner! I had never had chili until then and it was delicious.

A couple of funny work experiences: One was the day that we had a specimen sent from the US Navy located downtown to Pathology and labeled 'Uavy Oceanside'. It turned out to be a squid! We had a good laugh about that experience.

A second incident caused much amusement in the office. A leg (that's right; a leg) was sent over to Pathology. Well our ambulance driver had no idea he was carrying a leg under his arm, when I commented: "Oh great, our leg has arrived!" Well the driver went one way and the leg went on the floor!

One other thing I did remember before my hubby was there was that one of the guys from the lab had a tapeworm specimen and we unrolled it across the two hallways! It was so long and gave me the creeps!

Another particularly poignant memory for me came in 1963 when President Kennedy was shot. The entire lab staff went to the Chapel at the West Ruislip base for a remembrance service with many others. It was a very touching and humbling experience.

The biggest thing I learned from that job that we had to enjoy our life as best as possible as it can end so suddenly.

Edward initially got out of the service soon after we married and we left and went back to the University of Florida where he got his degree. Then went back into the service in the Army Medical Service Corp where taught the combat medics at Fort Sam Houston, Texas. We travelled all over the place — even overseas again and he retired as a Major and we retired in Washington State.

Little did I know when we left that a new adventure was around the corner — not only coming to a new country but having a job at the University of Florida with Dr Robert Cade who was experimenting with a new drink to help out the Gator football team, which became the well-known Gatorade. I was the original guinea pig for his concoction!

I have many wonderful memories to share with my two granddaughters of the memories their Nana has experienced.

Dalton & Iris White get a 'arch of bowling pins' on their wedding day. Photo courtesy Iris and Dalton White.

CHAPTER TWENTY

VIA RADAR TO WEST RUISLIP

Ted Clark

(MSgt) USAF
Ground Radar Technician
7568th Material Squadron
32nd Anti-Aircraft Artillery Brigade
Shellingford Air Station
May 1956 – December 1956
Vehicle Transporter
RAF West Ruislip
December 1956 – July 1957

It was a company notice board that put me on the road to England. I came from a small rural farming community in Georgia. If your family didn't own a farm or a business then there wasn't a not a lot of opportunity in the area.

I had finished high school and I didn't want to go back to school anytime soon so I took a job in the local department store. One day a week they would close early – at a half day so we could do some cleaning up. As we were doing so I stopped to look at a notice board. On it was a notice that a USAF recruiter was coming to the area. I thought maybe it was a way out. What the heck – I'll see what is on offer.

You think that a recruiter would be sell, sell, sell. But this one wasn't. Or maybe it was a reverse psychology 'sell'. "You have to take a test and you may not even qualify," he said. I took the exam and passed. Then there was a medical. I managed to pass that and so, on 20 June 1955, I enlisted and then it was off down the path well trodden by so many others to Lackland AFB in Texas for 10 to 12 weeks of Basic Training.

By the end of it I was ready to go. Except that I wasn't. As my fellow recruits received their orders and were posted, I could only watch. And wonder what to do with myself during the two days after the others had gone. How tough was that? Well I was volunteering to assist the trainers when my orders thankfully came through to order me to radar school at Keesler AFB in Biloxi, Mississippi.

Radar?

I had heard of it. I knew it was used with aircraft but I didn't really know the detail. Radar School sorted all that out.

Radar had really been developed in Germany and England but the English developed a version that used permanent magnets attached to the magnetron tube and that magnetic force kept the energy field concentrated effectively allowing it to produce more power which could be radiated out from the antenna.

The funny thing is that the equipment the school was using was developed shortly after WWII and my class was the last class to go through training on that equipment. It was then scheduled to be phased out of usage as improved radar was coming. Radar school training lasted about eight months and during that time we lived in basic WWII buildings – two-storey open barracks and one shower room for the lot of us. That was the 1950s.

Part of the training was to understand the use of early warning radar that could detect aircraft 200 miles away. It wasn't until I got overseas and started seeing some of realities of the real world that I started connecting the dots from the training.

And then my assignment came through. At that time they didn't give you a base name – they gave you an APO Number. Of course finding out what the number meant was task in itself – not helped by my fellow trainees. One of the guys took a look at it and said; I think you're going to France." Well that was sorted then. So it was no surprise when a little further searching identified my first duty as with the 7568th Material Squadron, at Shellingford AS (Air Station) in England.

It was exciting because it was the great unknown. I had relatives who had served in WWII and been in Europe. I had a little knowledge built up from a smattering of 'Europe' in High School history classes. I think the most comforting thing I knew was that they at least spoke English in England!

I flew to the UK on a military transport; possibly a C-54. We finally landed at Prestwick in Scotland and transferred to a place I recall as The Station House – it was like an inn with a train station underneath.

Unbeknownst to me there was a guy from Oklahoma on the same flight and staying at the same inn. It turned out he had gone through radar training with me at Keesler – only we had never crossed paths until we reached London where we checked in at the US military hotel at Douglas House. It was a TDY hotel and club so to speak.

From there it was a train ride to Shellingford AS. Now at that time the British had three 'classes' of trains: 1st, 2nd & 3rd. Third Class was the equivalent of 'cattle class'. And they didn't have a Second Class carriage on the train on which we were traveling. To make it more surreal when we got to our train station that was all there was. There was nothing else in the middle of nowhere. There was, however, a stationmaster present. So we asked him: "Is there a USAF base around here?" He looked at us as if we were crazy for asking the question. "Yes," he replied but that was all. "How do we get there?" we asked. "Well," he told us almost wearily, "I'll call 'em but it'll cost you sixpence."

We paid up the little silver coin. We were still trying to get to grips with the money. He called and then a little while afterwards a Vauxhall pick-up truck painted Air Force blue pulled up and a security policeman greeted us, helped us with our bags and took us back to the base and introduced us around. That was nicest security policeman I ever met!

And from that I point I was at the 7568th Material Squadron at RAF Shellingford in Oxfordshire.

The base was a small installation. Shellingford was really three different areas; The barracks were about 1/4 mile from the what I would call the 'social area', which contained the Dining Hall, Chapel, Recreation Center, NCO club, BX and Medical Facility (mainly first aid). The work area was about 1/2 mile from both areas. The mode of transportation was walking or biking. The unique aspect of this base, at least to me, was that there was a public highway that ran adjacent to the base and there were no gates or fences.

There was another surprise awaiting us – in the shape of the US Army.

Our joint mission was to provide Radar and Radio Maintenance on US Army equipment at the following USAF Bomber bases in England: Brize Norton, Greenham Common and Upper Heyford. While the USAF provided heavy bomber coverage for England and Europe, the US Army, via the 32nd AAA Brigade was tasked with providing air defense for the bomber bases. Needless to say, it was quite different from what I had expected. The men assigned this duty – both USAF and US Army personnel – worked together quite well.

There were about 124 Air Force personnel and around 50 US Army who were in two units. Ours was the 32nd AAA Brigade and then also there was the Smoke Generation and Signals Unit. They could 'make smoke' in order to shield us from enemy attack. We got along okay considering we were rival services.

We formed part of the 7568th Material Squadron and our Squadron was part of the 7500 Air Base Group at Denham Studio AB and, in turn, thereby part of the USAF 3rd Air Force which was headquartered at South Ruislip Air Base.

We had one radar set at Shellingford. It was the back-up unit. The operational equipment was at RAF Brize Norton, RAF Greenham Common and Upper Heyford. My own job was to provide repairs or replacement for the radars including the electronic system, the radar and the radar 'sail'. The work was relatively straightforward – complicated only by the users! But isn't that always the case!

We had a deuce-and-a-half (two-and-a-half ton) truck with an enclosed cabin at the back to carry the electronic bits. Someone in the military had at least figured out that the English weather and electronic parts didn't mix too well.

There were issues. And they were user issues. The Air Force would have one radar unit and operate it 24 hours a day with a Day, Evening and Midnight Crew operating radar in continuous mode. The Army operated differently. Each crew would operate the radar for eight hours, shut it down and go back to barracks leaving another crew to fire up the radar set.

This continual shutting down and starting up seemed to do more damage to their units and so we'd be out there a lot checking tubes in the system. Thankfully, we had removable sub-components and often had the spare parts in our van. We spent a lot of our time trouble-shooting and then soldering.

Given all this and the Cold War times we were in we didn't have any major alerts of exercises for ourselves that I could recall. The highest level of concern was when the Egyptians blockaded the Suez Canal and the British got involved. Tensions got pretty high and I can recall that the local British people were unhappy that the Yanks didn't go over and help their British comrades. But all those decisions were at a pay grade a lot higher than mine.

But change was in the air. By the fall of 1956 the Air Force no longer needed defense of the military bases by the Army and the role of the 32nd AAA was removed and the personnel reassigned. Once the 32nd AAA was deactivated then the mission of the 7568th was also going to end and, accordingly, we were assigned other tasks at installations throughout England.

I had orders transferring me to a new organization: 1st Motor Transportation Squadron. It was to be based at West Ruislip Air Station, England and that the personnel who were single would be housed in quarters at Denham Studios Air Station. That entire base was a part of the Denham Film Studios. The buildings were large open area bays where about eight to 10 bunks per one. We had a common shower and long connecting hallway. Everything that had to do with the movies had been removed and stored in other buildings. And I certainly didn't see any starlets!

Needless to say, I soon found out that I did not have any radar equipment to maintain at West Ruislip AS which we travelled to and from the three miles or so by bus. The mission and purpose of this unit was the repair/overhaul of heavy duty construction vehicles, namely dump trucks along with a scattering of other vehicle types.

The West Ruislip base wasn't open when we got there. It was turned over to the Americans by the British in the in early 1957 in an official ceremony. I was among about 85 of us in the unit on parade ready to take on the new base. In fact the event was covered by either the BBC or the independent television channel of the time. Amazingly for England in winter it was a good day that day – it wasn't freezing and the sun was shining.

We had a small Base Exchange shop at Denham Studios but we were so close to the Third Air Force Headquarters at South Ruislip we could hop in a vehicle and get right over there to the BX. South Ruislip was a well-laid out military installation and was neat and clean as a headquarters base should be. But going there was fraught with its own problems. That essentially meant far too many officers to salute. We tended stayed away from the high density of officers unless we had to go.

The entrance to the West Ruislip base came right off the main street running by the railway station The entrance comprised of a double gate made from wrought iron. Almost directly across the street from the base was a British teahouse. The base itself had a lot of WWII buildings on it. It went over the railroad and the tube, but the base itself was fairly open plan and it was mainly WWII-type storage buildings with narrow doors. The buildings had cross-hatched beams every 10 to 12 feet across and long open work areas.

So we didn't have any radar to work on but we did just about anything and everything else that needed to be done on base. Once the handover happened, they started bringing in British mechanics to work on all the vehicles that were being brought there for repair including dump trucks, fire trucks and the like. Lots of the vehicles came from Greenham Common some 45 miles away. There the runway was being upgraded from its wartime condition.

When they reached us the vehicles were often in poor condition. The USAFE would have the problem vehicle towed in and they would be placed in a parking lot containment area and the work order generated.

Our job, however, was not to fix them, but simply to get them to the mechanics and then order the parts they needed. Getting to the mechanics was a challenge – the garage /storage buildings that were being used at the time were not set up for maneuvering large sized vehicles in and we had to guide them through those narrow entrance doors pretty carefully. Remember these were vehicles that had to be towed or pushed.

One of the most interesting things that happened while I was there was they moved the offices of the Base Exchange (BX) network to West Ruislip. As part of that they brought in about 200-400 English ladies to work in the office. By that time there were about 20 of us GIs who were single – and there were all these ladies. Bliss!

We were young, highly impressionable and in a foreign land. It was heaven. We went to London (the London Underground had a station right next to the West Ruislip base). I went to Windsor. And in 1956, back when the USAF had base football teams, I went to see an American Football game played at Wembley Stadium. It was the country's main national station normally used for prestigious soccer matches. I can remember that it was a good game but that it was foggy and from where I was sitting so far up in the stadium I could barely see the players. But the stadium was pretty well packed.

Transport wasn't what we were trained for. And so in the summer of 1957 I was happy to find myself en-route to a new duty station, namely the 633rd Aircraft Control and Warning Squadron, Wheelus AB, Libya. I remained in this unit for about a year before returning to America for duty with the 727th Aircraft Control and Warning Squadron, Myrtle Beach AFB, South Carolina.

One note of interest; while assigned to the 633rd Aircraft Control and Warning Squadron, I did spend about six weeks on TDY to Incirlik AB, Adana, Turkey, during a border conflict with one of Turkey's neighboring countries.

Following a year of duty at Myrtle Beach AFB, South Carolina, I next traveled to the 908th Aircraft Control and Warning Squadron, Dobbins AFB, Marietta, GA. During this assignment I got married to Shirley L. Barber, a girl I had met and we had our first child, a son we named James Michael Clark. Then, in the Fall of 1962, I accepted orders for further assignment with the 615th Aircraft Control and Warning Squadron, Birkenfeld AS, Germany. It was an enjoyable two-year assignment before orders had me back to the US with duty with the 758th Radar Squadron, Makah AFS, Neah Bay, Washington, December 1964.

While on this assignment, our second child was born, a girl we named Gena Marie Clark. Then in late October or early November, we were transferred to a new unit at Eglin AFB, Florida. Here my duty was to help install and then maintain the new Space Track Radar system, the AN/FPS-85. This assignment was really neat and I enjoyed both the work and the living area; i.e., sunny Florida.

During this time of duty the conflict in Vietnam was heating up and in the summer of 1970 I was notified of my assignment to Vietnam, with duty being with Det. 9, 619th Aircraft Control and Warning Squadron. This was not one of my favorite assignments and in the Summer of 1971, I was on my way. England was already a distant, distant memory.

Above Left: Ted Clark outside the Denham Studios. Above Right: Ted in uniform operating a Forklift at Denham. Photos courtesy of Ted Clark.

CHAPTER TWENTY-ONE

THE HEART OF THE MATTER

Ellis Rolett, M.D.

Captain, USAF
June 1957 – June 1959

The Berry Plan, launched in 1954, would have been well known to anyone studying to be a doctor in the US in the late 1950s right up through until it was phased out in 1973. It was a deferment of US military service for those draftees in medical school. In my case it meant that I could complete my medical degree and one year of postgraduate training and chose the branch of service I wanted to enter.

I chose the US Air Force. And then I did the next best thing I could think of – I headed to USAF Headquarters in Washington DC and lobbied for an assignment to Continental Europe. I had previously travelled there and loved it. I knew that the bulk of USAF medical expertise was based in Wiesbaden, Germany and so I put a strong case forward as to why the USAF needed me there.

And that's why they sent me to England! But only after they deferred me for a further year – it turned out the USAF found it had more people than it needed. And so I went to Cornell Medical Centre in New York City to get another year of experience. It was to improve invaluable.

In the spring of 1957 I was given the rank of USAF Captain and my base assignment. "We're going to some place called "Rooislup" in England," I told my wife, Ginny. Then I found it on a map and said; "that's not so bad".

I was told to report Mitchel AFB on Long Island where I was given an allowance and told to use it to purchase summer and winter uniforms and then to leave a number where I could be reached at a moment's notice. I was given a date three weeks hence in mid-July to report to McGuire Air Force Base in New Jersey with my family for air transport to England.

For reasons I never understood I was not sent to Lackland AFB in Texas for basic military training. Instead I spent the three weeks on the beach with my wife and infant son. I had been told to report to McGuire in a summer uniform. I consulted the USAF Officer's Manual I had received at Mitchel and found that it contained photographs of the winter uniform but not the summer uniform.

With great care I pinned the various insignia, including my captain's bars, in the locations depicted for the winter uniform. As I was boarding the transport plane at McGuire, I was called aside by the Officer-in-Charge and informed that my insignia were not where they belonged on a summer uniform.

The transport plane was a four-engine prop job, seats facing to the rear, with overhead hammock for infants. The plane was filled with young officers like me and their wives and infant children. We were one of the last groups of officers sent to Europe with a two-year commitment to the Air Force for whom concurrent travel for dependents was approved. After August 1957 assignments to Europe required a three-year commitment.

I had little idea what to expect for my Air Force assignment except that I would be working at a base hospital as an internist. When I arrived at South Ruislip AFB my first impression was that the buildings looked like a factory – which it had been (I think I remember being told it had been a shoe factory).

I was surprised there was no airfield there! And I was surprised by how many senior officers were stationed there. My lack of basic training was immediately evident the first time I was saluted by an enlisted man and didn't know I was expected to salute back.

We – my wife Ginny, infant son Rod and I – found temporary housing at Mrs. Blood's Bed & Breakfast in Harrow-on-the-Hill. There was no on-base housing at South Ruislip and we wanted to stay on the economy anyway. From our temporary digs we moved to 25 Ladygate Lane (some of my colleagues had some fun with that name) in Ruislip.

It was a standalone bungalow with central heating (well, a water radiator) in the front hall, a larder in place of a fridge and coal-burning fireplaces in each room A cold water tap in the kitchen provided our drinking water. Water for the bathroom sink and bath-tub came from a cistern in the attic.

When I climbed into the attic I found the cistern was covered with nothing more than a piece of carpet. Worse yet there was a dead rat floating in the water. We were boiling our water after that!

We rented 25 Ladygate Lane from a British RAF officer, Squadron Leader Rees, who was based nearby. He would come by once a month to collect the rent which was £41 per month (£1 = $2.80 in 1959). We had nice neighbors such as the Websters next door. Mrs. Webster took Ginny under her wing – we had been married for just over a year when we moved to Ruislip.

Mrs. Webster recommended a housekeeper named Doris whom we hired for a shilling (about $.26 then) an hour. Doris taught my wife the ins and outs of English house-keeping such as polishing the front stoop and starching the parlour curtains. The most important rule seemed to be never put out the laundry to dry when the sun was shining for it was bound to rain!

The house was nice but we noticed the wallpaper in the sitting room was getting moldy. Why should this be? The carpeting in that room was also damp. I pried it up and shined a flashlight into the cracks between the floorboards. There was a lake of water six inches below!

It turned out that a neighbor had an orchid house and was watering endlessly. Mystery solved. Another thing that amused me was the house piping – it was on the outside of the house. I asked the owner about this and he said: That's in case the pipes freeze – they'll be easier to get to!"

It was Mrs. Webster who insisted on a midwife for Ginny who was by then pregnant with our second child, Barry. She contacted Nurse Adamson who had left the area to help run a B&B in a town called Swanage. Nurse Adamson agreed to become nanny for our second child, Barry.

In anticipation of having a second child we had come over with all the baby stuff – blankets and cloth diapers and baby clothes. Nurse Adamson looked over all the supplies and said: "I'll only need two of these nappies and you can put the rest of them away."

It turned out that she had also come with an enamel pot and every time she fed Barry she sat him on the pot! He never soiled a diaper! She had other quirks and we quickly learned to let Nurse Adamson have her way. Things turned out very well!

Things were also very interesting on the job front. There was a contingent of six of us with families who had been assigned for our two-year tour of duty to 7520th USAF Hospital in South Ruislip.

Because I had done an internship and residency in Internal Medicine I was assigned to the hospital's Internal Medicine department headed by a career USAF doctor, Major Martin Freedman. The third member of our team was Dr. Robert Spencer, a contract British doctor. Freedman and Spencer were older and more experienced but I had the latest training and so we made a good partnership.

The USAF South Ruislip Hospital was about 100 beds divided into wards, one or two of which housed patients under the care of our team. We rotated responsibilities for admitting patients, but as a British civilian, Dr. Spencer was not responsible for night call. The Air Force doctors rotated that responsibility as a dual call – one covering Obstetrics and another covering everything else, including Paediatrics.

My recollection of the hospital layout is fuzzy. The building had two storeys, mainly administrative upstairs. On the Ground Floor there was a long corridor, a few private patient rooms at one end; at the other end were the Paediatric Clinic and Radiology Suite. Arranged on opposite sides of the corridor were the Inpatient wards. Also on the ground floor were the Delivery Room and Newborn Nursery, Operating Room(s) and Cafeteria.

If I recall correctly, the Blood Chemistry, Microbiology and Pathology laboratories and the Morgue were at the nearby Eastcote base. (The Dental Clinic was also at Eastcote.)

I remember bussing over to the hospital from Ruislip and getting to work around 8 a.m. and then home in time for supper at 6 p.m. I also had a membership to the South Ruislip Officers Club. Back then the monthly dues for the club were just $3.00. I had an office on the second floor of the hospital, where I saw outpatients. I was an academic so I convinced the Adjutant's Office to give me a budget to order books, put in shelves and establish a library of medical textbooks that could be checked out. However nobody was really interested and so I had the books almost exclusively to myself!

During the course of the day we would see a lot of referrals from the other bases as well members of families of officers based at South Ruislip. I developed a small specialty practice for asthmatic and diabetic dependants. On several occasions I received commendations from Washington DC with letters of praise from grateful senior officers.

It was a great learning environment. At the hospital we had a small budget to bring a consultant out from London for one day a month. We brought out some of the top British doctors and surgeons to give lectures and consult on patients. Among them were Sir Russell Brain, the neurologist who became Lord Brain; Avery Jones, the gastroenterologist; Paul Wood from the National Heart Hospital and John Dacie, Chief of Haematology at Hammersmith Hospital.

Every Wednesday afternoon I had off for R&R and I would use that time to visit a London hospital. During my first year, I typically spent Wednesday afternoons at Hammersmith Hospital going to grand rounds and attending Haematology clinic.

The second year I went to the National Heart Hospital every Wednesday afternoon. I had been thinking about going into Cardiology and it was an unparalleled opportunity to be in a clinic with Paul Wood who was one of the giants of cardiology world at that time. Several of Paul Wood's registrars became prominent consultants in their own right.

The 7520th USAF Hospital South Ruislip was a pretty good hospital and in no small part because of a bright group of two-year doctors who came in under that Berry Plan. We were eager people who didn't mind working hard and staying late. People like radiologist Herman Shyken; James Rosen who became a neurologist and George Bauernschub in paediatrics.

When our second son, Barry, was born in the hospital Ginny was in USAF Ruislip Hospital for five days after the birth. The daily subsistence rate was $1.15 and the daily hospitalization rate was $0.60, for a grand total of $8.75. Of course in those days, whether or not you were a doctor, they wouldn't let you into the delivery area. It wasn't until a few hours after he was born that a nurse held Barry up to the window of the enclosed nursery for me to see him for the first time.

When I was working at New York Hospital, prior to entering the Air Force, a Puerto Rican teenage girl was brought to the Emergency Department and within minutes went into cardiac arrest and died. When the Chief Resident learned of the death he criticized me for not having performed emergent open-chest cardiac massage. It turned out at autopsy that the girl had an irreversible parasitic disease and nothing could have saved her. Nevertheless, the criticism lodged in my subconscious.

It was during my second year at USAF South Ruislip, while on night duty, that I admitted a Warrant Officer who was in his mid-30s because of an acute myocardial infarction. Those were the days before coronary care units and USAF South Ruislip had no cardiac monitoring equipment.

I had a premonition about this patient and before making night rounds in the hospital I put a scalpel in my breast pocket. I was in the next ward when a patient ran in and said that the Warrant Officer was gasping for air. I immediately returned to his bedside to find him unconscious and without a pulse.

Without thinking twice, I pulled out the scalpel and opened his chest and reached in and got my hand on his heart and started pumping it by hand. No gloves, no sterile technique. The heart was fibrillating. Within a minute or two the nurse on duty arrived and began mouth-to-mouth breathing.

No more than five minutes had passed when the patient regained consciousness. I put out a call for a surgeon and an anesthesiologist to return to the hospital. I must have performed cardiac massage for an hour with the patient's ribs squeezing my wrist before they arrived.

We moved the patient to the operating room, anesthetized him and were able to restore a normal heart rhythm by delivering a single electric shock directly to the heart. The chest was sewn up and, I'm delighted to say, the patient left the hospital in good condition. Over the next decade or more he would periodically send me a note that he was doing well and had no further heart problems.

Things being what they were at the time in England there was considerable resentment towards the American military presence. A Public Relations guy at the base thought this would be a good story and he called some of the London papers and told them what had happened.

The next couple of days were the worst of my life as press people camped out at my house trying to interview me about what it was like to hold another man's heart in my hands. All the London tabloid newspapers carried the story with front-page headlines like "Nurse Gives Kiss of Life." I didn't dare go home until Ginny said that the press had stopped camping out on the front lawn. I slept on a hospital gurney!

Several days later on my weekly appearance at the National Hospital, Paul Wood looked at me for a moment. Then he said words I never have forgotten. "I wouldn't have done that procedure if I had been in your shoes." It turned out that open chest cardiac resuscitation had never before been done in the U.K. It was, perhaps, a good example of one of those cultural differences between the American gun-ho approach and the more considered approach that the British had at the time. Certainly in the field of medicine, those differences have, to a large extent, disappeared.

Aside from looking after USAF personnel stationed in England we also were a way station for very sick service people from Germany and other bases in Europe on their way to the US. One day a colleague stationed in West Berlin contacted me about a patient who had acute liver and kidney failure and was being sent Stateside for care. He had apparently been cleaning engines with carbon tetrachloride in an enclosed hanger and the effects were deadly.

When he reached us he was at death's door and unable to make the long trip home back to the 'ZI' or 'Zone of the Interior' as we referred to the United States in those days. I contacted the Hammersmith Hospital where they had set up a specialty kidney dialysis unit – perhaps the only one in the UK at that time – and asked for their help. We transferred the airman by ambulance. There was no paperwork, no wrangling; it just happened. In the end however, the damage was too great and sadly the young man died.

It was not uncommon for Air Force doctors who were assigned to South Ruislip to be invited to join a weekend flight to someplace in Europe by a flight officer who needed flying time. RAF Northolt served as the local airfield. I recall a harrowing return to Northolt from a weekend flight to Oslo on the Base Commander's plane. Visibility was poor as we approached the landing field. I recall seeing rooftops less than a hundred meters below as we broke through the fog and within seconds were on the runway.

On another weekend flight, this time to West Berlin on a DC3, the plane's radio died as we were crossing the English Channel. The pilot said if we drifted into Soviet controlled airspace we probably would be forced down and no telling when we would be allowed to return home! But not to worry; he had flown the Berlin Airlift and thought he could navigate the official air corridor by sight.

He gave the passengers the choice of continuing or turning back. We chose the former and made it safely to Berlin. The Berlin Wall had not yet been constructed by the Soviets and a few of us dared to dash through the Brandenburg Gate into the Soviet Zone, have our picture taken and dash back before the Russian guards could apprehend us.

I was fortunate to be offered a TDY (Temporary Duty aka 'Temporary Duty Yonder') assignment by one of my patients to Cambridge University for a residential course for members of the USAF entitled 'The Great Powers in World Affairs." I still have my notebook from the course.

For newly-marrieds who had been living on $1,500 a year during my residency in New York, the Air Force salary with a hardship allowance for England gave us an opportunity to acquire our first household possessions and attend theatre and concerts in London. The best seats in the stalls could be had for one pound sterling.

The South Ruislip base Post Exchange (PX) was stocked with all kinds of goods from the UK, Scandinavia and Continental Europe. We still own china and flatware we purchased there at remarkably low prices. Selfridges, Fortnum & Mason, Simpsons and Burberry were among our favorite shops in London. I purchased a made-to-measure suit and a sports jacket at Jones, Chalk & Dawson. Years later when I revisited Jones, Chalk & Dawson it seemed as if only Saudi royalty could afford their goods.

And then it was over. The four of us returned to the US from RAF Station, Mildenhall, via Preswick and Gander to McGuire. I was mustered out of the Air Force at Fort Dix in time to return to residency training in Boston.

It was a memorable two years, both professionally and culturally. Our young family was welcomed into the neighborhood and we made many good friends. In the end, I could not imagine how an assignment to Wiesbaden would have equaled mine to "Rooislup."

Capt. Ellis Rolett in his 'formal' military pose.

A USAF DC-3 on the tarmac for a trip from RAF Northolt to Copenhagen in October 1957. Photo by Ellis Rolett.

Above: The 7520th Hospital Nursery at USAF South Ruislip in January 1959. Nurse Adamson with Barry, the son of Capt. Rolett. Below – A foggy Ruislip in 1957. Right Capt. Rolett with son Roderick in Ladygate Lane Ruislip 1958. Photos courtesy Ellis Rolett.

CHAPTER TWENTY-TWO

ICE STATION LONDON

Maurice "Mo" Gibbs

Commander, USN
US Navy Headquarters, London.
1957 – 1959
High Wycombe Air Station – 1968
US Navy Headquarters London 1969-1971

There's an old axiom in the military: *Hurry up and wait.*

That's exactly what happened I after I enlisted in the US Navy in Dec 1952.

I was born and brought up on Nantucket Island. It has always been my residence and I have over 100 relatives on the island. Remember in the 1950s the Cold War was in extreme mode. South Korea was invaded by the North and, after a while, the Chinese came into the war as well as (we now know) Russians operating covertly in North Korean aircraft.

The Draft was on and every able-bodied American male between 18 and 22 could expect to be drafted at any time. I chose to enlist to have some control over my career.

I had excelled in science in High School and had worked as a school boy at our local astronomical observatory. Whilst astronomy (then) was not an option, meteorology (and later oceanography) was. Hence I pointed myself in this direction.

Rather than being shipped out on the first bus, given a short back and sides and some sea legs faster than you can say "Yessir!" I was put on a waiting list for Boot Camp. Finally, in August 1953, I was summoned to Naval Training Centre Bainbridge, Maryland.

Then things started to move at a pace. By Fall I was transferred to an aviation prep school (Naval Air Station Norman, in Oklahoma, of all places) Finally, in the early part of 1954, I found myself attending primary weather school at NAS Lakehurst, New Jersey – the very site where the airship Hindenberg exploded in on 6 May 1937.

I quickly learned to distinguish my cloud formations and learned codes, codes, codes and more codes. Upon graduation I was dispatched to NAAS (Naval Auxiliary Air Station) Sanford, Florida, for my first "tour of duty".

They don't put the word "boring" in the job description, but I quickly became bored there and so volunteered for something that was a world away from the warmth and placidness of Florida: Operation DEEP FREEZE.

Usually the military has *Operation This* or *That* and sometimes those code names bear little relevance to the actual mission. But Operation DEEP FREEZE pretty much was what it said on the tin! We were heading to Antarctica on the first of what would become a number of different US and joint international exploration missions.

I soon found myself on the staff of Rear Admiral George Dufek, who commanded that task force, and already-renowned Antarctic expert Rear Admiral Richard E. Byrd. We were there as part of an international mission to investigate what was then one of the most unexplored regions on earth and report back on factors like the ice and the weather. The days were endlessly long but the work was most interesting.

It was while we were building Little America V base on the Ross Ice Shelf that I volunteered for London for the next tour of duty. My shipmates laughed at me and said;, "No one ever gets London." My answer: "If one doesn't ask, they'll not get it."

I asked and I received!

And I went in style. Even though I was only a Petty Officer First Class at that point I was shipped over on the *SS United States* and travelled First Class, if you will, in Cabin U41. What a shock I had when those tickets arrived. It was a great experience.

My first tour was an AG1 = Aerographer's Mate First Class (also known as Petty Officer First Class) to the Fleet Weather Facility, London co-located with and serving Commander-in-Chief, U. S. Naval Forces, Northeast Atlantic & Mediterranean (CINC-NELM) from 1957 – 1959.

I shipped out to England with, my wife, Francina S. (Reyes) Gibbs, 'Cina' for short, whom I had gone to school with in Naantucket. Cina and her parents had come home from Japanese occupied Philippines in 1945 to recuperate at her grandmother's home in Nantucket. They never left. Cina's father, Jose Formoso Reyes, a Philippine national had served at the Philippine Military Academy after receiving his Masters in Education from Harvard. He had married an American and they had returned to the P.I., where he became a Captain in the U.S. Army under General MacArthur who had created the Philippine Military Academy fashioned after West Point.

My father-in-law was head of the Department of Languages and Social Sciences until the war broke out. During the war, as the Japanese failed to learn he was U.S. Army, Cina her mother and brothers were not interned and Jose was in the underground after the loss of Bataan. At war's end they were liberated by U.S. Forces and were already en route Massachusetts when the war ended.

So Cina and I travelled over in style with our recently born daughter, Jacquelyn. I reported to the US Navy Headquarters, London, to Fleet Weather Facility and went to work under Commander William J. Kotsch, who had been the chief forecaster at Bikini for *Operation Crossroads* – the nuclear weapons tests in the Pacific. Kotsch was a great skipper, brilliant linguist and author.

My own duties were focused on being an enlisted forecaster, involved in collecting, analyzing and forecasting weather for U.S. Navy units serving in the European theater and in support of the Admiral's staff for their planning and execution of naval operations.

For both tours FleWeaFac London, as it was known, was located on the north end of the 5th Floor and we always entered through the 7 North Audley Street entrance. Only the Admiral and his senior staff officers were allowed to enter through the formal building entrance off Grovesnor Square.

This was 1957, of course, and Britain was still recovering from the war with rationing of coal, petrol and, if I remember correctly, eggs and butter, too. However we managed. We were fortunate enough to be able to use the military Base Exchange at the USAF headquarters at South Ruislip where we could purchase petrol, groceries or use the hospital for medical care. Our second child, Mark Edward Gibbs, was born in that Base Hospital on 1 May 1959.

My recollection of London at the time of my first tour was of many bombed out buildings yet to be rebuilt, but the city and the country was being cleaned up. I couldn't say the same for my visit to France a year later.

Away from work our first residence was 10 Alington Grove, Wallington, Surrey, just west of what was then Croydon Airport. It was being phased out but still had flying clubs and a few commercial flights. Our house came with housekeeper and her husband, the gardener. I fondly remember them: Mr. & Mrs. Leslie Garrett of Waddon, Surrey.

Mr. Garrett left me with one of the most memorable stories I have ever heard first-hand. He had been a Sergeant in the British Army unit that broke the gates down at Bergen-Belsen concentration camp.

I can to this day almost paint a picture of him relating the story to me as we stood together in a cold drizzle in the garden.

He related how the first sight they saw were approximately 9,000 unburied corpses and thousands of prisoners near death. It was later discovered to have been the place Anne Frank and her sister had died only weeks earlier. He related how many of the guards were Hungarian Nazis and were totally remorseless about the scene and balked at the orders to help bury the bodies.

Les told me that they took them around the building and shot them straight away. Something I've never forgotten and something that moved me into reading more and more. Incidentally, only recently I 'Googled' "Bergen-Belsen" to read the story over 53 years later. It was just as he described it on that rainy night.

Les Garrett was clearly scarred by this experience and I'll never forget him. At the time he was in his last years working for the power plant in Waddon. His young daughter, Wendy, became our babysitter and I often wonder where she is and how she fared in life. We lost touch after our second tour and very much regret that.

After a year of commuting (mostly by VW Beatle, as the trains didn't run early enough to allow me to reach Grosvenor Square by my 0600 hours reporting deadline), we moved to 54 Bucknalls Drive, Bricket Wood, Hertfordshire. Here I still had problems but could occasionally make it to work in time from the tube stop in Watford. But I still did many miles by my Beetle owing to the awful hours. Our neighbors were lovely and we remain in touch with some of them.

The travel through the thickest fog I've ever experienced driving home to Surrey one night in November 1957. That was the night a train ran through a signal and rear-ended another train under a bridge at Lewisham near London. The train knocked out the underpinnings of the overpass bridge and it fell on the wreckage below. It was a most tragic event with extremely high loss of life.

In my case I was crawling through Streatham Common in my little, black VW bug that evening. With it being left-hand drive, at one period I simply had idled along the curb so as to not lose my way. I also used a flashlight periodically to confirm I was still following the curb. It was like something out of an old movie where one could hear a horse clopping by, yet could not see it.

My first London assignment lasted just two years until 1959. I could have stayed another year, but as I couldn't achieve a promotion to Chief Petty Officer until I had graduated from Advance Forecaster's School, I opted to leave and head there.

The school was back in Lakehurst, New Jersey, and I began a series of tours that lead to a Commission.

Upon graduation, I had been retained as an instructor and very shortly after was promoted to Chief Petty Officer. A second son, Alan Benjamin Gibbs, was born at Fort Dix Army Hospital in New Jersey (just up the highway from NAS Lakehurst) on 16 Feb.

1961, whilst I was an instructor at Naval Air Technical Training Unit at Naval Air Station Lakehurst subsequent to that first London tour. Sadly we lost Alan at age 23 in 1985.

Then came a tour that was completely 'off the wall.' I was assigned to the State Department's Foreign Service Institute and then moved on to Pensacola, Florida, where I taught Geopolitics and World Marxist Strategy. I was a Meteorologist and totally out of my primary field.

Many warned me that this could only hurt my promotion chances. It didn't. I completed my tour at the CPO Leadership Academy with a commission and then went to the Philippines for a tour forecasting of both typhoons and for patrol flights into and around Vietnam. That tour was at Fleet Weather Facility, Sangley Point. It was memorable in another way because Cina, (who had survived the Japanese occupation of the Philippines), the children and I were now able to visit her many relatives throughout Luzon and tour like most Americans never could or would.

After that tour it was back to Antarctica as Lieutenant in charge of all meteorological and oceanographic services at the five U.S. bases. I wintered at McMurdo Station near the Kiwi's (New Zealander's) Scott Base. Following winter-over it was back to the Admiral's staff in Christchurch and, additionally, in charge of the Met Office co-located at Harewood Int'l Airport. All south-bound flights, regardless of nationality, were my responsibility. All flights to other destinations were served by the New Zealand civil Met personnel.

Our children – Jackie, Mark and Alan – were now in local schools. We thought we were in paradise. NZ is truly a veritable "little England."

At this juncture I was asked where I would like to go again and I put in for Britain. My shipmates again reminded me; "No one ever gets two tours in Britain." To which, I responded; "No one will unless they ask for it." I did and I did.

I received orders in the summer of 1968 to return to Grosvenor Square and the same unit at the building on Audley Street.

Except that I couldn't.

My second tour began in 1968 at USAF European Weather Centre at High Wycombe. I was a Lieutenant at this time, having been promoted to that on my earlier 3rd and 4th expeditions to Antarctica. I arrived back in London ready to serve only to learn that my clearance documents hadn't been finalized by Washington bureaucracy. Until clearance came I had to wait to serve in my London post. Remember the Cold War was at its height and working in the confines of the Naval Headquarters required exposure to very sensitive stuff.

At the USAF European Weather Central located at RAF High Wycombe I was able to serve two purposes: Be the US Navy contact for my London unit and, also, the entire Naval Weather Service Command as an on-scene liaison with the Air Weather Service (an arm of the USAF).

My duties were to provide forecast services to all USAF operations in Europe, Mediterranean and Africa. As an exchange officer (the only 'Navy Swab' in the unit) I was principally responsible for providing forecast services to all military operations in the Southern Hemisphere as well as northern Africa.

The first year we lived at 84 Shaftsbury Avenue, Kenton, Harrow, as our Czech landlord would be out of country for one year. This now made my commute perfect. I would ride from the Kenton Station to Baker Street Station without many of the timekeeping problems that I had on the first tour.

Once I returned to FleWeaFac London I became the Weather Services Officer and Weather Communications Officer. In the first, I was responsible for the forecast support to all ships operating in the NE Atlantic/Norwegian Sea. I used to routinely make ship visits/briefings to aircraft carrier groups in such ports as Plymouth, Portsmouth and Glasgow before they embarked on various US and/or NATO exercises.

In the second, as Weather Comm Officer, I was responsible for the circuitry and equipment through Europe, the Mediterranean and even as far as Asmara, Ethiopia. As we now had weather satellite readout throughout Europe, this unit in Ethiopia could relay polar orbiting weather satellite photographs in real time to us that covered the Middle East and western half of the northern Indian Ocean, an area that the Naval Headquarters vitally needed.

By this time the overall Commander's (the four-star Admiral's) title had changed from the above CINCNELM to CINCUSNAVEUR, meaning 'Commander-in-Chief, US Naval Forces, Europe'. Interestingly, the four-star (and there were two in succession on each tour) lived at Romany House which had been used by the US military since WW II. This lovely residence came with the job.

We additionally worked closely with the Royal Navy's Weather Centre at Northwood on this second tour. The station was known as HMS WARRIOR. There, when visiting, we were well underground. A profound memory of my time at Northwood was the "dining in" marking the anniversary of the great World War I sea battle, the "Battle of Jutland." Present as the guest of honor of the First Sea Lord that evening was retired Royal Navy Captain who had commanded the battle cruiser HMS WARRIOR for whom the station was named. He was in his mid-90's and ram-rod straight.

One should also note that all weather services of the world were linked and exchanged weather. Yes, both sides of the Iron Curtain cooperated in this. The matrices of circuitry, then and now, is little known by those outside the weather community. Even in the 1960s we had the equivalent of the internet.

At the flip of a switch in our London Weather Office I could talk on a dedicated circuit to our main computer site in Monterey, California. Thus, years later when we heard that Vice President Al Gore had 'invented the internet,' this became an inside joke. Back then I could send off a message; then called a 'wire note.' Today we call it an email. My how things evolve, but truly have stayed the same!

As I frequently hiked from my 5th Floor Office to the Command Center on the second floor, I can remember one thing. Unreliable, old elevators that required one to use the stairs (we in the Navy call them "ladders") if one was headed from 5th Floor to brief the Admiral or some other senior important senior in the command center area.

A short story about that fact indicates the problem:

When I checked in during late 1968, we had a logbook referred to as a 'PDL.' It stands for Pass Down the Line. The little things that wouldn't be incorporated in official regulations that became necessary info for all to either understand or execute.

In this case, as I checked in the entry, it simply said: "Use the ladder when going to brief the Admiral." When I questioned that, the answer certainly made sense and was the outcome of a most embarrassing incident.

It seems President Lyndon Johnson was concerned during one of the crises in Cyprus that the Turkish forces might actually land on the island. Of course where might just be weather-related, i.e. "Where is the surf up? Where not?"

In the wee hours (London time) the President called Admiral McCain (yes, Senator

McCain's father) who was the Commander-in-Chief. He turned to his aide and told him to "get the weather man down here." So a call within the building went to my mate, LT Karl Haake, who grabbed his charts and jumped in the elevator. You guessed it: At 2:30 a.m. the lift jammed halfway between floors. Naturally he pressed the alarm and was released sometime later by the Marine Guard force.

When he arrived at the Admiral's apartment next to the command center, the Admiral met him with a very "frosty" query. "Where @#!#@$% have you been?" We all had a later chuckle over this, but at that moment it was not a laughing matter.

Anyway, that one is a story that really did have a happy ending, albeit I've always thanked my lucky stars that it wasn't me. Actually when it occurred I was in my last week at the USAF European Weather Central, RAF High Wycombe. Owing to a glitch in my security clearance I, too, spent the better part of the first year at USAF European Weather Central, High Wycombe, as the exchange Lieutenant. A most pleasant time. After 11 months the 'Top Secret' clearance was again fulfilled and I transferred to Fleet Weather Facility, London and stayed on until the summer of 1971.

I was tapped to be one of the two permanent weather briefers who would each morning support the four-star admiral and his staff in the HQ's briefing theatre.

It was a room with seats rising gradually from front to back. The Admiral, other admirals and flag officers, plus the key staff department heads sat in the front row, the admiral directly centered in front of the podium. On each side of the podium were two large reverse screens – perhaps 12 feet square. Slides were projected from the rear onto the screens.

For most of my 1968-1971 tour, the Commander-in-Chief [CINC] was Admiral Waldemar F. A. Wendt, a tall, very soft-spoken gentleman and one whose courtesy I fondly remember. (I'll mention more on that later.)

Back to the procedures: In preparing our briefing of weather and oceanographic conditions we would have arrived at our Fleet Weather Facility very early. We would be privy to all naval operations, including the most sensitive material that was common in these days of Cold War operations.

Given the focus of the operations i.e.: What were the Soviets doing? What were we and our Allies doing? And what was our analysis of the met/oceano data? We would then photograph our various weather and oceanographic charts and develop a script to make them thoroughly pertinent to what we believed would be the interest of the admiral and his senior staff.

These were not the usual 35mm slides, but colored lantern slides of a special type – about 3 1/2 inches square. And, of course, we were always working under severe timelines to be ready for the briefing hour without fail.

As is the custom in most senior naval staffs the Meteorologist leads off the formal briefing. At CINCUSNAVEUR London this also meant the Met briefer would be given a script and provided appropriate slides to welcome any visiting VIPs. Prior to the arrival of the CINC I would be in my place, having run through the slides to ensure all were suitably focused on the operations that could/would be discussed. But it would start with welcome slides for any guests.

The level of sophistication of this process was remarkable. We would have the person's rank and name superimposed in Old English lettering along the bottom of the slide. The slide would usually be a beautiful country pub scene. However, we checked on the visitor's personal preferences. If it were known that he/she was a teetotaller, then a scene without the pub in the background would be used. If personal preference wasn't known, we'd take no chance and use some other landmark of importance.

What if it was a last minute visitor? Ah, we had that covered too. A slide of Buckingham Palace was inserted and the script would simply be read off by the Met briefer welcoming the guest on behalf of the Admiral (who would, of course, have escorted the personage in to his/her seat of honour anyway).

Room for error? It didn't seem so until one day the system broke down completely with Yours Truly at the podium.

With all officers in their place awaiting the arrival of the CINC (Admiral Wendt) a Lieutenant called through the curtain behind me to say the Admiral had a surprise visitor and there was no script yet prepared. Could I 'wing it'? Why, yes, of course. "Who is it?" The Lieutenant told me it was "Dr. Robert Walske, Member of the Atomic Energy Commission." I scribbled that on the side of my Met briefing script just as the Admiral and his guest walked in with all staff in their usual positions – rigid attention.

As the Admiral and his guest were seated I began what had been a very usual briefing. Calling up the emergency, last-minute slide, as the lights lowered I began what I had done hundreds of times; "Good Morning Admiral, Ladies & Gentlemen, we are honored to have with us today, Dr. Robert Walske, Member of the Atomic Energy Commission."

At that moment, Admiral Wendt sat bolt upright and corrected me in an unusually loud voice for him; "He is not." "Dr. Walske is Assistant Secretary of Defense for Nuclear Energy." Whew! At least I had his name correct! But I had committed a grievous error nonetheless.

Somewhat shaken, I continued with the briefing and all that part went well. However, upon arriving back at my 5th Floor office, the Deputy CINC, Rear Admiral Swanson, was already on the phone to my commanding officer, Captain Kotsch, making it clear that I had made a serious faux pas. And why?

The upshot of the investigation actually tracked back to Admiral Wendt's personal secretary who had called the Lieutenant in the room behind me. She had initiated the error and it was then passed to me by the unwitting 'LT' supporting the Graphics Department on the other side of the screens. The solution was that in the future, should I not have script already in hand, I would defer to the Admiral to introduce his guest.

It brings many chuckles now but in tension-filled times it was not funny.

Happy times in England. Cina was very active with weaving and spinning groups and we would be at many of the county fairs. I was an active member of the Weimaraner Club of Great Britain, doing both field trials and bench shows. We truly savored the country life and still keep in touch with people.

While in NZ in that tour after wintering over I became an active member of the RNZAF Gliding Club at Wigram, in Christchurch. When arriving in High Wycombe at the beginning of that second tour I again took up sail-planing and spent many a weekend flying from Booker Airfield over High Wycombe and the countryside. Alas, when I returned to the United States, the price was too high to continue and I surrendered to tamer things, sailing my gaff-rigged ketch, when it isn't "on the hard" in the back yard.

There is one other found memory worth mentioning when the CINC was Admiral Wendt, the perfect gentleman.

Early on in my tenure back in London from the High Wycombe stint I was directed by my skipper, Captain Kotsch, to prepare a point paper for Admiral Wendt concerning the feasibility of putting a weather satellite readout site at our Naval Communications

Station in Asmara, Ethiopia. As the CINC had responsibility for the western half of the Indian Ocean from the tip of India westward plus all of the Middle East, we lacked that 'footprint' with our polar orbiting weather satellites to provide him this coverage for daily briefings and/or emergency situations. As both the day and night passes over that area could be "seen" from Asmara, it was a technical and cost question as to whether this would solve the problem of our 'blind spot.'

Having carefully researched the issue, I determined that it would be feasible and that we had landline circuitry from Ethiopia through our switchboard in Moron, Spain, and into Q-Whitehall, London, to accomplish this photographic data transfer. All we would have to do is position the appropriate portable weather satellite readout van with a few trained personnel at an already-existing NavCommSta (Naval Communications Station) in Asmara.

I carefully addressed the issue with a series of appendices to the required single 8 X 10 1/2 cover sheet known as the "point paper." As I was his junior by many ranks, I signed the letter as Navy protocol dictated with the appropriate; 'Very respectfully, Lt. Maurice E. Gibbs, USN, Weather Services Officer.'

Two days later the cover letter was sent back to me with two red pencil notations. One read: 'Approved! Thank you Gibbs, fine report.' The second was a red pencil line crossing out "Very respectfully" and a note under it in the Admiral's hand saying, "Not necessary. All my officers are respectful." This left me both happy with the CINC's approval yet confused about the last red pencil entry.

It wasn't until 20-plus years later that the second entry was finally solved. The answer came when I read Thomas Buell's *Master of Seapower; the Biography of Fleet Admiral Ernest J. King*. During WWII King headed the US Navy and was a taskmaster. He absolutely abhorred wasted words. Ah-ha! A young Lieutenant, Waldemar F.A. Wendt was mentioned in this fine book as King's aide. Finally, over 20 years after that friendly correction by this fine gentleman admiral, I understood why.

As for me – the nickname "Mo" stuck for a couple of reasons. "Mo" wasn't just short for Maurice. It also was the initials for "Met Officer".

What began as an intended four-year commitment to the Navy ended up being 34 years in the making for me. I retired as a Commander, USN at the end of 1986 and completed my naval career on the staff of the Oceanographer of the Navy.

Interestingly, I began all this because of my interest and time at the Maria Mitchell Observatory on Nantucket in the late 1940s. Almost 40 years later I ended my career in the U.S. Naval Observatory in Washington.

Postscript: *Three years after retirement (1989) Maurice and Cina separated and divorced. He returned to his birthplace of Nantucket Island and married a childhood sweetheart, Millie, in 1991. They have been active in a number of the island community's nonprofit organizations. Maurice, as an active U.S. Coast Guard Auxilliarist, teaching boating safety and manning the CG station's command post, plus serving as organist of his Masonic lodge and a choir member in their church. Millie is a board member and volunteer of the Nantucket Cottage Hospital Thrift Shop.*

Left picture: Lt. Gibbs was present at when Rear Admiral Richard E. Byrd (Left) and Rear Admiral George Dufek (Right)held the commissioning ceremony for Little America V in December 1955. The first flag raised was the one for the National Geographic Society, shown here. Right picture: The "roster" photo of then Lt. Maurice Gibbs taken just after he had come off the "ice" in 1967 and a year before reporting to FWF London for a second tour of duty.

Maurice "Mo" Gibbs in 1955 on the ice. Petty Officer Gibbs, as he was then, can be seen with the flagship of RADM George Dufek, the Operation Deep Freeze Commander, in the background. "We were walking to the site where we were building Little America V, approximately two miles behind the photographer. We sailors had named the rough-ploughed roadway the "Broken Back Freeway"."

The USS STATEN ISLAND, an icebreaker, tied to the USS WYANDOT during arduous Weddell Sea operations in 1956-57. "Here we are almost over the exact spot where Sir Ernest Shackleton lost his ship 40 years earlier," Gibbs recalls.

In late Sept 1966 just after then Lieutenant Gibbs arrived at McMurdo Station to take up duties in charge of the weather office and other weather ops at the various U.S. Stations, he went up to "Arrival Heights" above the famous cross to Scott's very first casualty in Antarctica on his first expedition (seen by Gibbs' left shoulder) IMO Able Seaman Vincent. Observation Hill is behind. Photos courtesy Maurice Gibbs.

CHAPTER TWENTY-THREE

THE STRAW HAT

Ron Crowe

Dependent/Student
USAF South Ruislip, West Ruislip
Bushy Park & Bushey Hall
1957 – 1964

"Ban the Bomb".

That was the name and the chant of the 1960s group of people who protested against nuclear weapons. Their emblem became the peace sign that we all remember and know.

One weekend in 1963, as I recall, they organized a demonstration at USAF South Ruislip base. Naturally the base commander closed the base gates and put the military on alert. When the demonstrators arrived they chained themselves to the fence and plastered it with a few posters and 'ban the bomb' statements. A few of them tried to climb the fence and get on base but the Military Police quickly changed their minds... with batons!

Since I lived just up Victoria Road which ran in front of the base, several of us went to see the spectacle and stayed across the road while the demonstration was going on. In the end it was fairly tame except for lots of yelling "Ban the Bomb" or "Go Home Yanks".

I found myself in England because my parents had separated and my mother, Terry, was extremely ill with a stomach ulcer. Having been born in England she was able to have the operation done at Cliveden Hospital in Buckinghamshire. Also she told me that if she died my English grandparents, with whom we lived with in Slough, would raise my brother, Tony, and me.

She recovered and several years later, because she was an American citizen, she got a job in the Finance Office at the USAF South Ruislip base.

I had completed all my studies in English school, St. Joseph's in Slough and started attending London Central High School in September 1962 at Bushy Park near Teddington. I graduated in 1964 by which time LCHS has been relocated to Bushey Hall near Watford.

In the winter of 1962 the ITV television network came to the school to interview students about life in England for American students. The commentator interviewed Sam Hatfield, Senior Class President and a couple of other students. I was in one of the classes where the cameraman walked between the rows of desks and I made it onto TV.

My mother and grandparents thought I was the highlight of the show even though I was on camera all of three seconds. I still have that film footage and several others when television crews came to LCHS at Bushy Park and Bushey Hall.

One of the funniest episodes I remember from LCHS occurred in our teacher Jack Wernette's English class in 1964. Mike Shook, our superstar American football player, was sitting in the back of the class with his chair leaning against the fire exit door that had a crash bar across the middle, which opened the door. Mike either leaned too hard or fell asleep and the door crashed open. We all burst out laughing and Mr. Wernette simply smiled and said, "Mike, would you like to join us?"

Besides spending a lot of my time in the Snack Bar, Bowling Alley (4 lanes if I remember correctly) and the Airman's Club, the South Ruislip base was the place many high school students met before heading up to London to do whatever.

During the summer months a lot of us worked either at South Ruislip or the nearby base at West Ruislip. I never had the chance to work at South Ruislip but spent three summers at West Ruislip; two years in the loading area for soft drinks and for beer going to other bases.

We usually had two guys working each pallet stacking 10 cases per row and 10 rows high and we probably averaged 40 pallets per day. With 20 or so teenagers working each day, that was a lot of soft drinks and beer!

That was a long time ago so I can reveal this part of the story now. When we stacked the beer on the pallets the way the cases were stacked allowed for a square hole between the cases from the top row to the bottom. It was just wide enough for one of the skinnier guys to slide down the hole and cut into the side of the case and pull out a couple of beers or soft drinks.

Back then you had to open the can with a 'church key', a bottle opener on one end and a pointed can opener at the other end. I'm sure individuals unloading the pallets wondered how in the world did the case get cut open.

In the summer of 1964 I worked in the tire warehouse which was quite a backbreaking job. We had to load tires by size in floor to ceiling racks and when orders came in we had to unload them from the racks and roll them through the warehouse to waiting trucks.

One of the tricks we were taught to get the tires from one end of the warehouse to the other was to lay a heavy truck tire flat on the floor and throw the tire going on the truck at the outer edge of the tire lying down. With the right amount of force and hitting the floor tire just right, you could roll tires a good 100 yards or so. Often at lunch we would have competitions with other workers putting in a dime each into the pot and whoever got the longest distance of the lunch break got all the money. Back then West Ruislip was run by A.P.E.X (American Post Exchange) that later became A.A.F.E.S (American Armed Forces Exchange Services).

During the summer most of us would finish work and meet at South Ruislip Snack Bar before heading home or heading out for other things. Some of us would go over to the Airman's Club and shot pool or play Ping-Pong for money. Back then I was making 75 cents an hour and it would burn holes in my pockets! Anyway, it helped me get my pool game in good enough shape to let me manage a pool hall in Texas when I went off to college.

At the club we'd play the GIs for a dollar a game of eight-ball or a dime a ball on nine-ball with the nine ball worth a dollar. If we got real brave we'd go up to five dollars on eight ball and a dollar per ball in nine-ball with the nine worth $5. In the course of several hours a lot of money could be won or lost. Ping-Pong wasn't my game but one of my high school compatriots, Larry Patterson, learned at that club and when he later joined the military he went on to become USAF champion.

On Saturdays we had basketball leagues at the base. To my mind these were on-court fiascoes. None of us were good enough to make anything above Junior Varsity level basketball. But we were in awe of some of the GIs. I remember one black airman who had long fingernails. He'd pop shots from the baseline and you could hear his nails scrape the ball as it left his hands.

In 1963 the US Olympic basketball team was touring Air Force bases in Europe and they came to South Ruislip to play an exhibition game against the base team. I don't remember the score but the South Ruislip team hammered them!

And during that same year one of the most memorable and tragic moments for me occurred on Friday, November 22, the day John F. Kennedy was assassinated. The first announcement came on the BBC that he was shot in Arizona, I believe, and corrected a few minutes later that he was shot in Dallas and was dead.

The commentator said in respect the BBC would close its broadcast for the remainder of the night. My mother received a call from South Ruislip and was told to report to the base since all personnel were placed on high alert. She left and headed to the base.

That year I was President of the Teen Club at West Ruislip and I knew several of my classmates would be heading to the base and not be aware of what had happened or wanted to be close to friends. When I got to the base the Air Policeman on the guard would not let me in along with about a dozen other kids who were there. The American manager of the Teen Club, Mac (I don't remember his last name), told the 'AP' to let us in and any other kids that showed up, as long as they showed their ID card.

We stayed in the Club for about 30 minutes until someone suggested we catch the tube and go to the American Embassy in Grosvenor Square. About 50 of us went to the Embassy that evening and a huge throng of people had already arrived and were in line to sign a letter of condolence that would eventually be sent to Washington D.C.

English people were extremely sympathetic; hundreds were crying and acknowledged us as Americans. We never got a chance to sign the letter since the line almost completely filled Grosvenor Square.

A few days later a funeral ceremony was held at West Ruislip. I was with a couple of my classmates including Ted Puryear whose father, Major General Romulus Puryear, Commander of Third Air Force in Europe and base commander at South Ruislip, stood rigid with tears streaming down his face. I don't think there was a dry eye in the crowd.

We lived in amazing times and there were some other amazing moments. One summer Sammy Davis Jr. came to perform in the base gym. However three of us – Roy Hopkins, my brother Tony and I, had a private audience with him. It just so happened that as we were on the way to the gym for the concert a white Rolls Royce came up heading the road for the gym.

We saw it and moved a barricade out of the way so the Rolls could get closer. Inside were Sammy Davis Jr. and Shecky Green, a comedian who toured with Sammy. When we let the car through it went right up to the entrance of the PX. Sammy and Shecky got out and went inside with some military PR person.

We hung outside until they came out and Sammy thanked us for moving the barricade and started walking towards the gym. He was talking to Shecky and saying he'd sure like something to eat. That's when I piped up and said I could get him a burger from the Snack Bar. He started laughing and said on hearing my accent he felt like he was back in New York – which was strange as I had never thought of it as being a New York accent.

We started following him inside but were stopped by the Military Police who were keeping guard on the Gym and keeping it closed before the performance. Sammy turned to them and said: "They're my special guests – make sure they get in and get great seats!" They obeyed and we were treated to a great performance. Sammy was known as a great singer but he was also an incredible gun twirler. He could juggle a pair of six guns and was a quick draw artist.

After the show we hung around hoping to see Sammy leave because we were standing right by the door to the changing room. I looked back over my shoulder and saw a straw hat sitting on a stool and I went back to get it. Sammy had been using it during the show and I wanted to keep it as a memento.

I'm not sure what it was, but I had a change of heart when I saw him coming towards the door. I yelled to him and raised the hat so he could see it. He came over with a towel draped around his shoulders and looked like he was ready to cry.

He told me the hat belonged to his father when they performed in Vaudeville. After his father died Sammy said it became his good luck charm. He gave me a hug, took the hat and kissed it as he walked out the door to the Rolls.

To this day I still get a little choked up when I see a straw hat.

CHAPTER TWENTY-FOUR
EMPIRE TEST PILOT

Jack Blackwood

Captain, US Navy
Empire Test Pilots' School
Farnborough, England
(January – December 4, 1958

The year was 1958 and I was then a Lieutenant in a US Navy Squadron at Cecil Field in Jacksonville, Florida, when I received a set of orders to the Empire Test Pilots' School in Farnborough, England.

My class number was to be No. 17.

The school had been set up by the British in 1943 to help try and reduce the number of fatalities during new aircraft testing. The first graduating class was in 1944. Its motto was *Learn to Test. Test to Learn.*

I was following in the footsteps of other Americans who had been part of the course since J.R. Muehlberg USAAF and H Snyder USAAF became the first American pilots to join Course No 2 from 1944 – 1945. Synder was later killed in a flying accident in 1947.

I flew to England and my wife, Mary Lou, followed by ship. While in transit Mary Lou met a couple from England. He was the Commanding Officer of a base in northern England that, I believe, was a logistic base. So she had some good 'Intel' on what base life was like at British bases.

I reported directly to the Naval Support Activity in London that was co-located with the U.S.Embassy. At Farborough I lived in the Officers Quarters until Mary Lou arrived. Man did I like living there. I had a 'Batman' (in the British military an airman was assigned to every officer as a personal servant) who shined my shoes and dusted my uniform every day!

When Mary Lou arrived we stayed for a short while at the American Forces hotel, the Columbia Club, in London for a short while. It was close to Hyde Park. Not long after we rented a duplex townhouse in Camberley, just off the A30 road. We had fireplaces in every room and boy did we use them! In addition I had a paraffin lamp that I used in the study. Our laundry facilities consisted of a small washer and a clothesline in the kitchen where we dried the laundry.

Overall we enjoyed the stay in Camberley. Mary Lou did all the shopping at the local stores and since she didn't understand the currency she held out her hand and they took what they needed. They were quite trustworthy and she had a ball with all the transactions. At the time Mary Lou was pregnant with our son, Duke, and the local folks treated her very well.

The flying at Farnborough was great. We flew the following aircraft during the course; Varsity, Seahawk, Meteor Mk7 and NF-11, Devon, MK 1 Hunter, Chipmunk, Dragonfly, Canberra, Gannet, Provost, Vampire and the Vulcan. I really enjoyed all of the aircraft but the best was the Hawker Hunter. It was a fighter and I am really a fighter pilot. Most all of our flying was done in southwest England over land.

All members of our class were military pilots. There were two US Navy lieutenants, myself and another fighter pilot who was once a Blue Angel. The Third Yank was an Air Force Captain and multi-engine pilot. The other pilots were from: India (2), Australia, Italy, New Zealand, Canada, Royal Navy(2) and the others were from the Royal Air Force.

We attended class and flew for about three months and then had 10 days vacation during which we toured Europe including Brussels, Germany, Austria and France including Paris. It was quite a trip for my lady since she was quite pregnant with Duke.

On the trip across the channel in a ferry we had a funny experience. We were having lunch and found the folks around us staring at us while we were eating! Turns out they were amazed that we would change hands to eat with the fork in the right hand!

One other surprise was the night we took a bath. Next morning the tour guide promptly asked; "Okay who took a bath last night?" We did not realize that you had to pay for that separately! We really got a laugh out of that one.

I particularly recall two flying incidents during the course that stood out for me. First, since I was a single engine jet pilot my instructor took me up for a familiarization (FAM) flight in the Varsity which was a twin-engine bomber and fairly big trainer. I checked out okay so I landed and dropped the instructor off. The engineer and I proceeded to the operating area.

We were out about 45 minutes and the engineer pointed out that the oil pressure in the left engine was dropping and, in fact, we had to shut it down. We advised the base and they promptly gave a radar vector home. All things considered the flight to base went smoothly. I told the engineer he would run the engine while I steered. The landing was good – just a little yaw – and the watching crowd (they always watched when there was a problem) cheered as this fighter jock was successful in his landing!

In the second incident a friend of mine and I flew a Canberra, a twin-engine high altitude bomber, over the English Channel on one of the course exercises. He was the pilot and I was co-pilot in the right seat working the throttles and taking data. We were up at about 30,000 ft. doing 'Yo-Yos' to hit certain altitudes and air speeds. As we approached the top of the third 'Yo-Yo' I moved the throttles forward and the engine temperatures went to maximum.

Being a jet pilot this was not unusual. Usually bringing back the throttles would drop the temperature and you would continue the mission. Not so this time. Both engines froze. We were just too slow and did not get good airflow to the engines. We let the base know and again they gave us vectors to the nearest base.

We landed safely without having seen the field until we were about 2,000 feet downwind. Visibility was great looking straight down but horizontally the cloud build-ups blocked the view and we had to fly around them. Classmates gave me some good ribbing on that flight and jokingly threatened to charge me for the engines.

One thing that I must mention here is the quality of the air controllers. Those folks knew where we were all the time. Since we were busy with test work and above a cloud layer most of the time we didn't know exactly where we were. They did and they kept us in the operating area and helped in emergencies like the one mentioned above. Anytime we wanted a radar vector home all we had to do was click our radio 'mic' and give our call sign i.e. "Base this is Tester 28, request vector home." When we had the engine emergency they directed us straight to the downwind runway. Truly a very professional group indeed!

I loved the one flight I had in the Vulcan. It was indeed like a big fighter. The flight I had was 4.8 hours off the coast of Scotland where we did some bomb trials. That was a real challenge for a fighter jock and a lot of fun. To say the least I was really impressed by the capability of the aircraft. I just wish I could have flown it more.

Several times in the year we had a Dining In Night. This was a formal dinner dressed in our dinner dress uniforms. We started the dinner in the Ante Room where we had our sherry. When dinner was ready the Mess Treasurer came to the Ante Room to announce to the President of the Mess: "Sir, Dinner is served" at which time all would proceed to the dinner table. The President sat at the head table and the Vice President sat at the far end of one of the tables that were connected to the head table in an E-shaped arrangement. There were five courses served.

In the most formal sense all were served but no one ate until the President started. When dinner was completed the tables were cleared and wine and fruit were served. The President started a series of toasts starting with: "Mr. Vice – To The Queen", then Mr. Vice says; "Gentlemen – To the Queen" and we all repeat: "The Queen" and then we drink. Then Mr. President says: "Mr Vice, To the Heads of All Allied and Friendly Nations." The same procedure is followed and other toasts as appropriate.

After a short time the President retires to the bar and the Vice President moves to the President's seat. This signified the end of the formal dinner and the raucous part of the evening starts. Telling jokes and throwing fruit were common happenings. After a short while the Vice announces that all would retire to the bar. People of all ranks then moved onto the floor in a tug of war, dart throwing, trying to drink a 'yard of ale!' among other activities.

We didn't really get involved with the US military community in the UK but eventually we had to go to the 3rd Air Force HQ at South Ruislip in Middlesex. Mary Lou was about to give birth to Duke. Actually she was at Ruislip a couple days before she gave birth so that worked out okay. We drove up there in a 1958 Morris Minor that I had purchased before Mary Lou had arrived. The only problem was that Duke was a stubborn mule and tried to come out feet first kicking and screaming! Things got turned around and she came home with him a couple of days later.

There was plenty of flying and many classes with many guest lectures on the course. I logged about 180 flying hours while I was there. Looking back it doesn't seem like many flying hours but there was sure a lot of paperwork and analysis that went with it. The year went quickly and December 1958 the class was over. We had a graduation ceremony and were presented with a certificate of completion.

On the trip home we sailed on the SS United States. A bit of rough weather and we were three days late arriving in New York. We were aboard for New Year's Eve and that was a thrill. We were doing fine on the dance floor until a huge wave hit. All of a sudden we found ourselves on the opposite side of the dance floor! A lot of rockin' and rollin'. Quite fun actually.

I retired from the US Navy in September 1975 with the rank of Captain after having served 23 years in the US Navy. During my career I commanded two squadrons. I was the skipper of the EA6B squadron which was the first Electronic Warfare Squadron built from the ground up for Electronic Warfare. I was pleased to be on a staff in the Tonkin Gulf of Vietnam when the first EA6B squadron that I had trained went into combat.

My last duty was as Commanding Officer of the Los Angeles Recruiting District. It was just at the start of the all-volunteer service and quite a challenge and really a lot of fun.

The Empire Test School course had given me a real sense of responsibility as to how to fly airplanes. In one sense an aircraft is an aircraft is an aircraft. Some are two engines, some are jet engines, some are 'props' and some are single engines.

But the main point is that Empire Test Pilot school gave us the experience that meant we could fly any one of them with confidence. And we learned from each other. And these were lessons about flying and capabilities that we could take back and share with our own respective forces.

We really enjoyed the time in England. Our son was born at South Ruislip and so England and that assignment will always be in our hearts.

Author's Note: Jack Blackwood passed away in April 2011. His wife Mary Lou passed away in January 2011.

Jack Blackwood and his fellow graduates from the 1958 Empire Test Pilot School held in England. Then Lt. Blackwood is pictured 4th from right in the second row from the front. With him in the class was one other US Navy pilot and one USAF pilot.

CHAPTER TWENTY-FIVE

BIRTH OF A CAREER

Dr. Lionel Balfour-Lynn

Consultant Paediatrician
USAF South Ruislip
1958 – 1970

I came to the USAF Hospital at the South Ruislip base by way of a chance job application. At the time I had been a Pediatric (in Britain it is spelled 'Paediatric' and the practitioner is a "Paediatrician") Registrar at Hillingdon Hospital in Middlesex – a few miles from the base. However, as was the requirement, I had to apply for a new posting. The consultant at Hillingdon had heard about this job and suggested that I apply for it. My application was accepted and they took me on.

I had served in the Royal Air Force from 1954-1956 so knew a little bit about military life. However, I also knew that I wanted to build up my own private practice and so I found someone to job share at South Ruislip with me. Dr. Dennis Cotton was a Pediatric Consultant at Great Ormond Street Hospital.

And that's what we did. I took five sessions of a half-day each allowing me to spend the rest of the day developing a private pediatric practice in London and Dennis took four sessions.

The hospital was fairly modern having opened a few years earlier. But it was much like any hospital. Pediatric units in hospitals are, thankfully, mostly about life. The added bonus for me was that I was working with a great team but the people were always changing.

The US Military gave me a base Identification Card and every day I'd do a half day four afternoons at the USAF Hospital with one morning clinic per week along with an on-call shift one or two times a week.

In many ways it was like any other hospital – pretty standard. Deliveries ran about 1,500 babies a year and we had a busy Pediatric Unit because they weren't all military babies. At that time any Americans could use the hospital. And that's how I also became a Pediatrician to the US Embassy.

There was one way that stands out in my mind in which the American system was different to the UK one: In the UK as a pediatrician you only saw the babies who were sick when they were referred by the General Practitioner (GP). In the US system as a Pediatrician you followed the babies up on a 'well-baby' basis and looked after all babies and children from the moment the obstetrician handed them over. Thus we checked them regularly in the Well-Baby Clinics where they received all their injections and we treated all their injections up to the age of 15.

There was one other big difference as I recall and it had to do with Christmas. In a British hospital at that time Christmas was a big deal. Then there would be a party on the wards and they would have a turkey wheeled in and a doctor would do the honours of cutting it up and serving it up on plates that were then carried around by nurses to the patients. There always were some celebratory drinks. Not, however, at USAF South Ruislip. There was none of that and there was no drinking allowed. It made for a quiet Christmas as I recall.

One night I had just gotten home from a shift when I got a call from the base. There was a delivery that had complications and could I return to the hospital to assist? So I headed right back out and drove from my home in North London through Colindale towards South Ruislip at, shall we say, *a pace*. A policeman on a moped spotted me but couldn't keep up. However he radioed his police chums in a car and they did pull me up. I explained the situation and they immediately sent me racing on my way.

However, as I was closing in on South Ruislip base, the same police patrol pulled me over again. The officer strode up and I made ready with my 'doctor on delivery call' speech. But he cut me off.

"Go home," he told me. "There's no emergency anymore!" It turned out that the hospital had called my wife, June, shortly after I left and told her there was no longer an emergency. My wife had sensibly called the police and explained to them that I might be 'haring' to the hospital and, if they spotted me, could they stop me and send me home? And so I headed back...but with much less foot on the accelerator after promising to keep to the speed limit.

Some other funny things also happened. For instance, it was about this time that the links between smoking and cancer were first coming to the fore and we all had a lecture about this from the President of the US Cancer Association. Well most of us smoked but we refrained during this presentation. However, one of our number, our radiologist at the time, was a chain smoker. He took the speech to heart and actually gave up smoking.

Four days later all the other doctors were begging him to start smoking again. The reason? His hands were shaking so badly when he held the x-rays that it made them near impossible to read!

Another funny thing – though it was disappointing for me at the time – was my near encounter with Marlene Dietrich. The acclaimed actress was in the UK and she had contacted the hospital about her grandchild who was ill. At the time the USAF had a policy of not allowing their military doctors to go out on house calls. And so they asked me if I would go. I naturally said "yes". But it was not to be. When they called her to tell her they were sending a consultant she asked if I was British and when she heard that I was she said she only wanted an American doctor. I don't know if it was the hospital or the base commander but one of them authorized one of the American military doctors to go. I was terribly disappointed.

On another occasion Cassius Clay – as he was (Muhammad Ali as he became) – came to visit the hospital. It was after his fight with Henry Cooper here in London in 1966 at the old Arsenal football ground in North London. Clay had won giving Cooper a cut to his eye. Anyway, the American boxer wandered around the hospital and was very charming and I got to have a conversation with him.

The other great thing about South Ruislip was the chance for us to work with some wonderful doctors who also became life-long friends. Dr. Al Meneely, who came from Harrisburg, Pennsylvania, was one of them. Al had worked his way through college – spending eight hours a day in a Pathology Lab to pay for college, then eight hours in school and eight hours sleeping.

At South Ruislip he started a Pediatric Allergy Clinic and when he rotated on to his next assignment I was the only one prepared to pick it up. I didn't know much about allergies when I started but I ran the Paed. Allergy Clinic for 10 years. It was only then that I went to the Brompton Hospital to study it properly and make it the basis of my life-long research into paediatric asthma.

I remember taking him to Stratford-Upon-Avon to formally introduce him and his wife, Millie, to Shakespeare and that was that. From then on anytime anything new that came on stage at Stratford he was there for it. As he consistently said to me: "I'm never going to get to see this in Harrisburg."

Others included:

Dr. Schwartz who was known for working hard and late. He felt he could make money back in the States by collecting old grandfather clocks and selling them on his return. I can recall that there were at least 12 of them stored under beds and desks by the time he left. I always seemed to be coming across them!

Dr. George Bauernshub was a delightful pediatrician who has recently retired from practice in Ellicott City, Maryland.

Dr. Tom Roe has, I believe, retired from practice in Eugene, Oregon. He went on from South Ruislip to a career that saw him become a President of the American Pediatric Association.

Dr. Pascorosa became friends with the Rolling Stones band members and attended all their gigs. I later looked after all of the Rolling Stones' kids.

Dr. Herman Shyken in St. Louis & Dr. Gus Voyagis in Los Angeles both had very successful Radiology practices.

Of course the doctors and nurses would all come and then go usually on two or three-year assignments. I got to meet a lot of people that way and there was always new thinking and different approaches which was an education in itself.

Many expectant mothers at that time will probably recall Dr. Macnee. He was a small, spry Scotsman in the Obstetrics and Gynecology Department and was generally popular.

And one of the great things about the USAF Hospital was the desire and ability to develop considerable professional links. So all departments in the hospital had developed links with the London teaching hospitals. If we had a very sick child then through Dennis Cotton we had a direct link to Great Ormond Street Hospital. And so it went with other departments.

Many of the doctors I met remain friends. George Bauernshub, Al Meneely, & Herman Shyken have been life long friends and still holiday together.

In 1970 they decided to close the hospital as part of the base restructuring for Third Air Force. I left to go to the Brompton Hospital for four years as a Lecturer and then moved to the Hammersmith Hospital where I eventually became a Paediatric Consultant. as well as developing a large private practice in Harley Street.

However the 12 years I spent working at USAF South Ruislip gave me the chance to build up what would become at the time the busiest pediatric practice in London.

CHAPTER TWENTY-SIX

TOWN PATROL

Don Campbell

A2/C,
USAF Air Police
7500th Air Base Group
1958 – 1961

I arrived in England straight out of Basic Training at Lackland AFB in San Antonio, Texas. And both places were a long way from my home in Ocoee in Central Florida.

Essentially I was an Air Policeman stationed at bases around London – mostly Bushy Park. But sometimes we would get the Town Patrol for London.

The Town Patrol for London came out of Bushy Park. We were dispatched to the Douglas House then in Mayfair and then went out on Town Patrol from there. We worked mainly around the West End. I did get to work that duty and loved it. I learned a lot. At that time we did not have a holding cell; just a small room and desk. If any of the people we picked up needed to go to jail, then the British 'Bobbies' took them.

For most of the bases right around London unless it was a critical area we did not carry guns but on base we would carry .45 caliber pistols. That was because we could not be seen with guns if the Bobbies did not carry them.

When I first started Town Patrol I was still pretty new to the Air Force. I was 17 when I was sent to England and barely 18 when I started working Town Patrol. I needed training so the older guys would look for someone to pick up. The first one person found was a GI wearing khaki trousers. They were civilian trousers and it was against regulation (at that time) to wear them in town. So I got the honor of writing him up and completely ruining his day.

Another time we picked up a GI who had been in fight and was looking a lot worse than he was. He was put in the back seat with me and I asked what had happened. He said: "I was walking along minding my own business when two Air Policemen came along and started beating me. I really feel sorry for them." We rode along in silence for about three or four minutes when he said: "You know what? I don't feel sorry for them, I feel sorry for myself."

Most of the 'customers' we picked up were for being drunk. It was hectic when one of the US Navy fleets was in town. Those Navy boys loved to drink and party. We had picked one sailor and brought him down to Douglas House. When we would bring someone in we always checked everything they had on their person. We checked his wallet and he was completely broke. We asked him about his money and he stated that he had started out with nearly $500.00. We said; "You now have none." He said, with a big smile; "Man I must have had one hell of a time!"

One of my buddies ended up in a major court martial after being caught stealing something from the West Ruislip base. I remember it was a big deal back then.

However that paled in comparison to the big murder case during this period. It took place on June 9, 1958 when an American Master Sgt. (Marcus Marymount) who had been stationed at the USAF base at Sculthorpe had poisoned his wife because he fell in love with a younger British woman.

He was caught and brought to the stockade at Bushy Park. We all got to know him pretty well while he was waiting for his court martial (which took place at the Denham base in December 1958). He was found guilty and sentenced to life at Leavenworth Stockade in Kansas. But he stayed on at Bushy Park while going through the appeals process. He did seem somewhat remorseful but that just didn't come up much. He seemed a really nice guy.

At one point the Air Police did an Honor Detail at Cambridge American Cemetery. But the really interesting detail was for a few days at RAF Northolt during 1960. For about two days several of us Air Police personnel from South Ruislip were sent to RAF Northolt. We were not told why – just it was a guard detail. We were guarding a plane but that doesn't usually call for a special detail.

Of course we were able to figure it out. At night with the help of the plane crew we saw that it contained a rather large boat and a very nice one, too. Then the crew said it was going to Russia as a special gift from President Eisenhower to the Soviet Premier Nikita Khrushchev; a sort of peace offering.

But before it could get on its way, Francis Gary Powers took off from a base in Turkey in his U2 spy plane (it would fly on to Peshawar Air Station in Pakistan and then operate its final mission from there) and, while flying over Russia, was shot down and captured. This caused a very big uproar so the boat went back to Washington. *(Author's Note: The 18-foot-long, 183 horsepower-engine boat been made by Turbocraft in Indianapolis and eventually went to Africa to be used by American Christian missionaries from West Allis (Wisconsin) Presbyterian Church.)*

We knew the U2 plane had been at Incirlik, Turkey, because some of us had been sent there in 1958 for two months during the Lebanon Crisis and saw the plane taking off (secretly) every day. We were never told about what the plane was for. We called it the "Useless 2".

Later, of course, we figured it out.

CHAPTER TWENTY-SEVEN

TRANSPORTED

OLIVER CORK
A3C /A2C

1st Motor Transport Sqdn., (retitled 7500 Transportation Sqdn.)
West Ruislip (May 1959 – December 1960)
7532nd Material Squadron,
December 1960 – May 1962
Denham Studios
December 1960 – May 1962

I was 19 1/2 when I arrived at the USAF's Mildenhall AB, via USAF Transport aircraft from McGuire AFB, NJ. I had joined the USAF in Jan. 1959 and England was my first assignment. I was pretty excited about going to England as that is where my Grandfather was from.

He had been born in Audley in Staffordshire near Stoke-on-Trent in 1875 and his parents and siblings had immigrated to Canada at age eight then, at age 22, came to the USA and lived in Lowell, Massachusetts.

First off we went through Customs which was a breeze. Then we went to the terminal bank teller to exchange our money. That's when things got a little more complicated.

At that time in May 1959 all US Forces in the UK were using MPC Script, MPC was Military Payment Certificates (MPC), similar to Monopoly money. The money was all paper in denominations of 5 cents, 10 cents, 25 cents, 50 cents, $1, $5 and $10. Ten dollars was our largest denomination as military pay was not very high in 1959.

(Note: MPC was used by the military in various locations from 1946 to 1973. Post World War II all US Military personnel overseas were paid in MPC so that currency control could be maintained.)

At the bank teller's window I exchanged half for MPC and half for Sterling. I had no idea what MPC was or what it was for until I went in the terminal snackbar and ordered a hamburger and a coke, then finding the prices were in USD, I was shook up because I didn't realize the MPC I had was the equivalent of the USD used in the USA. The courteous British cashier kindly explained it to me.

In August 1959 all MPC was exchanged for US currency and coins except for the penny. We were not allowed to have US pennies as they were the same size as the old British sixpence ('6d' in old, pre-decimal British money) and could be used in the local vending machines to considerable advantage. All prices at the BX and Commissary were rounded up to the next nickel (5 cents).

I remember it took some of the British cashiers at the stores on base a while to learn our money system as all the dollar denominations are all the same color: G-R-E-E-N!!! What a day that was to remember when we got to hold American 'greenbacks' in our hands for the first time and spend them on base.

Anyway after sorting out the money I took a bus from Mildenhall terminal to the Douglas House in London near Grosvenor Square where two of us (we met on the flight over from the States) got a taxi out to Denham Air Station about 5 miles north of Uxbridge.

Denham was a rather small base but had huge supply warehouses, nice two-man barracks rooms and a couple of open-bay barracks for Motor Pool GIs and one for the Postal guys.

The Dining Hall at Denham served excellent food and the British Dairy that supplied us with fresh milk every day was superb. The small, half-pint bottles were nearly half cream as it was pasteurized but not homogenized. Delicious to drink and you could float the cream on your coffee if you had stirred in a spoonful of sugar first. (That was a trick I learned in high school back in Florida).

Denham also had a nice snack bar, movie theater, a barber shop, BX and a small library which was run by a British lady named Miss Jones.

From there I was assigned to work at 1st Motor Transport Squadron at base at West Ruislip. There was a work bus that took us young troops from Denham to West Ruislip every work day and brought us back each evening.

The Motor Pool Squadron was a close-knit bunch and I enjoyed my assignment at 'W/R' for 18 months. I started working in Unit Supply for a year then was transferred to Maintenance Control/vehicle parts and tools. This involved keeping mechanics' toolboxes up to date and ordering vehicle parts from local vendors like General Motors and Ford Motor Co.

But work was only half the experience for me. We lived for the off-duty time. Those of us in the barracks had a regular routine for the weekends. Saturday morning we would usually go to the BX at the main Headquarters base at South Ruislip – just a couple of miles further on.

In the afternoons during summer we would go to the Ruislip Lido where we would drink beer and meet the lovely girls who were there enjoying the afternoon sun. I was told that the summer of 1959 was one of the best summers in the London area for many years.

Saturday night was spent at the Tithe Farms Pub near South Ruislip, Then Sunday, in the early afternoon, we would head to the hotel and club for enlisted forces; Douglas House in London. We'd have a steak dinner and dance or watch floor shows.

During the time I was there the base at Denham housed the 7500 Air Base Group, 7532nd Supply Sq., the London Area Military Salvage facility for military disposable equipment and personal vehicles beyond repair, etc. At this time it also housed a Finance Office, automotive hobby shop and a service club for young enlisted troops and a NCO club for the Sergeants. The NCO club finally let all ranks join and use the facility.

West Ruislip was home to the 1st Motor Transport Sq, later renamed 7500th Transportation Squadron, a Detachment of the Defense Mapping Agency and an Aerial Photography Flight at the time.

Furthermore West Ruislip was also the main office and warehouse for the Army – Air Force Exchange Service (AAFEX/AAFES). The office had approximately 300 young women working there and we used to lay out on the grass along the main base road and watch the parade of young ladies going to lunch and returning to work. It was a parade to behold.

We used to eat lunch in a café across the street from the main gate until a BX snack bar was built behind the NCO club. Since there was no Air Force mess hall at West Ruislip we drew per-diem for our daily lunches which added to our meagre payday at the end of the month. I would often have a hamburger, fries and a Coke.

Saturday or Sunday afternoons at the Ruislip Lido were always a lot of fun; swimming, girl-watching and drinking beer. Ah!!! Those lovely British lasses in their bikinis were soooo easy on the eyes.

One Saturday night the NCO Club was having open house and we airmen were invited in. The NCOs got drunk and passed out or went home and we took all the iced-down beer and dumped it in the trunk of Hank Plough's '53 Chevy and took off. We drank out of that trunk for a week. Hank was an airman assigned to the maintenance shop.

The following Saturday was a work detail setting up the bleachers for the London Rockets USAF South Ruislip football games. They were a team that played American football and West Ruislip was their home field. We would carry a piece of bleacher framework to the site, stop by Hank's car, have a quick beer and then go get another piece of bleacher. It went like that all morning.

By noon the bleachers were finished and we left. The Commanding Officer had promised free beer after the bleachers were finished, but we didn't want any. He couldn't believe his airmen didn't want any beer! I guess if he ever saw this he'd now know why.

Some of us frequented the Three Tuns Pub in Uxbridge or the Green Parrot Pub in Denham. There was another pub in Denham that was a Watney's Pub. I had two German Beer Steins that I 'liberated' from an Oktoberfest Beer Hall in Munich in 1961 and would take a stein to the Watney's pub with me. It held exactly two British pints.

Lots of times late at night we would get a car full of GIs and drive up to the 'Busy Bee' truck stop north of Watford for sausage, egg and chips and a cup of hot tea with milk. The sausage and chips were great, with some sort of sauce poured over them. The tea was a lot different from what I was used to, but I learned to like the English tea with milk and sugar.

There was also a pizza parlor that opened near the base at South Ruislip called the Silver Dollar and their pizzas were excellent.

During my three years in England I owned two cars: A 1934 Austin which finally ended in the Denham Salvage yard and a '52 Morris Minor which ended in the Brunt-ingthorpe Salvage Yard. The latter car didn't quite make a curve near Leicester and hit a large wooden gatepost right next to the pavement. Too bad too, because I really liked that car. It ran real good and got exceptional fuel economy.

I would have to say that my years in the UK were great. The only thing that would have made it better would be having higher rank and making more money. The Brits gave me the nickname 'Corky' while I was there and it has stuck with me ever since.

I found the British people to be hard working, friendly and honest and I had many British friends while I was stationed there. I was a faithful reader of the British comic strip Andy Capp that appeared in the *Daily Mirror* newspaper, played the horses and did the football pools.

The countryside in England was beautiful; the weather when sunny was nice, but that London Fog was a killer. The base would be closed down for "fog call", then the base club would open for those who lived in the barracks. Rather amazingly to me the people who lived off base had 30 minutes to get their cars off base and then the gates were shut.

When I was in England I dated Jill, a girl from Bushey Heath. She worked for ITV and could get tickets to stage shows and TV live productions. I went to see several performances with her. I saw the stage play *West Side Story* and a performance by the jazz artist Eartha Kitt at London Palladium.

The A40 ran parallel to RAF Northolt flightline base and there were roundabouts at the East end and at the top of the hill well past the west end of the runway at Northolt. My drinking buddy was driving my Morris one night coming back from London. He ploughed through the concrete rail at the east roundabout, across the grass and came out the other side of the roundabout and kept going. The whole left front frame and yoke was bent but a British body shop fixed it perfect – just like new.

The portion of the A40 road beyond the roundabout at the top of the hill that crosses a ravine and, I think, a railroad track was only a two-lane road when I arrived. During my three years there, they completed the four-lane, then closed off the old portion and rebuilt that. The work was not completed before I left.

After I left in May, 1962 I heard that Denham was closed and the 7532nd moved to West Ruislip.

There was a lot going on to my mind while I was stationed in England. But one incident I vividly remember was a Pan Am 707 passenger plane coming from the continent to London's Heathrow Airport on 25th October 1960. The pilot thought he was on final into Heathrow and set the big thing down at Northolt. I don't know how he stopped the plane before crashing into the trees at the west end of the runway but he did.

Word had spread so quickly around the US bases that a 707 had landed at Northolt and we couldn't believe it.

All passengers, baggage, seats, etc. were removed and only minimum fuel to get to Heathrow was in the tanks. The pilot took off and the plane made it but you could hear it at West Ruislip. I saw it going straight up like a rocket ship. The runway was only 5,525' long. The pilot was very fortunate to have landed on such a short runway with no injury to passengers or damage to the aircraft. Most people won't believe that a 707 could land and stop on such a short distance. (See picture earlier in the book.)

(*Author's Note. In passing conversation with Oliver Cork the author heard him mention his old girlfriend, Jill and pondered whatever had happened to her. It was clear he was particularly remorseful about their parting. Following a few weeks of investigation driven out of journalistic curiosity (and carefully done with healthy dose of discreet diplomacy), one Jill Smart and one Oliver Cork were reconnected by the author. Both had travelled their roads with separate lives and loves and not a little heartbreak. So it's particularly pleasing to report that they rekindled their love of half a century previously. And furthermore they kindly invited the author and his wife to their wedding at St Mary's Church, Aspenden, at the very end of the road in the village of Buntingford, Hertfordshire, on 11 September 2010. They now live happily in Florida).*

Oliver and his West Ruislip buddies circa 1960. From Left to Right: Art Turgeon, assigned to the Photo Flight, Dave Cawley, 1st Motor Transport Squadron. Body Shop Maintenance; Oliver Cork, 1st Motor Transport Sq, Unit Supply Section. Next two were British acquaintances, a father and daughter – and left end is Bill Catlin, (who is holding the new baby son of a British host). Caitlin was in 1st Motor Transport Sq, Vehicle Maintenance welding shop.

Oliver Cork & his wife, Jill, on their wedding day in 2010. The pair were reconnected by the author.

CHAPTER TWENTY-EIGHT

DAYS AT DENHAM, SOUTH RUISLIP, & WEST RUISLIP

Rich (Rick) Silver (A2C)

7520th USAF Hospital, Denham
(Dec. 61 to Oct. 62)
West Ruislip
(Oct. 62 to Dec. 64)

It still seems so recent to me. It was December 1961. I was 18 years old.

Bill Hoffard, Jim Bickerton and I had finished two months of training at Medical Supply school at Gunter Air force Base, Montgomery, Alabama, a couple of weeks earlier and we met at McGuire Air Force Base in New Jersey for our assignment to Middlesex, England.

As a kid for some reason I wanted to go to England, so I got my wish. And the great thing – I knew I wouldn't need to learn another language!

We had to stay at McGuire an extra day so we took a bus ride to Manhattan and went to the Peppermint Lounge, a popular twist club on West 45th Street. We hung out there a few hours then took the bus back.

I didn't do much research about Middlesex but I knew London was near. The plane ride from McGuire was pleasant and we arrived at an Air Base called Mildenhall late in the morning. We had something to eat, then got in a van for a long drive to the enlisted military hotel Douglas House in London and then to South Ruislip Air Force Base in Middlesex where the headquarters of the Third Air Force was.

We got there about 4 p.m. and were taken to the administrative office where an airman named Marv signed us in and took us to the barracks.

My first sight of those barracks evoked a reaction of horror. Essentially it comprised a big warehouse-type building with separators that had a total of six people in each section. I don't think it had windows. It was a huge building but I don't know exactly how many people were in there. It had to be at least 100.

So we settled in for the night and told we would be met by someone in Medical Supply in the morning and be driven to a place called "Denham." This was where the supply warehouse for the Hospital and Medical Centre that was located since the base at South Ruislip. While home to the major 7520th Hospital, South Ruislip was not large enough for a warehouse. In end we had very little to do with the 7520th Hospital directly aside from the daily supply of medicine and other essentials.

Denham

Denham turned out to be the location of Denham Studios. It had been converted from a movie studio and was now used for supply buildings, medical maintenance and, I believe, civilian businesses. Movies were not being made there anymore.

Our first day, we were driven there by Airman Shirley Turner in his military truck with right-hand drive. Of course it was a frightening experience. The day before on the drive from Mildenhall on the left side of the road and this drive wasn't any less scary.

The Air Force people I remember include: Gordon Smith, Major Maher, Frank Farnsworth, Jim Stover, Art Garrett, Sgt. Tams, Gordon Badgett and Cal Freistedt. There was also Griffin who was in linen control), Al S. and Sgt. Jackie Hitchens, as well as civilian employees Olivia Salsbury ('Rosebud'), Henry Falshaw, Janet Behmber as well as an interesting character – Mr. Drinkwater. They were people who arranged 'local purchase'.

At Denham we worked in this large warehouse-type building. There was security to get on the base. All the supplies were in shelves and the office was upstairs. At the time there were 12 people working Medical Supply (10 military, including one officer, two NCO's and two civilians). It was a large facility. Lots of times we brought our lunch and they had a cafeteria we could eat at also.

At first I worked in the warehouse. Then, when Frank Farnsworth was getting ready to return to the States, I was picked by Major Maher and Sgt. Smith to replace him as local purchase clerk. We were always getting deliveries, stocking shelves and keeping inventory manually on stockcards as there were no computers back then.

We would get the orders from Central Supply at the hospital a few times a day, pick the items, deduct from inventory and deliver to the hospital. All the departments at the hospital would send their orders to Central Supply.

Turner always wanted to be the driver but we made sure we all got to drive. We had an old International Metro truck assigned to 9th Medical Supply. Remember Radar in M*A*S*H? Well, like him. we sometimes were able to trade and wheel deal supplies for something we needed like getting a motor pool guy to repair our cars. But there was very little drama at Denham.

A few months after we arrived Denham was closed and we moved to the base at West Ruislip and into a large warehouse-type building which was a short distance from the main gate.

It was bigger than the Denham facility. Everything was on a single floor, with Medical Maintenance at the left end of the building.

After a short time Captain Freistedt replaced Major Maher and Sgt. Hitchens replaced Sgt. Smith. Sgt. Hitchens was from New York and was the only female military person in Medical Supply. Her husband also was a Sergeant working at South Ruislip.

Bill Hoffard used to imitate her as she had a unique way calling us. It was very funny. She was a really lovely person. We all had a great time at West Ruislip.

In my job as Local Purchase Clerk I had to procure medicine in a hurry from the local (UK) market and go pick it up. I had to learn to type but I had a hard time. I was able to learn the keys but I used my two index fingers.

I got my military driver's license and became accustomed to driving left-hand drive cars and trucks on the left side of the road. I was on the road a couple days a week driving to drug companies all over London and some other counties. It was a great job and, because of it and the A to Z guide map, I became an expert knowing the streets of London.

I remember being sent to small companies in obscure parts of London. Sometimes it was very foggy and it was scary driving. Also, there were a few snowy days driving to work which were difficult. The furthest I would drive for a pickup was Reading.

About a year before I left, A3C Ken Detling joined our group and he replaced me as local purchase clerk a few weeks before I departed.

No barracks existed at West Ruislip so we were lucky enough to be allowed to live off base and I liked that a lot. The first apartment we lived in was at Harrow-On-The-Hill. Jim Bickerton, Bill Hoffard and I shared the place. We lived there about a year and when the lease was up, we split up.

Hoffard and I then shared a room a couple blocks from the South Ruislip base. That was awful, so we got Bill Sachs, Al Lewis, Bickerton, Hoffard and I together and rented another apartment at Harrow-On-The-Hill. I recall it was on Harrow Road and the entrance was around the back. It was a large place with four or five bedrooms. We lived there until I left England in 1964.

You can't believe how wonderful it was. We were out of those shitty barracks and were paid an allowance to live in local housing!

However the barracks did have one advantage – the showers. Although the apartment had a bathroom there was no shower – just a bathtub. And the hot water wouldn't last for five people. So, after work, we would stop at South Ruislip and shower. I'm sure there was resentment from those that had to live in the barracks.

LINEN CONTROL

Linen control was a part of Medical Supply. It required an airman to be stationed at the hospital and be responsible for the collection, cleaning and distribution of all sheets, hospital gowns, etc. Not one enlisted man wanted that job.

As mentioned when we got there an airman named Griffin was in Linen control. He was married and lived off base. We had very little contact with Linen Control. About a year before I left Griffin was leaving. It came down to which of the four of us, Hoffard, Bickerton, Detling or myself would be chosen to replace him.

None of us wanted to go there. It meant having to work at the hospital and move back to the barracks. Bickerton was the one chosen and we were relieved. He was upset. He had to accept the situation and he did. He had a girlfriend that he stayed with and he extended his tour.

RATION BOOKS

At the time, 1961-64, cigarettes, cigars and gasoline were rationed to us at a certain amount and price per month. There was never enough gas to last and if we ran out we would have to pay the local price, which was expensive. We became friendly with the ladies that pumped the gas at that time. They would somehow have extra coupons so we would trade cigarettes for gas coupons. That was a good deal for all!

DOUGLAS HOUSE

The Douglas House was an enlisted men's club in London. It was about 15 miles from Ruislip on the A40. We would always drive there and back. Sometimes we would take a different route to avoid boredom.

Douglas House was a hotel, dining room and a dance club. The restaurant was very good. Almost every night there was a live band and on weekends they had famous bands like Georgie Fame (before he made recordings) and Ted Heath and his full Orchestra.

You could drink at the table. Girls would wait in front of the building to get signed in by the guys. We met some of the most beautiful women that way. Some of us also became romantically involved.

Sidney, the club photographer, captured many key moments. He was not military. He was a local independent contractor photographer.

The place had a big impact on my life – I saved things like Douglas House Program cards. We also became friendly with the locals who worked there like the waiters, etc and we would invite them to our parties. We would get extra special service from them.

That was pretty much my experience at Ruislip. Work Monday to Friday and party Friday, Saturday and Sunday nights. An uneventful three years for me. Some of the guys we hung with got girls pregnant, didn't marry them and now have British children 45-50 years old or so.

A fond memory is the Silver Dollar Pizza restaurant in Pinner/Rayners Lane. A nurse at the hospital – I think she was a Major Patterson, opened it around 1963. She was very clever. Since there was little pizza available at the time it was a place we went to at least once a week. And the pizza was very good. I think she stayed there for awhile after her discharge.

The time went very fast. I was fortunate to be there when the British rock music scene began. I remember the first TV appearances of the Beatles, Rolling Stones and all the rest. We lived an almost normal life. We used the Commissary and AFEX a lot. Also ate a lot of meals in the South Ruislip Cafeteria and Mess Hall. Except for the uniform you wouldn't even think we were in the military.

I did travel to Dublin, Ireland and Copenhagen, Denmark. Took a lot of rides all over the UK. I really missed Ruislip when I came back to New York and was discharged in 1964. I kept in contact with people for awhile, then it stopped. I always had intentions of returning but never did.

Rich Silver ((second from left) with Friends at RAF West Ruislip in 1963 (*Above*) and pictured with his Morris Minor at the medical HQ (*Below*).

Rich Silver at RAF West Ruislip in 1963.

Douglas House was a popular hang out. Silver is seen here seated 2nd from left (*Above*)

CHAPTER TWENTY-NINE

DOWN BELOW

Robert "Red" Muir,

Airman 1st Class, USAF
458th Communications Squadron,
7th Air Division, Strategic Air Command
USAF High Wycombe Air Station
1962 – 1965

Travelling to England proved to be a bit of a time warp for me. Not stepping back in the sense of 'Merry Ole England' but simply in just getting there.

I had joined the US Air Force and gone through Basic Training and 'Tech School' in Texas and Mississippi respectively with two others who became long-term friends: Fellow Airmen 2nd Class-ranked Robert Moffett and Bill Brady.

In Mississippi we finished Tech School and got our orders in November of 1961. We were sent in various directions – Korea, Iceland, Thule Air Base in Greenland and Germany. The three of us had gone through the schools together and it was great that we all got assigned to the three-year stint at the US Air Force Base at Daws Hill, High Wycombe.

There weren't any choices about it. A bunch of guys from our class went to the other bases. When I think back it couldn't have been any better assignment but, at the time, we were three 21-year-olds and three years was an eternity when you are that age. Looking back now – well it couldn't have been any better.

As for the time warp, well the three of us flew out on of McGuire Air Force Base (now McGuire-Dix-Lakehurst) in New Jersey on 27 December 1961. However there was a snowstorm in England so we had to go through Prestwick in Scotland rather than getting a direct flight to a base closer to our destination.

And then, thanks to weather and logistics, we didn't fly any further. We stayed in Prestwick until the 29th December. When we departed it was via the bus for us to complete our journey to London. We headed for the US Forces hotel, Douglas House, in Lancaster Gate but it was full, so they sent us to another hotel.

And that's when we made our second mistake. We went into the bar next door to the hotel.

The wrong kind of bar!

It was our first encounter with a gay bar. We simply had been too tired to think about much except getting a drink. Even when the bartender said; "I really don't think you want to come in here!" we weren't initially clear what he was talking about. Then we looked. We got the picture and departed.

The next day was New Year's Eve and things started looking a lot brighter. We made our way out to the celebrations at Trafalgar Square where we met three beautiful girls – two Australians and one English girl – and went on to spend the evening with them.

After the general celebrations they invited us back to their flat. We sat in front of a gas heat fireplace and spent all night long getting to know each other better.

We were experiencing everything in life and what a wonderful introduction it was to the UK!

We knew only one thing about our new assignment. That it was about 30 miles from London. We thought it was an air base, so you can imagine our disappointment when we learned while in London that it wasn't a base but an Air Station and just a Communications HQ. So on January 2 1962 when we pulled up to the front gate it was drizzly and cold – just another miserable foggy damp day – the way it usually is in England. As I recall it was on a weekend and we weren't due to meet our Commander for orientation until the following day.

We were now part of the 485th Communications Squadron, 7th Air Division Headquarters, Strategic Air Command (SAC), USAF. The base then was strictly a communications facility supporting the UK 7th AD air bases including: Upper Heyford, Fairford, Greenham Common, Brize Norton, Lakenheath and Mildenhall.

The first thing I noticed on the drive onto the base was the antenna sprouting up and towering over everything. We were shown to the barracks, two three-storey buildings adjacent to each other at the far end of the base. I was given a room on the Second Floor with two other guys I didn't know. The barracks, which had red roofs, appeared packed. I was looking at these buildings and this place and thinking; "This is a prison sentence and it's never going to end."

The upside was that it was an Air Force base. And to my mind the Air Force was different from the Army & Navy in that it was the more 'relaxed' of the services. We were treated very well.

But while those antennae were crucial to our operation we hardly ever saw them bar coming and leaving the base – for the simple reason that we worked in the base's underground command bunker for most of our three plus years there.

Until I arrived at barracks I didn't have a clue there was an underground bunker there. And then that wasn't my first duty station. We received temporary jobs upon arrival until our permanent jobs became open. My temporary job was in the radio section in a building on the base and I was looking after the ground radio links.

We were part of Strategic Air Command at High Wycombe. In essence, the USAF kept a SAC command plane in the air 24 hours a day. Our job was to keep a radio link to it from the base as well as radio links to SAC Headquarters in Offutt, Nebraska.

So we did that and also started our training at TECH Control – that was the job we would be doing when we moved to the bunker. We were trained by the Non-Commissioned Officer (NCO) and he was with us at all times during the two months of training as we waited to move underground.

During the orientation they told us our job would be working in a very secret environment. We entered a small guarded building, showed our passes and walked down the steps to our assigned posts. The first time I went down there what got me was going down the stairs. It was about 50 feet underground.

It was quite impressive as a bunker because it was huge. Our Communications Centre included a Data Centre and Switchboard. There was also a Control Centre – where the officers were – on another level. And there was a Snack Bar for coffees and snacks. It was a 24/7 facility.

Over the time there I made, and got to work with, some great friends: Buddie Smith,

Stan Carlysle, Len Garcia, Ron Poremba, Dave Shortell, Ray Longi, Ron Schultz and Sgt. Otto Whittu all worked in the bunker together.

We were located in the Technical Control Unit with an open internal window that looked out on a big room that was filled with teletype machines. We had seven of the machines. Our job was to maintain the circuits for the machines and make sure they didn't go 'down'. We had two 'order wires' – essentially it was a jack plugged into different patch. Each machine had two lines.

People on the teletype would let us know when the machine wasn't working and we would mostly re-route circuits that went out of service.

Most of the failures were to do with the cable or radio. In all there were about 20-30 teletype operators and four of us. There was always pressure if something went wrong.

There was also a large command center in the bunker on a different floor with situation maps etc. There was also a reel-to-reel tape data center and a telephone switchboard.

We were supported by the local British GPO (Government Post Office) telephone company and by their cables and equipment. The GPO had a small telephone building above ground at the access road entrance to the bunker. A man named John Rand was the GPO man we dealt with.

The bunker always had guards. We had the highest clearance that we could have in the bunker *Crypto Clearance*.

And I was there during some exciting times.

I was in the bunker when President Kennedy was shot. I particularly recall it because I was a shift supervisor and it was relatively quiet. Then the bell started to ring on one of the order wires. As I recall it read: *Kennedy shot. Possibly dead. More to follow.* I left the message there.

Shortly after that we were notified that the base had gone to alert and confined to base and may be working extra shifts. I think President Kennedy's assassination left us all with a feeling of uncertainty for the future…especially being in the military and in Europe.

We didn't get any radio communications about the Cuban Missile Crisis that I recall. All we knew about that really was what we read in the *Stars & Stripes* newspaper and heard from television. Then, of course, there was Vietnam was just beginning. It only distantly touched us. It was a world away.

On one occasion we were ordered off base and down Daws Hill Lane and out onto the main hill road leading into the town of High Wycombe. We were lined up and down the road as an honour guard for a visit to High Wycombe by the Queen. She duly came by and gave us the Queen wave! It was an entertaining and educational time.

We worked rotating shifts. In essence, we didn't have a standard 8-5 job. We'd work three days 8 a.m. – 5 p.m. and then three days 4 p.m. – Midnight and then four days Midnight – 8 a.m. and then get four days off.

For me the greatest thing on the base was the Airmen's Club. That's where we would mostly go in our free time. We could bring girlfriends or guests up there. It's probably hard for some who came later to believe but the base had three different clubs then: the Airman's Club, the NCO Club and the Officers' Club. It was a fully-developed little base with a snack bar.

I recall Janie — one of the cooks who worked there at the time – she was hot! There Base Exchange and Movie Theater. I don't recall that it had a name back then but I heard it later became the Starlight Theater.

The base's Mess Hall was a blessing. The High Wycombe base had a wonderful Mess Hall. It was more akin to a restaurant. The left side was for Officers and NCOs and the right side was for the Airmen. The funny thing was that there was no difference in the food or the service. You got a table that had an artificial flower and a table cover.

The domestic help – the girls – would come over and ask you if you wanted tea. It was great. I believe it had been voted the best Mess Hall in the UK. And on holidays it was like being at home. You could bring guests.

The Mess Hall shut about midnight and reopened about 5 or 6 a.m. We'd go before it shut for our overnight shift and get them to cook eggs for egg sandwiches and we'd sneak it out and take it into the bunker.

Off base the three of us established relationships with girls in and around High Wycombe. I met my girl, Susan, on a Friday night at the Town Hall dances where they had live bands. She came from Hazelmere and would travel there on the bus. Once we got familiar I'd invite her on up to the base. And then we'd start going to London.

One very simple thing Susan and l enjoyed – the pub. Here we were young GIs coming into the pub. We loved pubs. I am not a fish person, but fish and chips, oh boy! Susan introduced them to me. The first time I saw people walking down the street holding a cone of newspaper and then could smell them. I knew they had to be good. Likewise I introduced her to cheeseburgers sand pizza. We had the latter at the Silver Dollar Pizza Parlour which had been set up by a US Air Force nurse who was an officer. Boy could she make pizza. So we were even!

I'm really pleased to say I've remained in contact with Susan ever since. She now lives in Dorset, England.

Some who were at the base at that time may recall "The Battle of Wycombe". It sounds pretty dramatic, however it was a fictitious battle that Bill Brady invented. Bill had broken his ankle in a car accident in or near town. Saying he had been involved in 'The Battle of Wycombe' sounded far more exciting and adventurous than saying that a beer can had rolled under the brake pedal of the car he was driving and he then hit a tree. We still fondly laugh about the 'war' story!

I left my assignment on 27 December 1964 and Bill Brady and I travelled home together. Robert Moffett managed to extend his stay for six more months. He had already married his wonderful girl, Jadzia, whom he had met there. (Bob passed away in September 2011).

I hadn't seen my family in three years. The Air Force asked me to re-enlist. Three years was enough. I had enjoyed it and I had been in a lovely place. It was a very formative time in my life. But now I wanted to get home.

As for the base itself it was, in my opinion, the best place any young man could have been stationed away from home. I was 21 and looking ahead to three years away from my home and family. But High Wycombe will always be my second favorite place to be. It was a wonderful base in a great location and, of course, there was a girl I should have married.

I often wonder what the inside of the bunker looks like today. Would I recognize anything? Probably not since it was totally redone back in the early 1980's I gather. I have used Google Earth to see a satellite view of the base.

Many of the buildings I recognized including the barracks I lived in for three years. Seeing it takes me back to a very different time. It was a very formative time in my life as well as one of the best times of my life.

Bob Muir preparing in his barracks room preparing to go on duty in 1963 and, inset, in civilian clothes just outside his barracks preparing for an equally important mission –on a date with his girlfriend, Susan. Many residence hall students from London Central High School, the Department of Defense Dependents School that was established at the base in the early 1970s may recognize both the furniture and the side-of-building walkway. Photos courtesy Bob Muir.

CHAPTER THIRTY

DOUGLAS HOUSE

Gerald A Collins,

Communications Technician Seaman (E-3)
Commander In Chief, United States Naval Forces Europe,
Naval Security Group Detachment Europe, London
August 1966 – September 1968

(Author's Note: Gerald "Jerry" Collins was stationed with the US Navy at the Head-quarters in London during what some might argue were some of the most exciting times in London life with the swinging 1960s closing out and protests against the Vietnam War becoming more common.

Collins was greatly taken with his time in England and went on to write one of the few books about US military life in England in the 1960s. Douglas House, published by 1st World Library in 2006, was written featuring his story as told through an alter ego but gave a fascinating insight into off-duty life and, as the name suggests, a good focus of it revolved around the eponymous enlisted men's club run by the USAF in London's Bayswater near Hyde Park. Collins' memories and experiences as a black US Navy sailor during that time in England make for fascinating reading.

Collins went on to become a Roman Catholic Deacon, ordained in the Archdiocese of Washington, DC, located in the Archdiocese of Atlanta and is on staff at a small Black Catholic Parish in Georgia. Collins also has also worked for the Federal Emergency Management Agency (FEMA) as an Emergency Management Program Specialist assisting communities to prepare for and respond to disasters.)

DOUGLAS HOUSE

I arrived in London in the early morning of Saturday, September 17th, 1966. At the time my rank was E-3 (Seaman), my rate was Communications Technician. I had just completed the Communications Technician basic school at the U.S. Naval Training Center, Bainbridge, Maryland.

HEATHROW

We were up most of the night after the transatlantic flight so George Pfeifer and I were extremely tired. I had met George at Bainbridge while in school. He and I received our orders together and agreed to meet at Kennedy Airport for the trip. Disembarking in London the morning was damp and dreary. Heathrow Airport was cold, barren-looking and very much like an old warehouse. The morning, when combined with the warehouse look, made a stark contrast to what we had left at JFK; ultra-modern jetways connected to terminals with walkways from terminals to docked aircraft. As we came down the truck-mounted stairway and touched the tarmac at Heathrow, my first introduction to London. It all seemed so somber, so dreary and depressing. The good thing was that there was a driver waiting for us. He called us by name without hesitation as if he knew exactly who we were. Of course we were easy to pick out because we were the only two individuals wearing Navy 'crackerjack' uniforms on the flight.

I don't remember much of the drive into downtown London; I slept most of the way. Although it was early morning in London, it was still late night in New York, which we had just left barely seven hours before.

ARRIVAL AT DOUGLAS HOUSE

The Douglas House was the terminus for all enlisted people arriving and departing London. From there buses, took the various travelers to their duty stations. In the case of George and me, there was no one waiting for us. So we were left to sit and wait. As it was Saturday morning the Douglas House looked like just another slightly dowdy hotel. We were not allowed in any of the upstairs areas of the hotel. The upstairs restaurant was not open. The downstairs restaurant was open but for a light breakfast and with a partial staff. I later learned that because the restaurant was open late night on Friday, it opened late on Saturday morning. There was coffee and pastries to be had but no real food. So the only place to sit and wait was in the oversized waiting area/vestibule just inside the hotel doorway.

Others who were going on to the Air Force bases that surrounded London were gathered in the same area. Most of them had come in through Gatwick or Heathrow. And still others were waiting to depart London going out the way we came in. Saturday morning at Douglas House was somewhat busy as the buses departed either for the airports or the bases. However nothing came for George and me. We were told that there was to be a car for us but no one could tell us why there was a delay. So we were told to wait.

After a couple of hours of waiting we were told there had been some kind of a mix-up and a car would be there soon to pick us up. Growing impatient George and I began to pepper the unofficial "dispatcher" with questions. Once pestered he placed a few calls and then announced that a car really was on the way. About an hour later a car marked US Navy arrived. We were dropped off in front of the Navy Facility (NAVFAC), London at the Keysign House on one of the side streets behind the US Navy Headquarters. We could not report into Commander In Chief U.S. Naval Forces Europe, 7 North Audley Street, because the building was a secure facility and although we were expected, we had no clearance to enter the secure site.

From NAVFAC we were then taken back to a rooming house not too far from the Douglas House.

As I state in the book, *Douglas House*, London in the early 1960's still had some of the residue of World War II. Although visibly there were no bombed out buildings, the impact could be seen by the fact that few homes had central heat and air-conditioning. While in-door plumbing and toilets (aka 'water closets') were not uncommon, many of the homes had water closets that had been added after the homes were built. At home, particularly in Washington DC, indoor plumbing and hot and cold running water were integral parts of the houses. This was not so in 1960's London. So I was taken aback by the fact that many in England did not find those add on "boxes" a bit unusual.

When I arrived in England in 1966 I had an image of London as the place of history, heritage, traditions and modernity. I knew little of the suffering endured during World War II and even less of the on-going long road to full recovery.

MY ASSIGNMENT IN LONDON

Essentially I was a clerk-typist. At Bainbridge I had attended a course that trained me to work in the classified Communications Centers aboard ships and at shore stations. We took the title Communications Technician from the communications centers where we were to work. For much of the latter part my tour of duty the Commander-in-Chief of the US Navy in Europe (CINCUSNAVEUR) was Admiral John S. McCain, the father of Sen. John McCain. My work included drafting, proof-reading and filing classified correspondence. I worked on the 7th floor of the North Audley Street Building in a secure area which was essentially a vault.

The Navy building, like so many along North Audley, had apparently been a hotel or apartment block, I think. Among its nooks and crannies the building was found to have large sealed up rooms that contained old bathrooms, huge walk-in closets and etc. The extensive presence of marble throughout the building gave the feel of a hotel. In those days the lobby where the marine sentry stood had the feel of a hotel check-in desk rather than a sentry post. The thing I liked most about the North Audley building was its central location. In less than a half hour you could walk to nearly every place you might want to go in the West End.

Restaurants like the Barley Mow located on one of the streets behind the building and the fish and chips shops along Oxford Street still conjure up images that make me salivate. Those images and the convenience of the other shops along Oxford Street, Carnaby Street, Baker Street and the nearness of Hyde Park Speakers' Corner, Buckingham Palace and of course, the Douglas House reinforces in my mind how special London was at that time. Anything I wanted to do or see was literally outside of the front door of my office

The group I worked with was a real team. Most of us were young, single and away from home. As a result the antics of young boys becoming men were not uncommon. Practical jokes were a routine part of the office environment. It was not uncommon for someone to blow up condoms to decorate an individual's desk in celebration of a birthday. The fact that also we wore no uniforms most days also tended to give the feel of a civilian office.

I had one heart-stopper of a moment as I entered the North Audley Street headquarters one afternoon. I had gone out for ice cream for my boss and some others. The US Marines provided the security for the Navy Headquarters. The front entrance was marble and sound reverberated everywhere. Just as I entered a Marine accidently discharged his weapon. I innocently and unexpectedly found myself at the business end of a number of weapons as the Marines locked, loaded and targeted me! A number of them had recently been in Vietnam and I recall that it took a considerable time to stand them down. The ice cream melted.

I'd like to think that Admiral McCain had a bit of a heart-stopper on the Saturday he apparently showed up in Civilian clothes and one of the seaman manning the Quarterdeck failed to recognize him and challenged him. The Admiral was irate and gave the sailor the sharp end of his tongue and told the rating that he had 48 hours to find out who he was and report back. However two days later, when their paths crossed at a briefing the Admiral, to his credit, commended him on his diligence in challenging him.

We had some miserable assignments. One of the worst was delivering classified documents to an incinerator in South London. And it was made doubly worse the day one of bags carrying classified documents fell off our truck. But the misery didn't end there. We had to twice *walk* the four-mile journey before we fortunately found the bag. Security services from both the US and UK had been notified and we had to open the bag in front of officials. What we found inside…well let's just say that the whole thing was a bit of nonsense and caused the US Navy in London to rewrite the rules in regards to disposal of so-called classified material.

And of course, out of work, Douglas House was our respite club offering a heady mix of drinks, music and women. During 'Happy Hour' (most every day from 3 p.m. – 7 p.m.) a rum and Coke could be had for 25 cents and a beer was two for 25 cents. With free munchies, sliced ham and cheese, 'Happy Hour' became the place for a free meal weeks between paydays when cash was perennially short.

On Thursday or Friday evenings between 5 p.m. and 6:30 p.m. the ladies would start coming to the club to meet their "dates." While many were legitimately meeting dates, some were there because of the reputation of the Douglas House as a place to party.

Club management would not allow the ladies beyond the huge double wooden doors that marked the entrance. Most could see inside the door, beyond the area of the receptionist who cleared them for entry. Often getting in for those without a "date" became a game of cat and mouse, motioning to service members whom they might have met to get them to sign them in. If they were caught making motions to service members both they and the member could be barred from entry not only for that evening but for future events.

The one time I was threatened with being barred from the club was because I had signed in a young woman I had met through a friend, a fellow sailor, Alfredo Dean. I passed her as I started into the club. As I passed by the receptionist, she called my name. At first I did not recognize her. So I went back out onto the porch at the entrance to the club. She pulled me to the side, out of sight of the receptionist. Then she pleaded with me to sign her in because, as she put it, she was supposed to meet Dean there. Hearing Dean's name I sort of remembered her so I signed her in. The receptionist whispered something to the manager as we passed. I paid no attention to it. Dean showed up about two hours later with his regular girlfriend, Marcia. By then the young woman had disappeared. And, I had gone onto something else.

As Dean came up the triple wide stairs from the first floor, the woman stepped out from beside one of the slot machines that sat in a gaming room that was set up on the landing between the first and second floors. As she came out she saw Marcia and lunged at her. Fortunately Dean moved between the two of them and caught the brunt of the lunge. Unfortunately an Air Policeman (AP) also saw what had happened. In short order he had other APs on the scene.

Soon it became obvious to all that the woman who had started the incident had been signed in by me. And, of course, without thinking, I admitted not knowing her. In turn, she, Dean and I were escorted to the front doors and put out. I was threatened with losing my club privileges. Dean lost his for two weeks.

A couple of weeks after they threatened to oust me from the club Ben E. King was to appear. Tickets were sold out way in advance. I had purchased my tickets long before the Dean incident. I had arranged a special date. I had even gone to Marks & Spencers and purchased a new suit! The talk of the office was about Ben E. King. In the midst of the talk about Ben E. King, John, my roommate, reminded me that my name might be on the list of people barred from the club. I laughed. But then it dawned on me that I really had not gone back into the club after happy hour in awhile. Usually they did not bar your entering during happy hour. But if you were barred from the club they would look for you after happy hour ended and ask you to leave.

I called the club. No one could answer my question. I asked them to check their list of suspended military personnel. They could not. I was told it was being updated and they would have to get back to me. Of course no one got back to me. Then came the Saturday evening of the King performance. I decided to take a chance. I met my date at the front door, presented my tickets and tried to enter. I was stopped. I began to protest. But before I could raise a real fuss I was pulled to the side and was asked to see my tickets again. Surprised, I gave them to the manager. He took a look at them and said; "I just wanted to double check your tickets. We have had several counterfeits. Yours appear to be okay." Then I said; "I had been told I had been suspended from the club." He looked at me and said, "Suspended? Hell, if that were the case, your command would have told you, not us. If you get suspended you are automatically subject to an Article 15 non-judicial punishment."

I entered the club feeling like a person who had just been liberated. My seats were up front near the stage. And when he sang "Stand By Me" I felt like the military had stood by me and not suspended me unfairly. I have never enjoyed a show as much as I enjoyed that Ben E. King performance.

It wasn't just the US Navy and US Air Force who could come into Douglas House – US Marines (who had their barracks in St. John's Wood), US Soldiers and an occasional Royal Forces member were also there. We all managed to get along bar the occasional inter-service rivalry coming out of the regional base football games. The Navy and the Army were never real competitors in those games. However the Air Force, because they lived on large military facilities, often had bragging rights. But Marines being Marines could never let the Airmen get away with just bragging. On one occasion a group of Marines challenged some airmen to a touch football game.

The game was to be held at one of the fields at Ruislip. The teams were put together, the field was reserved and everyone was to meet at the base. But as things worked out, the day it was to have taken place, the military Special Services Office (the office that booked tickets to shows, Battersea and other outings) offered some last minute discount tickets to Battersea. So the game fell apart because most of the guys went to the park rather than drive the near hour to Ruislip just to see a bunch of drunken military men "play at touch football."

It probably would have been better if they had played the game because for what seemed like an interminable amount of time (at least two months) the Air Force men who worked at Douglas House continued to brag about what they would have done to the Marines. Most of the Royal Forces personnel thought the thing hilarious because they saw American football as barbaric and without any real strategies.

The harshest memory I have of work and the office life was the day we had to don uniforms for a Captain's Mast. Captain's Mast is the equivalent of an Article 15. It is a form of non-judicial punishment that can have the look, feel and, in rare cases, the impact of judicial punishment.

In my book *Douglas House*, I speak about the consequences of one Captain's Mast. It involved a sailor who had married a young woman whose father had been a colonel in the Irish Republican Army (IRA). The sailor went to mast because he married the woman in spite of not having permission from the US Government. His marriage and his presence in England were considered to be an embarrassment to the United States. It was also seen by some as a possible interference in internal British affairs.

What I disliked about it was the fact that the entire command was made to wear uniforms for the mast. What I also disliked about it was the fact that the accused was allowed virtually no defense aside from his division officer. I saw no way for the accused to win because his division officer depended upon the commanding officer for the success of his career. The division officer would have been a fool to vigorously defend the accused against the man who was bringing charges. From that I learned that rules, though seemingly even, had different applications depending upon the circumstances. I relearned that lesson when I decided I wanted to get married.

When my girlfriend Marki and I decided to get married I saw how the rules had routinely been applied in the past. Senior Chief Petty Officer Thomas Shelton was the senior enlisted man who took it upon himself to defend my request to marry. Before Shelton took my task on getting permission to marry for most African Americans was a battle in futility.

Chief Shelton used the argument that my prospective bride was still a British subject who had right to citizenship and who had never demonstrably exercised her right to renounce her British citizenship by claiming Barbadian citizenship. To explain his strategy he met with Marki and me one evening after work vver a cup of coffee. To my surprise we won! However, his effort proved to be a rare victory. The threat of a Captain's Mast and possible demotion killed many relationships among African American sailors. Most whites were not scrutinized as closely unless they were attempting to marry someone like the IRA colonel's daughter. It was not an issue for them.

RACISM IN LONDON

When I arrived in London I was very aware that I had come from a place (the States) where race and racism played a major part of all relationships. While I did see the signs along the streets in Kilburn and Kensal Rise and in areas along the high roads in Kilburn, Shepherds Bush, Notting Hill Gate, Hammersmith, Kensal Rise and other places, it was something that was not a major part of my life.

It was not a major part of my life because other than George Pfeifer, who was Jewish, I had few people who were white whom I considered to be close friends, so my situations were basically race-free. For the most part I lived in a world occupied by people of color from the Caribbean, Africa, or Asia. Despite this I heard the stories and was reminded of the more ominous British form of racism as exhibited by the "young rowdies." I knew of the Skinheads, Teddy Boys, Mods and Rockers and the occasional "Paki bashing" but, for the most part, I had no encounters with them.

I saw more virulent signs of racism in the office in the disparate treatment (white sailors selected for assignments that black sailors were not given). I was also reminded of it because I could not get basic cosmetics like hair grooming products and shaving creme for ingrown facial hairs. There were few minority-oriented magazines or newspapers. When I met the woman I named Ivanna in the *Douglas House* book, I truly loved spending time with her and her son, Stephen. They provided me with the feel of family; the thing I missed most about home. They lived just off of Shepherds Bush High Road not far from Shepherd's Market. Like Shepherd's Bush, I loved going to Hammersmith Market and the area around Notting Hill Gate because of the diversity and the fact that so many people of color (Blacks, Indians, Pakistanis and others) lived in the area.

The issue was prevalent in the Commissary in the basement of one of the buildings in Shepherd's Bush and at the bases at South and West Ruislip and on the military bases generally, too. Product-wise the American military was blind to our needs. There were no newspapers, books, or magazines that focused on blacks and other minorities. However at that time the US bases would accept ads from white British landlords who would not rent or sell to blacks. If I wanted to read books or magazines about African Americans or about others from the African diasporas, I had to go to small shops operated by people of color in Ladbroke Grove, Hammersmith or Brixton.

Since writing *Douglas House* I had the opportunity to speak to a black woman who had been a naval officer stationed in London at the time I was there. In one afternoon she related to me how hellish her existence was. She could not get hair products; there was no place for her to get make-up. All make-up was for white women. I could at least get my hair cut off. She could not get her hair fixed. Relaxers were out of the question. Stockings were out of the question; they were too light or nothing at all.

She had to endure humiliation because of her sex and her race. This humiliation came from fellow officers and enlisted men both black and white. Because of this shunning she had married a 'Brit' and they both retreated to the relative isolation of a Pacific Island where they found the true diversity she wanted for herself and her children. When last I spoke to her she and her husband were still living there.

Without Ivanna and Marki, whom I later married, I would not have learned of products that I could use for grooming that transcended race. For example both Ivanna and Marki used Brylcreme for their hair. I had never thought of it as a product for African Americans. I usually had to wait for "care packages" from home to get hair products.

As a young African American man people had a tendency to respond to my presence before they responded to me. There were times when I went into to shops along Oxford Street and in the West End and would have people follow me, watch me and then ask to help me. But once I opened my mouth and they heard my accent they would relax and would want engage me in conversation to ascertain what "made me different."

Aside from that there was one other factor that got me – driving on the wrong side of the road. All US Navy seamen were supposed to be able to drive for chauffeur and delivery duties as and when. But despite having a US license I had to take a UK test. Let's just put it this way – the tale of my driving test with the British test official became the stuff of office legend and I was struck off the list for chauffeur duties!

I went to the Wembley area only once that I recall. It was because an office mate lived in the area and commuted to downtown London. He had a Christmas party and all were invited. I did like Blimpies a restaurant at Marble Arch. I liked Blimpies because it was near to what I was used to in terms of hamburgers. But, irony of ironies, most of us liked the grease and the "crunchiness" of eating fish and chips. There was a shop along Kensal High Road just north of the Kensal Rise Train station that had some of the best fish and chips in town. The only place I could think of that was better would have to be one of the chip shops – sometimes known as 'chippies' – in Shepherds Bush Market.

From the moment I left London that September morning in 1968 I knew that I had just completed a very important part of my life and that I had to write a book about it. The memories of London would literally haunt me for years until I finally wrote *Douglas House*. I would dream about going back and meeting some of the people I knew then. Aside from my wife, Marki, whom I have been married to for more than 40 years, there were two people who had a tremendous impact on my life. Both of them were Jamaican women who were exactly thirteen years older than I.

Aside from my wife, Sheilah and Ivanna touched me more than anyone. Sheilah was the consummate proper British woman. She had all of the affectations of a woman who was highly educated, motivated and ambitious. She took time to introduce me to things that she believed would keep me away from many of the traps that other GIs fell into. She enjoyed classical music, reading, walks along the Thames and teas in the afternoon. In many ways she was more British than many Brits I met.

She absolutely detested the Douglas House. She thought it a place full of rowdies and uncouth American GIs. We would argue about her notions of couth and uncouth. Her knowledge of history and the interconnectivity of social and political events fascinated me. She was one of the first people to speak to me about Marcus Garvey, Jomo Kenyatta, Mandela, Nana and the African liberation movements that were sweeping the British Empire in the 50's and 60's. In many ways Sheilah was very much the mother figure I needed while I was away from home. In hindsight, I know now – and accept more – the lessons she introduced me to when it came to staying focused on my goals.

In terms of Ivanna I wish things could have turned out differently. But she came along at the wrong time. Unlike Sheilah she seemed not to know where she was headed. She was very patient with me and my boyishness. I have missed her most because of Stephen. As I state in the *Douglas House* Stephen and my daughter looked so much alike that they could have been brother and sister. I would love to have played more of a part in Stephen's life. To say that I loved him as my own son is to state the obvious. Those are the only regrets I have. Otherwise there is nothing I would change about the people or that time.

Just after 9/11, when I first had the compulsion to write *Douglas House*, I let my wife read the first draft. Simultaneously, I sent a copy to Rachel (my girlfriend from the US who had come to visit me in England but with whom I broke up due in no small part to distance and whom I had managed to track down when I wrote it). Both she and my wife said essentially the same thing. That was; "I knew you were going through something. Now I have a better idea and it's okay." Then they both asked the same question, "What gives you the right to tell other people's secrets?" It was those two comments that made me decide that I should change the names of the characters in *Douglas House*.

The irony in changing the names is that in time, the different names and characters have helped to make the story new for me each time I read it. Each time it takes me back.

I've been fortunate to be able to physically return to this time of my youth. Originally in 1976, then in 1996, 2004 and even in 2010.

London – and the US Navy assignment – was the beginning of my finding me and what I wanted to do with my life. It was the place where I can say I learned to depend upon myself and be confident of my own abilities. London will always be my hometown. And like any hometown, it holds cherished memories where people remain young forever and are waiting for my return, if only I just look around the right corner.

Gerald Collins being promoted to Petty Officer 3rd Class at the US Navy Headquarters on North Audley Street in 1967. Captain John Alba Skinner, Collins' commanding officer, is awarding the promotion.

A picture of Gerald Collins, his wife and friends at Douglas House.

Chuck Steward and his wife Jean (*Left to Right*) were friends of Gerald Collins. "In the picture with Chuck were friends of his Tony and Maureen. The picture was taken in the mid-1960's while Chuck (who has sadly passed) was a manager of the Douglas House. They had stayed on in England. Photos courtesy of Gerald Collins.

CHAPTER THIRTY-ONE

THE WEST RUISLIP
WAREHOUSE WARS

Art Wallace

Dependent/Student
London Central High, Bushey Hall, Hertfordshire
(1967-1970)
Easton Village, Bentwaters
(1957-1960)

ATTENDING KINGS SCHOOL IN HARROW

From 1957-1962, when my father was General Manager at the South Ruislip Post Exchange (PX) store on the USAF base, we lived off-base nearby at 62 Melthorne Drive, South Ruislip.

It was near a huge dairy plant and the "158" double-decker bus line. At that time I attended a British boys grammar school called The Kings School on Harrow-on-the-Hill.

It was an experience I'll never forget.

First, wearing a uniform that included a blue blazer, blue cap, gray shorts and grey stockings. That took some adjusting after attending a co-ed primary school near the USAF's Bentwaters AFB, England (in a village near Wickham-Market). The uniformed schoolboy "pilgrimage" that began every morning on the buses and trains was a sight to behold. But after awhile you adapted because it was part of the normal landscape of life in England.

Second, adapting to disciplinary rules at school included respecting the role of the 'prefect', an upper-form student, usually 15-17 years of age, who wielded a level of discipline enforcement second only to the headmaster and school teachers of that time. Punishment was usually with a ruler but it was for disrespect to the teacher, being in the hallway without permission or just unacceptable behavior on their terms.

Third, learning how to play rugby, cricket and football (soccer) for your 'House' (mine was Glaston, with the St. George's red cross at its logo) were mandatory, so I was introduced early (age six) to learning new sports and the importance of team play and camaraderie in order to have any chance of being victorious in a match with other 'school houses.'

And last, the teachers who, while firm, were also very caring and expected only the best from each of us. From penmanship to a command of world geography and understanding Roman rule in the land of the early Anglos, Saxons and Picts, it was a stimulating school environment that prepared me well when I transferred to a DODDS school at Nouasseur AFB, Morocco, in 1962.

By the way, my most memorable friend while at the Kings School was a chap named John Eaton. His claim to fame at the time was his sister was an actress named Shirley

Eaton at Pinewood Studios. Her claim to fame (and shared by her brother with photographic evidence) was as the voluptuous lady who was suffocated by being painted gold in the opening scene of the 1960's James Bond movie, "Goldfinger".

A few of John's mates obtained "autographed" pictures of Shirley which probably are hidden in some dusty drawer at home in the villages of Yorkshire and Hampshire today!

MANAGING THE PX IN ENGLAND

My dad, Arthur Charles Wallace, got his start in the Armed Forces Exchange Service (AFEX) after the Korean War. At the time they were recruiting former military personnel into the junior management ranks of the growing PX system to run the post (and base) exchange stores in the growing number of overseas military locations in Europe and Asia.

My father's first store management position was at Ankara AFB, Turkey, where his biggest early challenge was teaching the new Turkish host nation employees on how to use the bathroom and flush the toilet! Cultural indoctrination was one of his biggest challenges back then!

Following that assignment he was assigned to England from 1955-1962 and then from 1967-1970 where he managed PXs at Bentwaters, South Ruislip and eventually as the deputy director over all the military exchanges in the United Kingdom.

As many of us recall the PX was the home of the *Stars & Stripes* magazine store where we feasted our eyes on the latest editions of our favorite comic books, *Mad* magazine, *Sixteen* magazine and other publications that connected us with our American homeland.

It was where the "Snack Bar", "Barber Shop" and "Beauty Shop", concessions and other hang-outs were located. Most of the PX employees were either family members of military sponsors or local British nationals. Only the management staff were regular PX career employees like my father.

The biggest challenge for the PX was deciding "what to sell" in the stores. Customer feedback, command guidance and PX merchandising via vendors and mainland US distributors determined what went on the shelf. Back in the 50's and 60's, if you didn't find it in the PX, you either looked in the catalogs from Sears, Montgomery Wards and JC Penney's (my favorite was always the Christmas toy catalogs!) or you ventured off-base to Marks & Spencer, Woolworths, Sainsbury's, W.H. Smith or other familiar British stores or magazine shops or supermarkets.

Of course the Internet was not a distraction or a temptation in those days. Everything was obtained either at the PX, off-base or via a catalog. And then you waited for the US military APO (which stood for Army Post Office) mail to deliver it....slowly!

As my father's career with AAFES (Army & Air Force Exchange System) came to an end in 1981, he realized the mission of the PX had changed along with service members' and family buying habits. On the mainland US young enlisted and officer families preferred the lower prices of Wal-Mart, Costco, Ross, Dollar General Stores, other discount outlets and, in more recent years, with Internet sites. The PX was primarily where you purchased toiletries, sports wear, unique import items and military clothing at the Uniform Stores. Retirees who also frequented the PX preferred high-end labels and manufacturers that did not appeal to younger buyers.

It's only in overseas areas where the PX remains popular because of its concessions and access to foreign merchandise at bulk discount rates. In the combat zones, the PX hooch and trailers are where you get your junk food, cigarettes, DVDs, candy and other items that brought you "close to home" in some strange connection.

What happens to the PX in the "joint service organizations" of the future remains to be seen. One final point to remember: Profits from PX/BX sales are returned to the installation Morale, Welfare and Recreation (MWR) budgets so their value as a source of revenue to keep the gymnasium, athletic teams and other community activities available at no cost. That cannot be overemphasized.

LIVING WITH AMERICA

I attended London Central High School (LCHS) in Watford, England 1967-1970. It was situated on an old RAF radar site and base converted into a DODDS junior high and high school campus with historical ambience and a wonderful wooded tree setting. It was secluded which fostered a unique sense of community among its military school children culture that none of us really appreciated until we left to go to large, sprawling, impersonal campuses in America.

In 1969 I was on the school track team with Dewey Bunnell whose Dad was an USAF Senior Master Sgt. Dewey and I ran the Low Hurdles so we became acquainted while trying to master the strategy of maximizing our speed before that next little fence blocked our route!

At that time he was learning to play the bass guitar on his own. It was a passion and influenced by the magical presence of British rock and roll music from the Beatles, Cream, Rolling Stones, Yardbirds and its growing influence on the Motown and bubble-gum rock (Tommy James, Paul Revere, Buffalo Springfield, etc.) scene in America.

Dewey was eager to join a new rock band that included LCHS senior Dan Peek, whose dad, Col. Peek, worked at the West Ruislip Armed Forces Exchange (now AAFES) headquarters with my Dad. Another band member was senior Gerry Beckley whose Dad was the West Ruislip Base Commander. Dan was the talented acoustic lead guitarist and Gerry played keyboards/piano. Whenever you heard them play, you had a suspicion that they were destined for "bigger and better things"....eventually!

To my recollection, the band experimented with names to include "Genesis" but that was a popular name that didn't last long. After graduation from LCHS in 1969 and playing some stints at the West Ruislip Teen Club, the boys started playing gigs at small clubs in London. By then, they had changed their name to America.

The producer for Apple Records (the same producer used by the Beatles in late 60's) was George Martin. He discovered America about the same time he discovered Mary Hopkins, the singer of *Those Were The Days My Friend* and was instrumental in guiding them to their eventual hit single *Horse With Know Name*.

It's ironic that I had lost contact with Dewey and Dan after they graduated but saw them on the cover of their debut album at the University Book Store at the University of Kentucky in 1970! It was a classic; "Hey, I know those guys" reaction!

After America became known nationally in the 1970's I had a chance to go 'backstage' to visit them after concerts in Virginia and Texas.

In those early tour days, they were receptive to "London Central High" alumni that they knew to being allowed access after the concert to briefly visit with them. I remember that they were friendly and enjoying seeing "familiar faces" from their high school days.

By the late 1970's/early 1980's Dan Peek (who passed away in July 2011) had parted from the band to join a religious community near St. Louis. America played on with Dewey, Gerry, Dave Atwood (on drums) and Doug Kenny as a stage manager. It was still a LCHS-influenced band. Over time, as new musicians joined the core Dewey and Gerry members, its attachments to England have, perhaps naturally, waned a little bit but I still remember it as a reflection of our experiences growing in the British music revolution of the late 1960's.

WEST RUISLIP WAREHOUSE WARS

Back in 1967-1968, one of the most sought after summer jobs in England was working minimum wage (then $1.65 per hour) at the West Ruislip PX warehouses, loading and unloading freight trucks and ship-land containers from the East Coast providing PX and commissary items for US forces in England.

Most of us who were hired were high school football players who weren't resistant to lifting large heavy boxes all day long while playing hide and seek and other 'mischief' that was made possible by the maze of large boxes and storage shelves!

To pass the time of day we would occasionally separate into teams and play 'war games' like "capture the case of Butterfingers" or some other innocent competition that involved a scavenger hunt and finding an item in different sections of the ware house. It was one way to pass time in between unloading containers and loading trucks that would transport the merchandise to other USAF bases at Upper Heyford, Lakenheath, Bentwaters, Chicksands, Mildenhall and South Ruislip.

Long days in the dusty warehouse at West Ruislip were spent unloading the large metal containers that arrived from Bayonne, New Jersey, with foodstuffs, toothpaste, clothes and countless other items that filled the shelves at the local PX or commissary.

What amazed us was the number of pilfered, damaged and missing items that were discovered when we did the inventory count to verify that "xx" number of boxes or items had arrived. It was obvious that there was a lucrative theft ring pilfering merchandise in Bayonne, New Jersey, before the security lock was placed on the container hinge lock.

As "summer" teenage labor we reported this to our civilian manager but we never knew if anything was done about it. It was just one of those discoveries you made when working in the warehouse and merchandise shipping culture.

COMMENCEMENT

The London Central graduating class of 1970 held its commencement ceremony not at the school in Watford but at Brent Town Hall in Wembley, Middlesex. The ceremony started at 14:30 hours with the processional – "Hukliguns March" and an invocation by Chaplain Robert E. Merrell. I was the Senior Class Speaker. Then I was followed by my classmate and class Valedictorian John Kidd who introduced the principal speaker — someone from Oxford University who would give the commencement address.

At the time – and in the blur of excitement that surrounds graduation — probably none of us gave much thought to who it was or what he said. But that speaker, who was a Rhodes Scholar Lecturer at Oxford, was then just Mr. William J. Clinton.

Yes. The future President of the United States came to Wembley and spoke to us!

Afterwards the Senior Chorus sang *Your Shining Eyes, All in the April Evening* and *Aquarius* before Brigadier General F.L. Gailer and 7500th Air Base Squadron Commander Colonel H.L. Porterfield joined the Schools Superintendent Clarence Kennedy and our own Principal Rolla Baumgartner for the presentation of diplomas.

It wasn't the only time I was to meet the President. My own career took me to the White House where I worked in the Medical Unit from 1989-1993 and I met President Clinton again in person when I administered his allergy injection at our West Wing medical annex.

I showed him the program from 1970 and he had vague recollections of that trip to Wembley. He did recall the wonderful experience he had while at Oxford.

Above Left: Toyland in the BX (Base Exchange) at USAF South Ruislip believe to be in the late 1960s. Above Right: Art Wallace's father Arthur Wallace (far right) along with staff and USAF officials at a presentation at the AFEX (Armed Forces Exchange) office at USAF South Ruislip. Below: Arthur Wallace (seen in bowler hat) joins staff for an Easter Event again believed to be in the late 1960s. Photos courtesy of Art Wallace.

CHAPTER THIRTY-TWO

ALONG-TERM RELATIONSHIP

Henry Farwell

Chief Master Sergeant
US ARMY/USAF
Boxted (May 1942 – September 1942)
Great Dunmow (1943 – 1944)
Burtonwood (October 1949 – February 1950)
USAF Alconbury (1959 – 1964)
RAF Upper Heyford (1952-1955)
USAF South Ruislip (1967 – 1970)

England became a long-term relationship for me. And my first 'date' with her was at World War II.

I was brought up in Milwaukee, Wisconsin, and I enlisted in the US Army in May of 1942 aged 20. I did my basic training and was assigned to the Army Air Corps (as it then was). I was sent to Scott Field, Illinois, where I became a Radio Operator Mechanic. From there I went to MacDill Field in Florida. I had gunnery training there.

In December of 1942 the 386th Bomb Group was activated on paper and consisted of four squadrons: 552nd, 553rd, 554th and 555th. We moved into it in January 1943 and so I was a charter member of the 554th Bomb Squadron. They flew the B-26 Maurauder, a twin-engine tactical bomber.

The commander of the 386th Bomb Group was a famous pilot – Colonel Lester Maitland who was born in Milwaukee. He was the first pilot in the US to fly faster than 200 mph and he also became the first person to fly a plane from California (San Francisco) to Hawaii in 1927. He would go on to fly 44 combat missions with us.

After McDill we transferred to Lake Charles, Louisiana, where we completed our flight training. We trained in low-level flight. We had four-package guns on the outside of the aircraft for strafing. As radio gunner I operated the radio. When on combat missions I manned the waist gun – a .50 calibre machine gun half way down the fuselage.

My fellow crew included:
Capt. Asa Hillis — Pilot
Lt. Jerry Soper — Co-Pilot
SSgt. EJ 'Mac' McDonnell — Tail Gunner
SSgt. George Cheadle — Engineer/Gunner

Of course nicknames were and still are very popular in the military. The 386TH were known as the *Crusaders*. Our plane, Serial Number 131622, was christened *Litljo*. That was because the last two numbers were 22 and in dice a throw of two and two made four – which is known in the dice world as 'Little Jo'e. The plane was special to us. We had taken delivery of it from the factory.

Somewhere along the way I picked up the nickname 'Fireball'. I think it was because Farwell was my name and I was always ready to go and do something. I was never known as 'Henry'.

We then deployed to Boxted airfield near Colchester, England. We flew over via Newfoundland, Greenland and Iceland, Prestwick and reached Boxted the end of May 1942. It took us about 12 days to fly across. We could only fly if the weather was clear at both ends of the flight. There was a lot of waiting.

The time we took in training and travelling to our combat base was invaluable. We had all just thrown together in the January. By May everybody on my crew had been together – except for the bombardier – since that point and we were a tight-knit team on the field. The only time that wasn't the case was when we went to town. Then everybody went their own ways. But once on the airfield and in the plane we were a team.

That was crucial. We had trained for low-level attack. But there was an immediate problem when the 386th came to operations and the first Marauder Group that came before us flew to a mission in Holland. Every airplane was badly shot up. Three days later they had a second low-level mission. One of the planes aborted mid-Channel.

It was the only plane that came back.

So we had a problem. There were our four groups coming to Europe and more being formed. Low-level was out; so was flying high-level up at 20,000 feet because we had no oxygen systems on board.

The only alternative was medium-level. We flew as high as we could without oxygen – 9,000 to 10,000 feet – and our losses were relatively low. This was because we always had fighter escort. RAF Spitfires in the beginning and, as our own Air Force built up, we were then escorted by American P-47s.

With all this in the background our own unit's first mission was 31 July in 1943. The target was Woensdrecht in Holland – it was a German airfield. We were attacked by German fighters and two of our airplanes were shot down. We had Spitfire escort and two or three Germans were also shot down.

On the base itself we settled into an operational routine. Personnel were dispersed by squadron in various areas. And I do mean dispersed. You'd certainly get a workout. It was three-quarters of a mile from my hut to the mission briefing room. Then the airplane was a mile from the briefing room. Most of us had bicycles.

We had few problems, bar one major one, with *Litljo*. It was an exceptional plane that eventually flew over 100 missions. Occasionally, however, we'd fly in a different plane or individually join a different crew for a mission if an operator was sick.

After the each mission we'd head to the briefing room with the Intelligence Officer for debriefing. Then we went to the Mess Hall.

One odd thing was the blue badge: If you had gone on combat missions you were able to wear a blue rectangle just under your wings. It wasn't an official thing – it happened just like the British RAF fighter pilots would wear the top button of their tunic undone. It was a message.

After about nine missions we transferred to Great Dunmow, Essex, on the 23rd of September 1943. I liked that base and I liked the local village. One thing about the place the particularly struck me was hearing about something called the Dunmow Flitch. Once every four years, in peacetime at least, married couples would try to convince a jury of six maidens and six bachelors that for a year and a day they had not wished themselves unmarried. The flitch of ham was the reward. Even though I never saw it take place there that really appealed as a concept.

I had my 22nd birthday in 1944 – it was just the right age to be in England. I did go to London and I went to the theatre and to music concerts. I learned to dance and had a good time there. At that time the Royal Opera House was a dance hall.

I got down to London and whatever happened, happened. I might see a movie or a show. I sometimes went to the Rainbow Club – the US Forces Club in London. I was interested in music and I went to Westminster Cathedral and was able to stand and pay respects at the grave of Henry Purcell. Another time I went to the British Museum and, of course, there was the Rosetta Stone.

I was a Tech Sgt – which was equivalent in pay at least to the rank of an English major at that time. That made a difference. I can recall that Lady Astor, the American-born woman turned British aristocrat and the first woman to sit in the Houses of Parliament, proposed that the American Forces pay be cut back to the level of the English. That didn't get very far!

So every flight we'd get up to altitude – usually 10,000 ft – then I would go back to the guns. On one mission we were going over to France and a lot of German fighters were up. I saw a plane going down and realized that it was my buddy Tech Sgt. CH Burdict's plane.

War does strange things. Rather than worry about him or mourn him – all I could think at that moment was: "Burdict, you bastard, you owed me £5!"

As it happened, Burdict was picked up by the French Resistance and they managed to get him back to London. And so I got my money back!

Of course we were all stereotypes. The Engineers were seen as dirty. The radio operators were lazy and the armory gunners were seen as the dumbest. If they had had any more smarts they would have gone to a Tech School!

There was always a bit of tension between radio operators and engineers. I used to get at George Cheadle, our engineer on *Litljo*. I had a pretty easy time on most of our flights – but he would worry all the time.

In those days, especially in the winter, if there was a mission in the morning they'd put you on notice late the night before. But if the skies were cloudy I'd come back in and tell him: "George, I saw a star – we might be going after all." Then, after a little while, I'd tell him: "Don't worry George, it's cloudy. And then after a little while more: "George I just saw two stars." This had him going. We did wind each other up.

And then it was D-Day.

You could say we had some clues. Everybody was restricted to base. And then, as I recall, late in the afternoon the day before they started painting black and white stripes on our planes. So everybody was speculating. That evening the Squadron Commander went around to each hut and said: "Get an early night tonight." Every hut asked him the same thing: "Is this the big one?" "Can't say," came the reply. "Just get a good night's sleep."

We started early that morning, about 3-4 a.m.. Speculation was at fever pitch in the briefing room as Col. Joe W. Kelly got up. There was a big blanket covering the wall map. "Men, this is it," he said. Then they pulled back the cover and there was a big map of Normandy. "Two hours ago the 82nd Airborne Division parachuted into France here and the 101st parachuted into France here and the British 6th Airborne parachuted into France here. Right now, as I speak, there are 15,000 men on the ground in France. Yesterday afternoon ships from all along the British coast headed out. By the end of the day there will be 100,000 men in Normandy."

He concluded with the rallying message: "The troops will be landing at 6:30 am. And we will be flying over and dropping our bombs on Utah Beach five minutes before." There was a dramatic pause and then he continued: "The greatest military operation in the history of man has started. The 386th will be right there in the front."

This was going to be my 63rd mission and we were all veterans in that room – and still he built us up. There was no cheering but you could feel the optimism. We went out to our airplanes but the weather was terrible. We wouldn't have flown on any other occasion.

As it happened the weather was to our advantage because of low clouds. We had to fly at 2,000 feet or less, which would increase our bombing accuracy. We flew over the Channel and below me were so many, many ships. We were on target and on time. The Germans were taken completely by surprise.

My memories of the day comprised seeing the great armada of ships below us and dropping our complement of bombs. They had given me a K-20 handheld camera to use that day. It was an aerial camera and took 4x4 images. I took a number of shots that day. Later when they were processed I recalled something else that been subliminal at the time but that registered through the photography – the countryside of Normandy and in particular, the hedgerows around the fields. They were so thick – so distinctive.

We landed and debriefed and then I went straight to the Mess Hall and the back to my hut and to bed. After all, you can't be a hero all the time! We had been together and part of something momentous: D-Day.

And we were also together on the day we got hit by flak.

It was on the 15th of June 1944 during the Normandy invasion. I was up in the nose serving as 'toggler' (at that point some shortages of bombardiers dictated that a gunner handled the bomb drop procedures from up in the nose of the aircraft: opening bomb bay doors, setting the intervalometer device for bomb spacing and then releasing the bombs on the signal of the lead plane).

All I can recall was a 'boom' and then shooting stars.

Our pilot, Capt. Hillis, got hit through the wrist. We lost an engine and the hydraulic system was shot up. That would be soon become an issue going home.

But that wasn't a cause of concern for me. I had other problems. Lady Luck didn't have her eye on me on that day because I got hit in three places. My flak helmet, which I was thankfully wearing, probably saved my life. It was knocked from my head and I received a deep cut in my scalp. I was also shot in the hip but thankfully it was a flesh wound. I was also hit in the groin.

We flew back to an emergency base on the English coast at Friston. It was the closest landing point when you crossed the Channel. The abiding rule was that when you lost an engine you landed as soon as you could.

As we prepared to land it became clear that the hydraulic system was a problem. The main gear came down through force of the wind speed but the nose wheel wouldn't. We had to land so we prepared as best as we could – or should I say my crewmates prepared. I wasn't in much of a position to do anything except be stay where I was.

Those moments before landing are an eternity and then bump – wheels down. And then bang – the nose wheel collapsed. We went skidding along on our nose and then came to a stop. Everybody but me got out and ran. They all feared the plane, which still had fuel on board, was going to blow.

In that scramble they managed to leave me behind! Everybody was standing around about 30 yards away when I managed to clamber down.

Of all those people – and there were plenty – only one man came up and helped me: George Cheadle my fellow crewmate and squabbler (radar operators and engineers always argue). I'll never forget this – he put his arm around my waist and helped me away.

They took Capt. Hillis and me to the 93rd General Army Hospital at Oxford. They sent a message to my family that I had been seriously wounded. I think it must have sounded worse than it was. It was serious short-term, but in the long term I made a full recovery. Some days later they took the stitches out.

There's something I didn't mention about that flight that should probably be mentioned. The day we got hit we flew with Mabel. She was the pilot's pet dog. Aircrews did this on occasion but it was the first time we had a pet along.

Mabel didn't fly with us again.

I had been lucky. After hospital I went back to my squadron and after one more week I was cleared to fly again and so I rejoined a crew (my own crew had dispersed) for another mission over Normandy.

It was to be my last.

Something happened. I'm not sure I can fully explain except that it was a beautifully clear day as we got over Normandy. The first 18 aircraft were ahead of us and it was then I saw flak exploding around them. And suddenly I changed my mind about wanting to be there that day... or any day. And then we were into the middle of the flak ourselves.

I made up my mind as we were leaving the coast of France and going back across the English Channel: When I get back I'm going to get a pass to London and I'm not going to fly any more missions.

We landed and I got back in time to catch the Noon train to London. By 1:30 p.m. I was sitting in a London restaurant, the Lyons Tea Room in Piccadilly, and contemplating the incongruity of it all: This morning I was up getting shot at and now I'm sitting here having a good meal.

And that was that. I had done more than required number of my missions. I returned to the US and went home to Milwaukee on leave.

It wasn't the end of my flying however. I was sent to Langley Field, Virginia and flew training missions in B-24s. And life was easy...until December.

It was on 7 December 1944 that I became a member of the notorious Caterpillar Club. We were up on a training mission when the one of the propellers ran away and the engine blew up. The pilot commanded: "Everybody out!" Now you don't need a second invite when you hear those words. I was out that door so fast that it was a blur.

It was night-time but it was clear and there was hardly any wind. My parachute opened and I just happened to land not far from a farmhouse in North Carolina. The navigator jumped right after me and so we landed close by together and the farmer took us into town so we could get picked up and taken back to base.

Why the Caterpillar Club? Because from caterpillars came the silk that made our parachutes. If your life is saved by a parachute jump you become a member. I was given a small pin in the shape of a caterpillar that I wore on my uniform.

Despite this I was getting tired of training missions and found myself yearning for action. I got put on a combat team that was due to go the Far East but then the plans changed as the war ended. I was actually relieved when I left Langley in June 1945.

And then in September 1945 WWII was over.

I had left the Air Force in August 1945 and got a job working in roofing insulation back in Milwaukee in October. Quite quickly it was made painfully obvious to me that the dizzying heights of roof contracting were just that. I was on a roof and the scaffolding gave way.

This time there was no parachute and I broke my ankle. It was as I was recuperating that I came to the conclusion – civilian life is not good for me and so I re-enlisted!

But it was the best thing I could have ever done.

It took me back across the world and it enabled me to meet my wife.

My new route through reenlistment took me to McCord Air Force Base, Washington and then to Fairfield (which became Travis Air Force Base) in California. I had no particular skills other than being a radio operator so I was assigned to work and learn Air Traffic Control in the control tower. I began my ATC training in 1947 and became qualified as a tower operator.

In 1948 I was sent to Germany and the Rhein Main Air Base. This was during the Berlin Airlift. It was a hectic eight months. Given the necessity to get supplies in as fast as possible there was a virtual non-stop take-off and landing operation. As air traffic controllers for the single, 6,000-foot-long runway we would say to every plane: "You are cleared for an immediate rolling take-off."

And from there I went on work at Wheelus AFB in Tripoli, Libya (for a few months) and that was a change of pace. In fact, on my first tower duty, I barked my usual: "You are cleared for an immediate rolling take-off." The pilot swiftly queried the reason for this and so I explained…and learned to slow down.

Before long I headed back to familiar territory – England. This time I went further north to Burtonwood, then a major US Air Force base halfway between Liverpool and Manchester. It was a major staging post for the inspection and repair of aircraft flying airlift goods to Berlin. The control tower there had two people per shift.

From Burtonwood it was a myriad of military assignments and relocations. I decided to re-enlist and took a Strategic Air Command (SAC) course at Keesler AFB and went to Wright Patterson AFB. Then back to Europe – to RAF Wyton and then to RAF Upper Heyford. That's where I met my wife, Lottie.

She was a nurse. It was a dance on a Saturday night where we met – though I always told everybody and my kids that we met at Bible Study class! I had a motorcycle in those days – it was a Triumph Thunderbird 650cc. I didn't have much money then but I did have a great bike!

We got married in 1952 and Susan, our first daughter, was born in Oxfordshire in 1953 at the Churchill Hospital in Oxford. It was the same hospital I had been at nine years earlier when recovering from my post D-Day injuries!

We stayed at Upper Heyford quite a while with me working in the Ground Control Approach. Our standby was a 16'x32' tent that was located about 550 feet from the main runway.

The purpose of the GCA post to maintain runway operations even in bad weather. I worked on the ground on the flight-line. When the weather was bad – as it frequently was in England – we guided in the planes if it was just a half-mile of visibility using ground radar and talking them down.

We had some interesting times. One day a plane came in with only enough fuel for one run and the weather was below minimums, but we got him down.

The most memorable story of my working time there was on a quiet Sunday. The tower called GCA: "There's a six-car convoy heading out your way." Sure enough six cars come racing up and lots of suits got out lead by 4-Star General Curtis E. LeMay, who had headed US operations for the Berlin Airlift and had gone on to command the Strategic Air Command.

He took one look at the tent were using as our standby billeting and asked: "Why is there a tent out here?"

"We have a standby hut on order," the MSgt replied. Gen LeMay looked at the Base Commander who confirmed this; "Yes, Sir. It's on order for the second quarter of the next Fiscal Year."

"Have it here by Wednesday," LeMay commanded. And it was as if the voice of God had spoken.

By Wednesday it was there!

I spent three years at Upper Heyford before we went to Abilene, Texas. In 1957 our second daughter, Claire, was born while I was stationed at McClellan AFB in California.

In 1959 I went back to England to RAF Chelveston – a satellite base of RAF Alconbury. Then we hopped back to a Strategic Air Command base – Dyess AFB in Texas.

We were there for about two years and then I was deployed to Bein Hoa Air Base in Vietnam as Chief Controller of GCA (Ground Control Approach). It was a busy place. Not only were planes coming in and out all the time but they were all kinds of planes from transport to fighters, helicopters and even the U2 spy plane. I was proud that in the year I was there not one aircraft accident or incident was attributed to ATC. It was a pretty intense place.

It was in 1968 we moved back to England and to the USAF Third Air Force Headquarters at South Ruislip. We bought our house in village of Ickenham nearby and moved in during January 1968.

My job at South Ruislip was to look after the Air Traffic Control for the US Air Force in England and for five years that's exactly what I did. From there we undertook and inspection and analysis of ATC units at based around the UK and we'd go to each base – Mildenhall, Lakenheath, Bentwaters and Woodbridge and make a report and indications of what needed to be done to improved in procedures and with the systems.

South Ruislip was a good base. It was small, so most people knew each other. They had an awfully good NCO club there and they had some fine entertainment. I always remember Lonnie Donnegan performed there. There was also Two Ton Tessie O'Shea and even Englebert Humperdink.

When the USAF transferred its Headquarters from South Ruislip rather than move me to the new HQ at Mildenhall in Suffolk they sent me to Goodfellow AFB in 1972 where I saw out the last nine months before my retirement alone while Lottie stayed in our home. On 1 June 1972 I was promoted to Chief Master Sergeant.

Retirement happened at Goodfellow on 1 April 1973. I had 30 years and a couple of months in the US military and about 18 of those had been in the United Kingdom.

I was coming home.

(Author's Note: Henry Meigs Farwell passed away on 3 September 2013 with a Service of Thanksgiving held at St. Giles Church in Ickenham held on 12 September. When he was alive his WWII uniform was given to Duxford Museum in England and his flak helmet with the shrapnel damage was donated to the Milwaukee Museum. It has since been transferred to the US Air Force Museum at Wright-Patterson Air Force Base in Dayton, Ohio.)

Above: Henry Farwell (*Back Row, Left*) and the crew of *Litljo*. Photos courtesy of Henry Farwell.

Farwell was in this flight of B-29s flying over the Channel on D-Day (6 June 1944).

Henry Farwell in
his official Air Force picture.

CHAPTER THIRTY-THREE

RELOCATING LONDON CENTRAL HIGH SCHOOL

Rolla Baumgartner

Principal,
London Central High School
Bushey Hall, Watford 1969-71
High Wycombe Air Station 1971-72

(EDITOR'S NOTE: London Central High School was the main High School for Ameri-can military dependents whose families were stationed at bases around London, the UK and even overseas. Originally opened at Bushy Park, Teddington, to the Southwest of Lon-don in 1951 it was relocated to Bushey Hall in Watford in 1962. Rolla Baumgartner was instrumental in what would be the school's final move west to High Wycombe Air Station (aka Daws Hill). LCHS, as it was also known, operated from High Wycombe Air Station from 1971 until 2007 when it was closed down as part of the general drawdown of US Forces in Europe.)

I became the Principal of London Central High School at its Bushey Hall campus in Watford in the fall of 1969 coming from a staff position in the headquarters of the United States Dependents Schools, European Area (commonly known as USDESEA) in Karlsruhe, Germany.

The school enrolled about 750 students in Grades 7 through 12 at that time. I was 32 years old, one of the youngest members on the staff and was the fourth principal there in four years! It was a challenging assignment!

Near the end of that first year, in the spring of 1970, I was told of a proposal being floated to convert the Air Force Base at Daws Hill, High Wycombe, into a high school to replace London Central High School's facility at Bushey Hall, Watford. I was certainly open to any idea for facility improvement as the Bushey Hall facility was composed of many temporary World War II buildings with classrooms which were difficult to keep clean and maintain and were far from ideal for educational use. We could probably only improve our facility by any move.

High Wycombe was an American Air Force base which was unique in that the other installations were actually RAF installations, but High Wycombe was American from the beginning. However it was, at that time, no longer to be used by the U.S. Air Force as an operational base. The last base commander at High Wycombe, Colonel Nave, had been a strong school supporter and, as best as I remember, came up with the notion of using High Wycombe for a high school. It was a marriage of the need for a better facility and the availability of an entire base. I joined "the brass" for many tours of both facilities with the final tour being with the four star commander of U.S. Air Force, Europe.

Prior to a final decision rumors were rampant within the community and the teaching staff. At one time I recall being told by the Air Force that the project had been turned down. When asked about the situation at a teachers staff meeting, I told them that it looked like a move was dead. A short time after that we learned that the move was on again. We were going to move and to move fast! By the time the decision was firm we had only a year to plan, renovate and move!

Although the new facility was to be a good distance from central London, there was never any discussion of changing the school's name from "London Central High School". After all Watford hadn't exactly been in the center of London either. The school at High Wycombe would remain London Central.

Colonel Nave was appointed as the senior project officer. He was easy to work with and was very supportive of the project but was not involved on a day-to-day basis. Unfortunately the project engineer with whom I was to work on a daily basis didn't work for Colonel Nave but reported to the South Ruislip base commander, a Colonel Porterfield. At times Colonel Porterfield did not appear to share our priorities which made things difficult.

I soon learned that I was to spend half time as LCHS principal over the next year and half time planning for the conversion of High Wycombe into a high school. An additional administrator, Walt Radford, would be moved from Lakenheath to London to help out in the planning of the move into the new facility and would become the Deputy Principal of the new school.

An Air Force Major was appointed as the project manager and we began to make decisions about how to use each building. We got off to a slow start. Each time I would suggest a use for a building the Major would give me several reasons why it wouldn't work. When Maj. General Bell, 3rd Air Force Commander, subsequently visited High Wycombe he asked me about progress and I told him that we were having difficulty putting together my suggestions with his engineer's ideas. That was on a Friday and on Monday morning a new project manager arrived.

The new project manager was Captain Larry Harrison. They could not have selected a more skilled guy or one who was any easier to work with. He was a "can do" guy and we worked over the next nine months making decisions about building use and supervising the work that had to be done. Captain Harrison (now Colonel Retired) and I have remained good friends to this day.

What I did not know until later was that on that Friday afternoon Captain Harrison was called by Colonel Steele who was the Air Force Regional Civil Engineer. He gave Captain Harrison verbal orders to report to High Wycombe Monday morning where he would head up a design team and work with the British who would do the renovation of the buildings. This involved converting 29 buildings of an Air Force Base into a high school with two dormitories.

There were lots of decisions to be made and a short time to make them as the school was to open the following fall. On my very first visit to High Wycombe the buildings were empty with little or no indication of their previous use as military facilities. My task was to determine how the facility could best be used for school purposes. Captain Harrison and his work force made the modifications to the buildings to make them work.

There were many meetings with teachers to determine their classroom and other requirements. Like most projects there was not time or money to provide everything that everyone wanted. Although there was no official budget for the project, Captain Harrison asked for (and was granted) additional funds on four separate occasions. He and I worked many extra hours determining what facilities had to be provided and how best to present these requirements to faculty and to the Air Force.

The planning was pretty informal as I look back on it. I had my list of facility requirements based on projected enrollment and attempted to find available buildings which met those requirements. I didn't have to work from a budget as at that time the school system didn't fund facilities. The Air Force paid for the entire renovation to include the cost of the move. I know that Captain Harrison had a budget but I don't recall his ever telling me "We can't afford that!" regarding anything that was really important for the school.

It is also surprising to recall that there was very little paperwork from the school side of the project. Things had to be done quickly and decisions were made on the spot without a written record. I'm sure there was a paper trail on the Air Force side. The strong working relationship between Captain Harrison and myself allowed decisions without lots of paperwork. Today we'd probably have to build an extra building just to store acres of paper to protect ourselves. But we were young!

Despite a lack of formal paperwork we had a great deal of trust in each other and knew that the things we had agreed upon would be carried out. Some of these agreements ended up being real challenges since American requirements and designs were not always acceptable to the British builders. This was worked out on a case-by-case basis and resulted in getting the required facilities.

We were to open with about 1,200 students, approximately 300 of them dormitory students. The Base Chapel was made into the Library – a nice library I might add. The snack bar became our cafeteria for day and dorm students. The base gym needed no renovation and became an outstanding high school gym.

We remodeled what had been a USAF Weather Station into a Girls' Gym and made several small office buildings into classroom buildings. There was one existing large brick building, Building 900, which comprised a series of wings and became our largest classroom building.

A couple of the classrooms in the 900 Building were even wood-paneled, with one room having at one time been the Eighth Air Force Commander's office. It had a small rest room with shower and a chinning bar. Another room adjacent to the Commander's office was also wood paneled, having been used as a map room with sliding map boards inset into the walls.

The old base command building, at the front of the base, became our administration and guidance offices with the old Officers' Club becoming an AYA and student recreation area. The airmen's barracks became our dorms – one for the boys and one for the girls. My recollection, which could be faulty, is that we planned on two beds per room in the dormitories and I think we had two per room in many cases. The number of boys and girls was assumed to be pretty equal. The bath facilities for the girls had to be converted from gang-type showers, a conversion which was complex.

The existing fire system and hose cabinets were left in these dorms and, on one occasion after we opened, the students turned on the hoses and had a water fight on the top floor. This caused water damage and all the repairs had to be done immediately since there was no place to temporarily relocate the students.

We made an office building into a Medical Aide Station that the Air Force staffed with a nurse for the exclusive use of the dorm students. At some time after I left LCHS I'm told the dorms were named (Mansfield & Trinity Halls). They were just the boy's and girl's dorms when they opened.

There was an elementary school on the other side of the gym which had been used for some time. When the high school opened, the elementary school, which had its own principal, came under my supervision.

I don't recall that there were buildings that we didn't want to use and I think we used pretty much all but one of them. The one building and area that was "off limits" was the base's underground command center that had been constructed during World War II and was, I gather, a miniature version of the one in Colorado's Cheyenne Mountain.

Not many people at that time knew this facility existed or had seen it. The rooms were constructed on 'springs' and many other advanced construction techniques for that time. I never went in the building and I don't think that others did during the two years I was working with/on the base. Naturally, I always had a curiosity about it but never got there.

I don't recall other areas that were "off limits" for the general school population other than the base housing which was on two separate sites on the base. When the school opened I was given what had been the Base Commander's quarters. The quarters were very nice but resulted in my being "on the job" for very long hours and available at all times. A mixed blessing.

The year of planning and renovating went very quickly. The planning and execution of the move during the summer of 1971 was a massive project. Everything from chemicals to basketballs was moved and had to be in place in the new school.

We opened in September of 1971 although there were still many things yet to be done. I recall a few ditches for heating and gas pipes with wooden makeshift bridges over them. Classroom pencil sharpeners, maps and movie screens weren't up. It was a chaotic time as we welcomed students and faculty to the new facility.

The start of the school year was not made easier when we learned that a North Central Association (NCA) accreditation team would be arriving shortly after school started. The visit was within the normal cycle of accreditation visits but couldn't have come at a worse time for us. With the confusion caused by the move and the large increase in student body we were hardly well prepared. We passed but it isn't a good memory.

We welcomed a great group of young people to the new facility who were living their teen years through the tempestuous years of protests, drugs and a great deal of anti-American feeling as a result of the Vietnam war. They were an unusually bright group of students with lots of leadership skills.

I would have to think that students from Bushey Hall really liked the new campus as it was a significant upgrade from what they had. The new dorm students were mostly coming from Lakenheath where they had been in that dorm. The campus setting and feeling of the base was probably appealing to them as well.

A memory I have of that first year at High Wycombe includes the strike by electrical workers in the UK causing us to lose electricity at various times during the day. We had a back-up generator and you could hear students all over the area counting "ten, nine, eight, etc" with the lights coming back at the count of "one".

I also recall an assistant principal having to spend time roving in the woods behind the school herding back students who had opted out of a particular class. We had a detention hall after school in which students had to make up double the time missed due to 'skipping' class. Crowds in the detention hall diminished as the year went on as it wasn't an enjoyable hour for students. During the move I had Margaret Fuller as my Secretary, but after the move Mr. Fred Denton moved from the school at Eastcote to become my secretary. Both were British employees.

I enjoyed my three years in London but was ready for change. That period of the late 1960s and early 1970s was not an easy time to be working in a high school. The Director of the European Schools at the time said that he would not leave the principal of a large high school in place for more than three years unless the principal requested to stay. I left LCHS after the first year in High Wycombe, third at LCHS and headed to a new assignment as Assistant Superintendent in Naples, Italy. Following that assignment, I spent two years as a Deputy Superintendent in Stuttgart and 19 years as a Superintendent in the DOD schools in the United States.

In the summer of 2002 I had the opportunity to visit the school at High Wycombe – the first time I had seen it since 1972. Many of the buildings we had used were no longer used as school buildings. The Chapel had, I believe, sometime after I departed reverted back to being a proper Chapel and, then, in the ensuing years, reverted again to become as the school music/performance centre.

I was told that there had been a bit of controversy about that building. It seemed that there was an Air Force regulation which decreed that no Chapel could be abandoned for other purposes without the approval of the Air Force Chief of Chaplains. We had not known of the regulation and had not sought such approval. I was told at the time – probably in the mid-1970s – that the Chaplain Corps took it back and that it served those who lived on the post.

When I visited in 2002 all of the buildings near the gate were no longer used by the school. A multi-storey bespoke classroom building had been built at the back of the base which housed the better part of the school. The enrollment had gone from the 1,200 plus at opening to fewer than 300 students. For me the chapter was closed. It was hard to imagine the facility as seen in 2002 as the vibrant, alive and exciting place it was in 1972.

CHAPTER THIRTY-FOUR

PROGRAMMED TO RETURN

Steven J. Guenther ACS, CL, SMSgt USAF (Ret)

South Ruislip
Jan 1970 – May 1972
RAF Uxbridge
Sept 1985 – Sept 1988

I had never heard of the Cumberland Hotel next to London's Marble Arch when I started talking to the red-haired, hazel-eyed Irish girl that night at the Oak Rooms in Lancaster Gate. The hotel itself really didn't interest me but Ann O'Neill, who worked there as a chambermaid, did.

I was a USAF Airman First Class (A1C) with the Third Air Force Headquarters Intelligence Directorate, stationed at USAF South Ruislip. I never knew an awful lot about the hotel, even though I did visit it a couple of times.

But I did learn a lot about Ann, Ireland and the village of Newmarket-on-Fergus where she was from.

In fact, it wasn't long before my 'intelligence gathering' was complete and I had the answer to my big question. I knew I couldn't live without her – we were married a year later!

My tour in the London area was totally driven by luck. I started out my time in England assigned to the 79th Tactical Fighter Squadron at RAF Woodbridge, near Ipswich in Suffolk. My unit was selected to refit to the then new FB-111 aircraft and during the transition the Intelligence branch would not be needed. I was reassigned to an open slot at South Ruislip. So 11 months after arriving in-country I was on my way to London.

I arrived at South Ruislip on 20 January 1970. The base was small, almost postage stamp-sized compared to what I was used to. There was one main street going in from the front gate and the first large building on the left was the headquarters of the Third Air Force.

A couple of hundred feet further in on the left was the BX (Base Exchange) and Cafeteria, on the right side of the street, the base Photo Lab (a place where a lot of my work actually was done). Going one street in on the left brought one to the barracks, where I was to live for the next few months. I was told it was a converted shoe factory although I never knew if it was true or not.

My office was on the first floor of the headquarters building, down a long corridor to the right of the entryway, beyond the front desk where the armed Air Police (as the Security Police were known) guard sat. It was in a vaulted room behind a big, thick, green, steel door that was locked at night.

There were four or five rooms in the vault and I worked in one of them with my supervisor, Captain Bonita H. Markison. I'd say there were about six of us working in there, although I never knew much about the others. They were part of the Air Force Security Service... very hush-hush.

My job as an Intelligence Specialist was to maintain Order of Battle maps and lists, sort through documents and prepare projection slides that Captain Markison used to brief the General in command of Third Air Force each morning. It wasn't exciting but it certainly was enlightening.

Shortly after my arrival there I was given permission to move off base and found a bedsit near the underground station in South Ruislip, about a half-mile down the road at 19 the Fairway. I don't remember the name of the family that owned the home but they were quite gracious and I remember several times having 'Tuna on Toast' (cheese on top) with the husband in the kitchen.

A few months later one of my co-workers (Harry MacKay) and I rented a house together in Harrow Wealdstone at 3 Belmont Road. That didn't last long because on August 21st of 1971 Ann and I were married at the St Gregory the Great Church in South Ruislip. Ann bought the bridesmaid dresses and veils at the Marjorie Spence store in Hammersmith.

We had our reception at the NCO Club on base (we didn't have much money then and that was the best way to go). We moved into an apartment on Rydal Way in South Ruislip that we rented from a Royal Air Force Colonel whom we never met.

We had many fond memories of our time there. The nights down at 'Bangers' on Moscow Road in Bayswater were historic. We would scorch our names or initials on the ceiling with candles while drinking Lowenbrau and sing songs to the music of an accordion played by a fellow sitting on a stool up front by the bar who wore lederhosen and one of those caps with a feather in it.

There was a memorable snowball fight in a snowstorm at midnight on Christmas Eve 1971 in the middle of Queensway. There were eight glorious days in Benidorm for our honeymoon and a five-day trip to Rome.

We didn't have much in the way of interaction (or at least I didn't) with anyone off the base except as we met them at, or going to and from, parties. There was a mini-cab service we used a lot and the driver we always seemed to get was a wonderful fellow who worked too much. I remember a couple of times we were afraid he'd fall asleep at the wheel and we had to nudge him.

My mother, father and little sister came over from Florida for the wedding. They stayed with us in the apartment as well as while we were on honeymoon and did some traveling on their own (Stonehenge, Stratford upon Avon, even Scotland. The Cultural difference between their lifestyle and the local one in which I now found myself became obvious to me (sometimes almost painfully so) after my even limited exposure to British living.

One particular time was when mother opened the kitchen window and shouted out my sister's name a couple of times to call her in for lunch. I winced when I heard that N her calling out; "Trudy... Trudy... lunch!" I could only hope the neighbors either didn't hear or couldn't tell from where the shattering of the peace came.

This would not be an uncommon occurrence back home in the US but now it seemed far from normal in the quieter society of the London suburbs. No matter. On that visit we did a lot of sightseeing together going to the Tower, Covent Gardens, St James Park, Hyde Park, Buckingham Palace and so on.

There was also a Christmas Dinner Dance at the Cumberland Hotel. Ann brought me as her guest and I got to meet her boss, Mrs. Hancox. She was a very proper, stern lady, always in charge and keeping her "girls" in line. Still, there was a twinkle in her eye. Ann told me later that she was quite impressed with me. I naturally replied… "Of course!"

I remember buying bunches of 'Daffs' for a pound or less at the South Ruislip tube station to bring home to Ann after work and buying 'the messages' (Irish slang for 'groceries') at the shops on Victoria Road.

There were many, many sightseeing trips all over London as well. And the unforgettable trips in Harry MacKay's old mini, crammed full of people as we went looking for parties in London and over in Pinner (the US Navy guys who lived there had great parties). There was a hole in the floor of the back seat of that old clunker and we could look down and see the road flying by as we went from party to party!

I remember Sergeant Bing Young, a fellow that worked for 3rd AF personnel and Colonel Lester McCloud the Deputy Chief of Staff Intelligence for 3rd Air Force, Captains Markison and Bill Woodin (my bosses while there) and co-worker Technical Sergeant George "Joe" Shaw, the Best Man at our wedding.

A great memory also were the all-night card games in the South Ruislip dorm. The dorm was divided up into rooms that had cinder block walls but no ceilings. If you locked yourself out all you had to do was scale the wall. Any small valuables had to be locked up, since anyone else could scale the walls as well.

I remember Staff Sergeant Richard "Herbie" Braun, my first supervisor there. He had married a British girl and was a silent partner with someone named "Nat" in a men's clothing store in Ruislip. I spent a lot of my spare dollars there.

But it was my visits to buddies who worked in Data Processing that changed my world. They showed me a computer punch-card sorter and I was so enthralled that I got out of Intelligence and into the computer field.

After leaving RAF South Ruislip and HQs 3rd Air Force in 1972 my career carried me to West Texas for a short period. I was no longer in Air Force Intelligence but had retrained as a Computer Programmer. This was the dawn of the computer age and I loved it. In fact, I'm still in the computer field now as a Systems Analyst/Programmer.

So imagine my surprise when, after a year and a half, I was dragged back into Intelligence and assigned to Nakhon Phanom Royal Thai Air Force Base in Thailand. My one-year tour there ended with my being reassigned to an Intelligence unit at Offutt Air Force Base in Omaha, Nebraska.

It took me four years to get back into the programming field and, after two more years, I was reassigned to Washington DC to work with the Air Force Office of Special Investigations (AFOSI), again, in the computer field.

When you are in the military you never really say 'goodbye' because there's a good chance you'll be back. And so it was with me.

After three years in Washington DC, I was reassigned to work with an AFOSI District office – District 62, located at RAF Uxbridge back in the UK. The base was actually an RAF Headquarters and it was located just a few miles from USAF bases at Eastcote and West Ruislip. By this time South Ruislip was gone. The 3rd Air Force Headquarters had relocated to RAF Mildenhall in Suffolk.

In September 1985 my wife, Ann, son Michael, daughter Patricia and I arrived at London Heathrow to begin what would be a three-year tour of duty at RAF Uxbridge. We were met at the Airport by the man who would be my supervisor and co-worker, Robert

L. Thornton, a retired Chief Master Sergeant AFOSI Special Agent who was in charge of computer operations for the OSI District. There were just the two of us and the area we supported included all OSI units in Europe.

Our mission was mostly training agents in how to use the newly adopted PCs in their jobs. I covered the UK, Spain and Germany and Bob covered the rest. My family and I lived in the Holiday Inn at Heathrow for three months waiting for quarters to be available for us at RAF Daws Hill, High Wycombe. Did we ever get sick of hotel food! We used to drive the children to a bus stop to go to school.

It seems strange now I have hundreds of pages of records and orders covering my twenty-two years in the Air Force, but virtually nothing covering my arrival back in the UK. I was, by this time, a Senior Master Sergeant (E-8, the second highest enlisted rank in the Air Force).

We lived in the enlisted quarters section of base housing at High Wycombe and I drove the 20 or so miles to and from Uxbridge every day. After a while I got into a car pool and four of us rotated the driving one week at a time.

Our house at Daws Hill was on Alabama Circle. I don't remember the number, but do remember there was a public swing set in a grassy patch to the left of the house when facing it. I know the house was on the outer edge of the base, because when looking through the board fence in the back yard we could see British civilian housing.

On the day that we first saw what would be our house (it was under renovation at the time) we also took the first step to adding a new addition to the family. As we walked up to the front door we discovered the lady in the house on the other side of the swing set had a litter of Cocker Spaniel puppies. Since we had promised our children a dog when we got to England, we bought one.

We had been bombarded with "Please!" and "You promised!" comments. Ann and I never had a chance. He was a pure bred and we registered him with the American Kennel Club.

He was the runt of the litter but grew pretty big. Very fluffy and fuzzy and the Security Police used to jokingly ask if I wanted them "to shoot the black bear" that was romping out on the front lawn as they walked by on patrol. We named him Black Prince of Daws Hill (Blackie for short). He was with us for 13 years and it broke our hearts when we had to have him put to sleep.

I was never very involved on the base. I wasn't very sociable and didn't even belong to the NCO Club. In any event anyone working for OSI was not too popular. The fact that I wasn't assigned as an agent but in a support role probably wouldn't have made much difference.

There were two families that we became close friends with; Dave and June Watkins (Dave was a US Army Recruiter who worked the England and Ireland areas) and Dennis and Cathy Glass (Dennis was the senior NCO at the base medical Dispensary).

My OSI duties involved getting a computer room built for District 62 and travelling around to the different OSI Detachments in the UK, Spain and Germany to teach Computer Literacy.

The computer room we had installed housed a Wang VS-100 and the PCs in use were Zenith Z100 and Z150s. Looking back on it, one could almost say there was "much ado about nothing" in that all the system was used for was word processing. We had a centralized system with a lot of 'dumb' terminals hooked in. All of the offices were wired and the computer room had a huge 'patch panel' where they all came together to connect to the main CPU. These units were independent of the PCs that the agents used and so we had to teach in two different computer environments.

This was the time when the PC was first coming into its own. There was no easily accessible Internet yet. I worked on a few simple programs some database applications and standardized forms. We managed to get a very basic network up and running in the form of telephone modems that were used by the District units to send and retrieve reports. In an organization that was heavy into reports and paperwork it made a big difference in daily operations and efficiency.

Uxbridge itself was an RAF base and when I was there OSI occupied a lone building down a tree-lined street. I don't even remember exactly where on the base it was, except that we used to turn left inside the front gate. There was a courtyard in the back with a long, low outbuilding e used for storage and a workout room with exercise equipment.

In family time we used to drive to USAF base at RAF Upper Heyford every two weeks to go shopping at the Commissary and there was the occasional medical appointment. Ann went on lots of shopping trips with other wives to buy Wedgewood pottery and so on. She was a member of the Wives Club there and went to a lot of their activities.

We also used to go to the Black Knight store at Upper Heyford and built up quite a collection of David Winter Cottages. My mother came to live with us for six months, too. Our children went into DoDDS schools. Patricia attended the elementary school at West Ruislip and Michael attended London Central High School there at High Wycombe. When we were there for about a year or so, Ann started helping the lady that did the washing for the resident students who lived in the dorms at London Central High School and soon was the sole "Laundry Lady" for all of the students. She worked hard at it and did a good job.

There was so much going and coming between the two dorms there that she eventually managed to contract a bad case of pneumonia and had to spend a few days in the local Wycombe General hospital there in High Wycombe. They wouldn't even transport her to USAF hospital Upper Heyford as they said the trip would be too much for her. It took a good few months for her to recover. After that she found another one of the wives on base there that helped her out. That lady even took over the job when we left in September of 1988.

We loved going to the local town market there in High Wycombe. British supermarkets were also part of our shopping trips including Asda and Tesco. We used to get fruits and veggies from a local vendor who came around the base once a week in his van and we'd rent VCR tapes from another vendor who likewise made the rounds.

Our milk was delivered daily and we watched the Armed Forces TV, the British soap *Crossroads* and of course, the other major British soap *Coronation Street*. The TV shows I enjoyed were *Callan* and *Special Branch* and a particularly British comedy, *Steptoe and Son*, was also big at the time.

I couldn't say there was any 'drama' in our lives there except for one incident at a fast-food restaurant in the town of High Wycombe itself. We stopped for lunch and some local toughs came in and started to harass the customers. One of them picked up a chair and made to hit someone with it.

I grabbed him from behind and told him not to hurt anyone. It was really rather comical. I was certainly no hero and would not have lasted two minutes in a tussle with any of them. However, my action was so incongruous that it seemed to break the mood and everyone was back to munching their burgers a few minutes later. I think I was really in more danger from an angry wife who told me I shouldn't have gotten involved than from the bully boys.

The only mysterious part of our life there at High Wycombe was towards the end of my tour. All I knew was that a very new, super secret, underground installation was being built. I never knew anything more about it; don't know if it was ever finished or if it even ever really existed. I didn't have what was called "a need to know". Actually…I didn't want to know.

Shortly after arriving back in England, I bought an Austin Allegro car to get us around. It was a heap.

I was really taken to the cleaners with that purchase and one of the local High Wycombe garages made a lot of money from it! However Ann used to drive it on shopping trips to Upper Heyford with Cathy Glass. Somehow they managed to cram groceries into the boot and back seat along with Cathy's two babies. We finally sold it and bought a Pontiac Grand Am through the BX (Base Exchange) at Upper Heyford. I had to take a ferry ride to Belgium to pick it up.

That last tour there at Uxbridge should have lasted longer. I had put in for a one-year extension and it was supposed to have been approved, but I found out that my position had been "eliminated". Imagine that! I was made… redundant! As a family we were very disappointed. But, in light of later events it was probably the start of a USAF personnel drawdown because, a few months after our leaving, it was announced that OSI District 62 would be merged in with the other European District (I think it was District 10). I don't know what happened to the facilities or the people after that. I was gone.

On 2 August 1988 we were on the way to what would be our last assignment before retiring: Reese Air Force Base in Lubbock, Texas – back in the USA.

Looking back on my five years and five months in the greater London area, I have visions of great fun, great sights and great friends. When I left the US for my first tour in the UK, my mother said to me; "You never know, you might come back with an Irish bride". How prophetic that turned out to be. I came to England that first time a single bloke, unknowingly looking for a soul-mate (most single men are) and found one. I came away much richer. Our family even found a treasure the second time around (our pet Blackie). For all those good times and memories, I raise a glass in toast to the UK and London.

Steve Guenther on his wedding day in front of a church in South Ruislip close to the base. Pictured from Left to Right: George and Gertrude Guenther, Steve Guenther and Ann, Ann's mother (Maureen) and Ann's cousin's husband (Joe Tibbels) who stood in for her father.

CHAPTER THIRTY-FIVE

THE LAST HURRAH FOR EASTCOTE ELEMENTARY SCHOOL

Ken Zebrowski

Principal
Eastcote Elementary School
1971-1972

The first I had heard of Eastcote Elementary School was in May 1971 while I was located at the Stephen Decatur Elementary School on the Italian island of Sicily. By then I had been in the dependent schools system for 10 years having joined with the Dependent Education Group of the United States Army, Europe in the summer of 1961.

A call from the Mediterranean Superintendent's Office in Naples informed me of a message from the Directorate, USDESEA (United States Dependent Schools European Area) as the school system was known by at that time) headquartered in Karlsruhe, Germany. They were offering me the Principalship of Eastcote in England for the School Year 1971-72.

A simple few lines of orders from the Director of USDESEA to the Superintendent of Schools for the North Sea District followed by Telex orders at that time simply saying: *OFFER MR ZEBROWSKI POSITION AS PRINCIPAL, EASTCOTE ES, EFFECTIVE SY 71/72.*

The opportunity for our family to relocate focused on our daughter, Cay, who was then attending Vincenza High School in Northern Italy as a sophomore dormitory student. Much as we loved Sicily it had been frustrating experience for us as a family because she had to go away to school. So the opportunity of a reassignment meant that she could live at home during her last two years of High School.

Having a bit of stability was important. After Kindergarten in Wisconsin from age 4 to 5, Cay's schooling involved 10 different schools in six countries including Germany, CONUS (Continental United States), Okinawa, Turkey, Italy and France (where I had performed my military service in the US Army from 1956-57)). The transfer would also place our son, Jon and our other daughter Ann, in Grade 5 and Kindergarten respectfully.

Confirmation of the Eastcote administrative position was soon received and we planned for our departure from what had been a wonderful hillside home overlooking the Mediterranean Sea.

We arrived in London in August 1971. I went through the routine of processing through the Civilian Personnel Office and then checking in at the USDEASEA North Sea District Office then at South Ruislip, the HQ of the Third Air Force.

After meeting the Superintendent's staff we left and drove a couple miles via the Field End Road to Eastcote to visit the school.

I was surprised and pleased with the route traveled through the quiet British locale. I found the surrounding neighborhood provided the school and playground area with a fit setting for the pupils.

A walk through the school's sloping central corridor with its highly polished tile floors made a good initial impression. The classrooms, library, office and storage areas were all secured for the summer recess. I looked forward to checking my office and reviewing the files in preparation for the new school year.

Important to my family were the other local schools and the home we found on Rectory Way, Ickenham, was convenient for nearby (about half a mile) West Ruislip Elementary School (a second elementary school for military dependent children in the area) and for London Central High School (about 17 miles away) accordingly.

The August orientation for that 71-72 School Year gave me the first chance to meet the staff of 10 teachers and support personnel and outline plans for the 225 pupils in grades 1-6. Preparing their classrooms and attending faculty meetings were familiar routines for the veteran educators. A reduction in the pupil population required the transfer of a teacher from the previous year's staffing levels.

My check with the Civilian Personnel Office identified the individual with the least seniority for the personnel adjustment. However, when I approached her to explain the need to transfer due to staff reduction she was quick to note her already numerous years with USDESEA. While concurring with her in regards to her considerable longevity in the schools system, it was with regrets I had to point out that her colleagues had even longer periods of service and that she was the 'junior member' of the faculty at Eastcote.

The School Year commenced as scheduled and minor adjustments to classroom activities were coordinated by the Principal and faculty. Two of the things I remember were the playground recess schedules needed changes and the routine fire drills shifted exits.

However a surprising turn of events took place in November 1971 when a USDESEA North Sea District office staff member transferred. The Director of Administration position was temporarily given to me based upon my administrative background in the Offices of the Superintendent in France, Okinawa and Turkey.

To accommodate the school and District duties, I split my duties as Eastcote Principal to mornings and then drove to South Ruislip and assumed the multi-school tasks at the USDESEA office for the afternoons.

The Eastcote school staff adjusted well to the situation of a half-time principal, but further news of change was to follow.

In Spring it was announced that Eastcote Elementary School would be closing in June 1972. The announcement was made formally by Colonel H.L Porterfield, then Commander of the 7500th Air Base Squadron and Base Commander of the Ruislip Complex.

The school had opened in 1953 and had 10 classroom and other special purpose rooms that had once formed part of a women's prison according to the *Overseas Schools* newsletter of the time. (*Author's Note: There is no other confirmation that this was the intention for the base as far as the author could discover. Everything points to it being intended as a D-Day Hospital. See entry earlier in the book about the Eastcote base site.*)

The closure required the teachers to complete transfer paperwork for new school assignments. The June departure of the faculty was coordinated and arrangements made for the clearance of the furniture, school supplies and miscellaneous maintenance and support equipment. It was a major undertaking that went well as the final phase of closure was accomplished on time. The Eastcote Elementary School doors were locked and the keys turned into the base office for transfer in July to the British Ministry of Defense.

It was during my final walk through the building that my thoughts centered on the veteran faculty and their dedicated educational endeavors that provided the Eastcote pupils quality learning.

At the time I was quoted in the local paper in the area as saying; "t was a really happy school with a marvelous atmosphere."

The closure of the Eastcote School was cited as another adjustment for the shifting of military personal in the UK and on the Continent. Departing the British compound on the final day rendered a professional resurgence that focused on the future of our worldwide school organization.

My Superintendent's Office position became a full-time assignment and I assisted our District Office staff in the move from South Ruislip to Daws Hill, High Wycombe that summer. I stayed there until 1975 before departing to Belgium to become Principal of SHAPE Elementary School in 1975. It was during this time that I received my Doctorate in Education from the University of Southern California.

CHAPTER THIRTY-SIX

'SIR' HARRY

Harry John Wynn

Engineer
Public Services Agency
USAF South Ruislip
1965 – 1971
Carpenders Park Military Housing Area
1971-1994

(Author's Note: I am indebted in particular to Neil Hamilton, a retired British police constable who had the Carpenders Park Military Family Housing Area as part of his 'beat' from 1988 when he was posted to Oxhey Police Station).

Mr. Hamilton has managed to maintain a website dedicated to the memory of the Carpenders Park which can be traced back to the 1700s when a house was built there by one Samuel Moody. Mr. Hamilton was there when the very last American family left and the site was closed.

Mr. Hamilton's website details the history of the 70-unit married quarters estate made up of a combination of three and four-bedroom, two-storey brick houses along with large single-storey homes (known in Britain as 'bungalows') for senior officers. The homes were fitted with both 110 and 240-volt electricity supply included polished marble flooring.

The 11-acre site was first purchased by the Ministry of Defence (MOD) for US military housing in 1960 for £65,000. Eventually when the post-Cold War US drawdown in Europe took hold, the housing stock was emptied and the facility was returned to the MoD. Time and nature then nibbled away at the buildings and site.

For a while it became both a location for a number of British television shows including "Silent Witness". British police also used it for police dog and police firearms training. In 1997 the MoD sold the site to a private developer for £9.5 million. The development of private homes by housebuilders Laing's and Barretts there is known as "Fairfield". Mr. Hamilton's contribution to the Carpenders Park history can be seen at: http://www.london-central.org/photos/carpenders.htm

It was Mr. Hamilton who was able to put me in touch with Mr. Harry Wynn and his wife Peggy. This is Harry's story.)

You could say that 1965 didn't go to well to start with. I was made redundant from ACI, an air conditioning engineering firm at the top end of Victoria Road in South Ruislip.

A friend, Sid Pantry, had a job nearby at USAF South Ruislip, the Third Air Force Headquarters where he worked maintaining the base boilers. It was he who got me the job at the base. I finished work on a Friday at ACI and started work the following Monday at USAF South Ruislip.

I started as a hot water fitter. That mean that any time there was a steam leak they would dispatch me to work on it and sort it out. The primary area to work on based was the usually with the Base Hospital. If there was a breakdown on the hot water system then we got out and repaired it.

If the heating went out in the hospital that was a priority. In the roof space at the hospital there were also a lot of air conditioning units that needed their filters changed regularly. It was a great job and I was made welcome by all of the boys there. There was a staff of 10 on maintenance for the base.

I guess you could say that my private life was also a bit steamy. I had been living close to the base but I had left my wife and kids. It was what it was. Peggy had lived a couple of doors down. She lost her husband at a very young age but had two daughters, Sandra & Shirley. Anyway, she asked me to repaint her bathroom. And here I am some 40 years later still intending to do it!

The USAF left South Ruislip in 1972 for new headquarters at RAF Mildenhall in Suffolk. But the base was still there and I stayed on providing maintenance to what became an increasingly quiet and sad place until 1976 when it was closed completely. It was then that I was offered a job at the USAF housing quarters at Carpenders Park and started work there on 30 September 1976 as a General Handyman working for the British Property Services Agency (PSA) and reporting to a US Air Force Civil Engineer.

It broke my heart that they were closing down South Ruislip but little did I know that it would open up the way for what I still consider to be a job in a million.

The Carpenders Park job was truly a pleasure. It was a 70-home estate located on a gently sloping estate near Watford. In a way it was like my own *Brigadoon* – the mythical village in the movie of the same name. It was a magical place to work and a magical job for me.

The main entrance to Carpenders Park was opposite the dairy on Oxhey Lane running from Hatch End. It was a brick pillar affair and to each side was a high wall of brickwork – over 6 feet high. As you drove onto the estate you could turn left or right. The majority of residents seemed to turn left – that was the American way to go! The buildings started with 1 Block – a block of four homes – and then there were bungalow houses from there: Numbers 2 – 9. Those were mainly quarters for Colonels or other senior officers. Then there was a back entrance to the estate beside the last bungalow which you could use to go down the road to the shopping centre at the bottom of the hill and the Carpenders Park railway stop.

Then it came to Block 10 – a block of four more homes (A,B,C & D) From there you would be heading up a corner and up a slight hill. Then there was 11,12 and 13 and you turned the corner and went further up the hill where you would find Block 14. Then you would come down to my shop, make a left and there would be other block (15). Two more blocks (18 & 19) and two large individual houses (20 A& B) were close to the back entrance. I recall in one of them lived the Jackson family who resided there for many years.

In the centre of all this was a playground, a club hall and the main site office where I worked. The main office was about 16' x 12' and then there was a workshop side and a kitchen. In the workshop we kept various fittings; gas fittings, water taps, spare parts for heaters and other bits for the homes. We also had 110 Volt and 220 Volt power points – the homes had both.

We would do all the repairs to the housing to a point but for major repairs such as roofs we would have to get in specialist contractors. Also the gardening was contracted out to a private provider. If it was a big job then I would phone the base at Daws Hill, High Wycombe where the PSA had its head office and they would send a team down. And then there were the general facilities upgrade which would also be undertaken from time to time with contractors coming in to do the work.

There were seven colleagues when I started at Carpenders Park. Phil was the electrician and he knew the place inside out. Louis was the carpenter while I was there for a year. I recall a lamppost slid off a truck and struck him while they were about to install it – he was out for a couple of weeks but recovered. John was the painter. The employee numbers quickly reduced however and eventually, 18 years later, I was the only one employed.

The incredible thing to me about the community was the politeness and the hospitality. Everywhere I went the wives would open up a window or door and call out to me: "Mr Harry, would you like a cup of tea?" It was lovely but it made it hard to get around the estate to carry out all the various jobs when you didn't want to turn down another cup of tea.

The working day was usually 8 a.m. – 5 p.m. But I adored the job. I was usually there by 7 a.m. – an hour early. It made good sense to me – I could catch people before they left for the day and they could tell me what was wrong. By the time they reported something through normal channels it would have taken two days to receive orders to repair it.

But sometimes we also worked out-of – hours. In one case a resident and her husband were having a party and the hot water went out. We were able to come back and strip down the boiler and install a new thermostat. I also was a key-holder and would have to go back when people locked themselves out of their homes. Sometimes we were also invited back to a meal or a party.

Carpenders Park was seldom empty. When one family moved out there were usually moving trucks there waiting to move the next family in. During the day, of course, the mostly fathers would head off to work and the kids would head off to school. The USAF police would patrol occasionally but our local British police would also stop by and patrol once a day. After about 4 p.m. the kids would start arriving home and then later on the parents who had been out working and then things would be jumping a bit before dinner.

I used to go to the old scrap yard near the old South Ruislip base and pick up wheels from prams and other bits to make go-carts for the kids. But who used to get on them and want to use them? Only the fathers themselves! They'd get on them up at the top of the hill and then come flying down the road and around the corner.

And it really was about community. There was an outdoor swimming pool just before the bridge near the housing area. On occasion I would skive off work and tell Phil the carpenter to hold the fort and I would go with a group of kids and teach them to swim. Later on I took the kids to what was then the new swimming pool at Highgrove. I loved swimming – I had started myself at Ruislip Lido. I think over the time I helped maybe around 100 American children learn to swim. And this was in addition to teaching my own wonderful children: Fran, Goff, Janet & Marie.

It was always fun and a real community with events like Easter and Halloween but on the 4th of July – Independence Day – it was something else. The families would bring things – barbeques, tables, food, games and drinks out into the centre of the park and have a real party. I would eat whatever was going. In my case though I'd drink just water or a Coke. But they didn't' do fireworks. For some reason the military authorities wouldn't allow them to have fireworks there.

One of the most memorable days for me was when USAF South Ruislip had a visit from the Queen Mother in about 1970. The base was always pretty but the boss had the local garden centre come out put small potted trees in a line. The visit went well but when they came back to pick up the plants I'm not sure if they knew that they went back one short. I believe one of my former colleagues may have had away with it.

Carpenders Park closed in 1994. I can recall that the Kathy Allen and her family were one of the last military families to leave.

And that also marked the end of my time there. I retired in 1994 and received a number of Letters of Appreciation and Commendation from various military and US civilian groups and organisations. On 3 May 1994 I also received the Imperial Service Medal. A nice note accompanied it from the Registrar of the Imperial Service Order: *I am commended to award the Imperial Service Medal which Her Majesty The Queen has been graciously pleased to award you in recognition of the meritorious service which you have rendered. I have the honour to be, Sir, your Obedient Servant.* It was signed by a British Lt. Colonel. The award itself was presented by a USAF Colonel.

It was a great honour and it also resulted in a funny case of mistaken identity thanks to the use of the word 'Sir'. I had to go to hospital in June 2009. They had to get an ambulance in for me. Unbeknownst to me at the time the ambulance personnel saw the note from the US Embassy – thanking me for my work – and Imperial Medal certificate up on the wall – with the aforementioned words.

However one of them read the word 'Sir' and mistakenly read into it that I had been knighted. And from that moment on I was suddenly 'Sir Harry Wynn'. I wasn't in great shape to argue but I tried. Even when I got to hospital and looked above my bed it was written "Sir Harry Wynn". No matter how much I protested to the nurses they were having none of it. It was 'Sir Harry' and when Peggy came 'Lady Margaret'.

To be honest it's not a bad mistake to make!

The main entrance to Carpenders Park from Oxhey Lane. The gates that can be seen across the road were only put in once demolition started. Photo courtesy of Neil Hamilton.

Harry Wynn visiting Carpenders Park just after demolition on the site started. Photo courtesy of Neil Hamilton.

CHAPTER THIRTY-SEVEN
STEGGY & OTHER STORIES

Edward J. Brennan, Ed. D.

Administrator at West Ruislip Elementary School (WRES)
(1978-1982)
Teacher, WRES
(1982-1984)
Math/Social Studies Coordinator,
DODEA Atlantic Region
Eastcote, UK
(1984-1989)
Teacher, WRES
(1989-2002)

During the 1950's and 1960's the USAF's Third Air Force Headquarters was located at a large compound in South Ruislip (until it closed in 1975) with support bases at Eastcote (closed in 2005), West Ruislip and High Wycombe (closed in 2007).

In the late 1950's plans were developed for the construction of a new elementary school at West Ruislip, a place best known for being at the end of the Central Line on the London Underground. The area had long-established links with the military going back to WWI.

The school was to consist of 20 classrooms in the shape of the letter 'H' with the the horizontal line being the link corridor to the classes on the vertical parts of the H. There were to be twelve classrooms on the west side and eight on the east side along with the Administrative Offices, Cafeteria and main entrance.

During the 1980's the main entrance was moved to the north side of the link corridor as the old entrance was blocked off and replaced by new offices. A multi-purpose room was built as a separated facility that was used for physical education classes, assemblies and community gatherings.

The playground was extensive surrounding the multi-purpose room including a large area north of the building known as the 'hardtop'. It should be noted that Mr. Clarence Kennedy was the Superintendent of the schools at the time of the construction and he was based in South Ruislip with a small office staff and one secretary, a far cry to what it consists of today in the Office of the Superintendent.

West Ruislip Elementary School (WRES) was officially opened in the fall of 1962 and it was to be another 44 years – to June 2006 – before the school officially closed it doors for good. The RAF base at West Ruislip was also closed and sold off for housing development. As of this writing the school building still exists and the building is being used by the London Borough of Hillingdon for child care and has been considered for use as a general school for British students.

During its 44-year history the principals in charge of WRES were: k

Mrs. Eunice Matthews	1962-1969
Dr. William Kilty	1969-1973
Ms. Gertrude Riley	1973-1974
Mrs. Lois Robertson	1974-1980
Dr. Edward Brennan	1980-1982
Dr. Karla Stark	1982-1988
Mr. James Burns	1988-1993
Dr. Raymond Smith	1993-1994
Ms. Judith Mayo	1994-1995
Ms. Cathy Magni	1995-1999
Ms. Martha Parsons	1999-2006

Over that time WRES (Grades K-6) also saw several thousand students pass through its hallowed doors. During the first 13 years the school had enrollments of 400 plus students a year, thereby relieving the nearby Eastcote Elementary School of its heavy student enrollment.

However in 1975, when the base at South Ruislip closed, the school's population decreased significantly to approximately 140 students and it primarily serviced the families of students who lived in the Hendon and Carpenter's Park housing complexes and the villages near West Ruislip.

There was also an elementary school at High Wycombe which serviced about 100 students which was adjacent to the London High School campus. In 1980 WRES and High Wycombe Elementary Schools consolidated at West Ruislip and for the next 26 years student enrollment ranged between 200-300 students – only decreasing rapidly in the final three years as the US Navy, who had taken over from the USAF as the primary military command authority, drew down and redeployed to Naples.

It was Mrs. Lois Robertson and myself who took on this consolidation project and I opened the combined schools at West Ruislip in 1980. There were many challenges regarding this change as many at High Wycombe military members did not want to send their children to WRES for obvious reasons. The distance between the two bases was about 20 miles and many weren't keen on their youngsters being schooled far away. However, this potential parent problem dissipated quickly over the course of the first year.

West Ruislip ES was a very active place where a myriad of educational experiences were offered to the students and some of the highlights include the following:

WEEKLY ASSEMBLY

Up until the late 1980's there were weekly assembly programs every Friday at 9 a.m. where various presentations were organized by staff members. All staff members took a turn in organizing a short program (10-30 minutes) in which the students and staff shared their learning. It included drama and movement presentations, activities around the various holidays, multi-cultural presentations, music events, talent days and guest speakers. One of the school's famous assemblies was a visit by a team of American astronauts during the 1990's.

During the final 10 years of the West Rusilip Elementary School's history, the school sponsored a Talent Day which included student presentations of all sorts and at times the staff would do something as well. Indeed, in the 1990's I became a member of the "Four Raisons" where myself, Dr. Maria Ramirez, Ms. Sally Wingate and Mr. Chuck Dudley got on in front of the student body (supposedly in disguise), dancing to Credence Clearwater Revival's version of *I Heard it Through the Grapevine*. The place went mad. We were disguised as raisons but some one jokingly told us we look like dried up dates!

MINI COURSES

During the 1970's and 1980's WRES conducted a mini-course program where students could take special classes which included: Drama, Board Games, Bridge, Chess, Tumbling, Gardening, Cookery, Folk Dancing, Knitting, Macramé, Recorder, Cardboard Carpentry, Choir, Flag Football and Library Club.

The classes would usually last for eight weeks, two times a week for one hour usually conducted during the last hour of the school session. It was well received by the students and parent involvement was strong. Parents taught the classes as well and the more there were the smaller the groups would be. The average group size was approximately 5-15 students.

HOST NATION

One of the key programs at WRES was the Host Nation program where the instructor was a qualified British teacher. This resource specialist offered classes on British culture to the students and organized a myriad of study trips and experiences for the students both in school and away from the school.

Day trips away from the school took in Hampton Court, the Unicorn and Wimbledon children's theaters, the British Museum, Imperial War Museum, Museum of London, Tower Bridge, Tower of London, Greenwich (Cutty Sark), National Portrait Gallery, the Tate Gallery, the Museum of London, local farms, Shakespeare at Coventry, visits to British schools and to the Hillingdon Outdoor Activities Center. This is just a sample of the many that were taken.

One of the most notable trips I was involved with was taking a group of students to London in the spring of 1979. The city of London was celebrating the International Year of the Child in Hyde Park for two days and WRES was fortunate to received 25 tickets for the event.

In order to go they had to perform in one of the pavilions and it was decided they could do some folk dances from other countries. We took extra time and did lots of dance practice for this memorable even. Some of the students had the good fortune to meet the Queen and the Duke of Edinburgh while another group met the then newly-installed Prime Minister, Margaret Thatcher.

OVERNIGHT STUDY TRIPS

For many years there were a number of extended overnight study trips that would last from three to five days. These included trips to the Isle of Wight, Kent, York, Bath, Cirencester, the Slimbridge Wildfowl Trust, Strafford-upon-Avon, Wales and an outdoor education week on the Devon seaside. When I was teaching I took my students on two five-day overnight study trips to Kent and to the Isle of Wight.

The reason for the two trips is that some years I would have Grade 5 and take the students to the Isle of Wight and then, when I moved to Grade 6 the following year, I would have many of the students again thereby a need to organize another trip. Many people thought I was totally mad but I believed strongly that this kind of learning experience was important for the children and that, in the long run, the payoff would something that could not be measured.

The trips covered all segments of the curriculum. It was life. I had the assistance of the Host Nation resource teacher, Mrs. Mary Andrews, who was very helpful in making the trip go smoothly. The five-day trip to Kent included visits to Chatham Navy Dockyard, Canterbury, Dover, Chartwell, Leeds and Rochester Cathedral.

On one of the Kent trips the students visited the National Trust site at Ightham Mote and participated in a drama with two other British schools. It was a wonderful experience for all.

The Isle of Wight trip included Beaulieu Motor Museum, Shanklin Zoo, Osborne House, Carisbrooke Castle, Portsmouth (HMS Victory and Mary Rose), the Spitbank Fort, the Needles Battery Fort, Portchester Castle and Winchester Cathedral.

There were extensive preparations before the study trips and upon their return the students spent a great deal of time in completing a scrapbook on their experiences. The students shared their projects with parents during an open evening during the first week of June.

Parent support for this activity was strong. Several classes that participated with these trips took it upon themselves to raise money to reduce the cost of the trip. They conducted fundraising activities such as bagging at the Base Commissary on a Saturday or Sunday, bake sales, book fairs and holiday catalogue sales.

One year one class which went to the Isle of Wight for five days raised over $4,300 and most of the students did not have to pay a penny for the experience! Over the years I took two trips to Kent and 12 trips to the Isle of Wight.

PARENT SUPPORT

Parent support at WRES was consistently strong and this was done through the Parent Teacher Group (PTG). This group raised funds to support the school and this was done by sponsoring bake sales, evening dinners at the school, book fairs, a Halloween fair, holiday catalogue sales (Santa's Workshop) and individual/family and class photo sales.

Over the years the PTG made tens of thousands of dollars to support school events that normally could not be funded by the DODEA school system. One of the most popular school activities was conducted at the end of the school year and that was "Fun Day" when the students would come together and have all sorts of team games and many times the staff/parents became involved. It was a lot a fun and a grand way to end the school year.

LIBRARY/MEDIA CENTRE BOOK FAIRS

During the year the Media Center (as the Library became rebranded) sponsored successful book fairs (both American and British) where students could spend their pennies on books. Also there were special visits by well-known children's authors who, over the years included Roald Dahl, Betsy Byars and Paula Danziger. These events were well received and supported by the community.

My class also sponsored a book fair for many years and the profits went towards the overnight study trip and the students also received many free books as well as having books for the classroom library. The class would have sets of one book so that they could read the books together in an organized format called Book Club.

WRES STUDENT COUNCIL

During the 1970's the Student Council was organized which was considered innovative for elementary schools at the time. Students were allowed to provide input in the school's programs and discuss issues that were important to them. While I was Principal, in the 1980/1981 School Year, the Student Council assisted in selecting the school colors and mascot which were to last until its closing in 2006.

The chosen colors were blue/light blue and the school mascot was a stegosaurus named Steggy Spike; in short – Steggy. During the summer of 1980 I purchased two stuffed animals (a dinosaur which was to become 'Steggy' and, also a walrus) at the Smithsonian in Washington, DC and brought them back to WRES as possible options for a school mascot. It was the year we consolidated as children moved from the elementary school at High Wycombe – it was a good time to decide.

Steggy became very popular and received with much affection from the students. During the 1990's the school's cafeteria was renamed the Steggy Café.

SCHOOL IMPROVEMENT WORKS

Over the years there were many projects that significantly improved the facility at WRES. Examples include lowering the ceilings in the classrooms, carpeting in all classrooms and telephones in all classrooms. A new and more efficient heating system was installed, the side cladding was replaced for the entire building, extension of the Media Center and development of the Computer Lab undertaken. Storage space was also added in the halls while there was construction of several offices to accommodate more education specialists and the Cafeteria was renovated.

One of the most extensive projects biggest took place during the summer of 1986 when asbestos was removed from the building; one would have thought the building was being dismantled. Finally, the traditional chalkboards were removed in 1993 and replaced by white boards, hence no more chalk dust for which my sinuses were most grateful! The building look in 2005 was vastly different from when it was first constructed in the early 1960's.

SCHOOL YEARBOOKS

In 1981 I took it upon myself to organize the first yearbook project for WRES. For the entire year I took heaps of pictures of the school's activity and then with the assistance of a US Navy printer from London we managed to publish the first yearbook which was given to all the students free of charge. After this first endeavor the yearbook was published by a professional company in the States – a process which would last until the school's closing in 2006.

SPECIAL SERVICES PROGRAMMES

During the late 1980's and 1990's there was significant extension of special services programs for the students which included music, physical education, art, gifted and talented, speech, pre-school handicap, computer science, compensatory education and health services.

This was in addition to the already present programs in special education, media and Host Nation. In the last 10 years extensive work was done in Information Technology where students did all sorts of projects using the internet and other media resources.

The school had it own Computer Lab with 30 computers and the individual classrooms had two computers with internet access. The late 1990's was a much different time than the years before that with regard to learning experiences for the students. Finally, by the late 1990's, a full time Kindergarten was instituted at the school.

THE WRES CAMPUS

Some of the highlights of the WRES campus include a memorial garden which was constructed in 1978 in memory of one its long-standing teachers, Mrs. Louise Pinneo, who passed away the previous year. There was also a hardtop play area where there was a giant outline map of the United States with all the names of the state capitals. Additionally there was a painted as well as a times table (and in later years a large computer keyboard) and a tetherball area.

At the time of writing this there are still located somewhere within the grassed areas between the buildings at least two buried time capsules.

On the west side of the primary wing an extensive play area was developed in the 1990's. Over the years two of the most popular games on the hardtop were Four Square and Tetherball; it was amazing how the students sorted out the play as the rules could easily be argued or fought over. That rarely happened.

My tenure at West Ruislip was a total of 19 years where I was an administrator for four years and a classroom teacher for 15 years. There was never a dull moment in either position I can assure you. While an administrator I was involved with teaching classes, organizing the mini courses and assembly programs, sponsoring the Student Council, providing in-service training on classroom management, yearbook project, organizing the North Central Association schools inspection authority visits plus a myriad of other activities.

While teaching I generally had a Fifth or Sixth Grade and several times it was a combination of the two. In SY 1983/1984 I had a family group of Fourth, Fifth and Sixth graders, a total of 26 students. A typical routine day in my classroom would include the following:

0825-0900: Administrative and opening activities and writing in journals.

0915-1035: Language Arts which including reading, writing, speaking and spelling. Also the students were scheduled for a special class during this time for 30-35 minutes in Music, Physical Education, Host Nation and Art. Spelling test with dictation was always on Friday and usually I was administrating tests at three levels.

1035-1050: Morning recess and snack break upon their return. Many students preferred staying in the class during this time so I allowed them to do this, two days a week. A limited number of students were all allowed to go to the Library.

1055-1145: Math and for most students this become their favorite class. This was my favorite class as well and we did all sorts of things from mental warm-ups, the lesson itself, non-routine activities, problem solving. My students loved math when they left this classroom.

1145-1225: Lunch recess.

1230-1300: I read to the students everyday and if I dared try getting out of it, I was in for trouble. They loved being read to and it was something that was done everyday of the school year unless we were on a study trip. On average I read 12 books a year.

At times, I would also read a book that most of the class had already read as the students were insistent I should read it to them, for example, The *Philosopher's Stone* by J. K. Rowling or *Mathida* by Roald Dahl.

One of my favorite books that I read to the children was *Great Expectations* by Charles Dickens. It was rewritten so that the children could follow the storyline and when I completed reading the book I would show the original 1940's black and white film version of the story. It may be hard to believe but they loved it.

1300-1445: We did all the other subjects including Social Studies, Science and Health. During the last quarter of the year I taught a program in Sex Education that was usually a topic that my colleagues would avoid and at times the parents would question. With the assistance of the school nurse and the base medical facility we developed an outstanding program. There was little complaint after it was over.

Also as part of the health program my students' participated in the D. A. R. E. (Drug Education) program which would last for 16 weeks and was taught by the local military security police.

I also took my students on a myriad of day study trips over the years and some of them included: British Museum; Annual Coin Fair in London and seeing a matinee play; the Imperial War Museum; The National Portrait Gallery; seeing a ballet or opera the Royal Opera House, London; Bodiam Castle, Battle Abbey (site of the Battle of Hastings, 1066 A.D.), Stratfield Saye (country home of the Duke of Wellington), Hampton Court, Windsor Castle, the local wildlife part and the Hillingdon Outdoor Activity Centre, Harefield.

During the last week of the school year we would have a fun day with bowling and a barbecue to follow. On one occasion we managed to go to Thorpe Park – an adventure park about 20 miles away in Surrey. These are just a few examples of some of the outside school experiences the students had.

WRES was the center of community activity for over 40 years. Thousands of students have passed through its doors of learning and feedback from many of them has been more than positive which can be easily validated. Over the past 25 years I have received letters and emails from former students who remember their WRES days as wonderful and positive experiences; ones they will not forget.

I have also had many students return and visit me in my home and now as adults have shared and reminisced their WRES experiences. The number one thing the students remember were the study trips they experienced.

In 2005 I traveled to Maine to participate in a wedding of a former student, Jim Hassard, (Grade 6, SY 1982/1983). While visiting he talked about his good times at WRES and he still has his scrapbook of his trip to the Isle of Wight. He was on the first Isle of Wight trip in 1982.

I'm pleased to say that Steggy still exists as it was given to me when the school closed in 2006. In turn I gave it to two of my former students who spent many years at WRES;

Jesse and Adam Tate. Hopefully they will pass Steggy on to their children telling them of the wonderful experiences they had at WRES. Jesse who went on to graduate from LCHS (Class of 2005) and Adam (who went on to another base school – Alconbury High School – Class of 2008) have both graduated from college (Tennessee State).

While the school buildings may be standing for the moment – there is some hope that the local council will reutilize it as a school (and indeed a childcare company had, at last hearing, leased about one-third of it). Our WRES is no longer and it now rests in the hearts and minds of the students and those who worked in the school to make learning fun for its students. WRES was my home for many years and the rewards from the work were plentiful. These successes need to be celebrated.

The author and Ed Brennan stand in front of the school on a visit to the site in 2008. The school building has been largely left untouched and is owned by the London Borough of Hillingdon. As of 2013 the school is presently home to the Growing Tree Nursery which can cater for 104 children. The Nursery's address remains as 'The West Ruislip Elementary School' and it operates 51 weeks a year for children from 3 months to 5 years old.

CHAPTER THIRTY-EIGHT

CHECK LIST

Mark Woodbridge

British Civilian
Bank of Fort Sam Houston, American Express, Bank of America, Nations Bank
(1985 – 2003)
Security Guard (Mitie Company), Daws Hill
(2007 – 2010)

It wasn't that I was a British civilian on an American base that made my job a bit unusual, although that in itself probably would have qualified.

Rather it was the fact that each night, as I sat in the office in a remote corner of RAF Daws Hill in High Wycombe, I would usually hold about $2 million in my hands.

It was in 1985 that I was recruited as a British civilian by Maxine Jones, an American lady who was the wife of a base commander, to the Accounting Department at the Headquarters Bank of Fort Sam Houston. At that time they were the local banking provider for American service personnel and their families in the UK and Iceland.

I had been born and grew up in High Wycombe and had been a trainee accountant. I believe the advertisement for the post had been in the local paper, the *Bucks Free Press*. I had known a little bit about the base but had never been on it.

And when I did go on for my interview and got past the armed officers at the gate where was the first place they took me? Only right to a building located adjacent to the base's secret underground bunker entrance.

And so I was hired and worked as a Check (or in British: 'Cheque') Processor/Account Manager for the bank which had the contract to service the US military in the UK. The office was about 50 yards long by about 20 yards wide and it was the nerve centre for the banking operations.

At various times during my time the contract changed from Bank of Fort Sam Houston to American Express and the Nations Bank. The contract would come up every few years.

My first day was overwhelming – like any job. I drove to the base – I had ID which would be checked by US Air Force Security Police at the gate. And then I drove down through the main base and then beyond it down the long road to the bunker area.

My job was to work nights processing the checks from the day that couriers would have picked up from between 13-15 bases for which we had the contract. The number of bases varied from year to year.

All the checks written by personnel on the UK bases that arrived each night were processed so they could then be sent out to the US the following morning. So, for instance, if you cashed a check at the American base at RAF Bentwaters in Suffolk one day it would be in America the next morning – with the time delay it might even be the same day on some occasions. We used to get the checks from US base in Keflavik, Iceland a day late.

I usually didn't eat at night and didn't chat too much as we did the work in the machine room. We did have a music player. We would encode each check and then when all the encoding was done we would go to the machine and run the checks.

We used to run the checks through a number of passes in the machine so that it would then sort them into number order so that they could be posted back to the account holders.

It wasn't dangerous work but it had its moments! We'd get finger cuts from the checks at times while processing them. We got an upgrade and the new programme meant we could sort them into order in four passes, but if you had an error message in those four passes you had to start all over again.

After the bank stopped posting the checks back we used to keep them and send them down for storage in a space in Desborough Road, High Wycombe.

Also about in the room were modems linking to the US and to Germany.

At first we had an IBM 3694 and then an NCR 780. The IBM machines were wicked for giving you electric shocks. Whenever I had to change the bulb or fuse I'd always get an electric shock. I was glad when they went!

There were three of us at night (as part of a total banking headquarters team of eight) and we had two sorter machines from IBM. The first machines we had were slow – we actually punched the check details in on the machine. By the time I left we were using the NCR 780 and then things sped up.

We used to process about 3,000 checks an hour. The problem was that the checks also still had to be stamped – otherwise we could have done 10,000 an hour. At that time the bank was sending the checks back to customers along with their statements.

During the week it might average out to 7,000 checks a day. But Christmas tended to stand everything before on its head in terms of statistics.

The US Forces would have to post gifts early to get things home and so check-processing numbers started climbing in early November. By December we probably had double the number of checks and would be processing something like $1-2 million in value a day.

And on some exceptional nights that figure was a mere pittance in monetary terms. There were a few occasions when savings bonds came or tax refunds came through and we would literally process through $600 million dollars!

And I once recall a truly exceptional amount – I processed through a US Government check through for close to $1 billion – for what I don't know. I think it was payable to the British Government. That was amazing.

I can still recall the codes we had for each of the bases:

75	Woodbridge
76	Chicksands
77	Greenham Common
78	Harrogate
79	Lakenheath
80	Mildenhall
81	High Wycombe
82	Upper Heyford
83	Croughton

84   Keflavik

85   London Navy & Brawdy

86   West Ruislip

87   Fairford

88   Wethersfield

Any codes with a 90 on them meant that they were UK Sterling Accounts which were available to the US Forces bank customers. These checks were processed through Barclays Bank.

In the early days the checks themselves were pretty boring – yellow paper with blue printing for some of the Sterling checks as I recall. But in later years as more personalized checks were introduced you'd see different designs and that made things a little more interesting.

In November 1984, I think it was, there was a fire in the Headquarters that damaged the whole of the Banking Operations building. I'm not sure they ever resolved how it started.

All I knew was that the building had to be demolished and we couldn't process documents for several months. When that happened we moved what we had in operations to a local hotel and then got the operation going again.

Eventually we worked in a 'pre-fab' place for about six months close to the Bowling Alley. Then a permanent two-storey building was erected at Daws Hill and our machines had a separate room which was where we stayed until the entire operation was transferred to Germany.

Over the years I was lucky to work with a number of nice people including Enid Gibbs, Angela Christopher, Yan Lada and Frank Popovich, among others.

Towards the end we were handling only about 3,000 checks a night. There were two reasons for this: The bases were drawing down but, also, people were using their credit cards and opening up more shop accounts.

I didn't know really about the rest of the base. I simply drove to the bunker and later, to the new building and back. There were some days I couldn't; not because of security but because of snow. It made the drive down and back up too difficult.

Only once did I ever see any real major activity at the bunker. And that was only when I encountered an armed Security Policeman who was dressed up in Nuclear Biological Gear. He looked a little bit like an alien!

At night the area around the bunker was eerie. There were bats about and you'd hear various animal noises. In a way that prepared me for what was to come – but more of that later.

In 2003 with declining military presence and personnel in the UK I was made redundant. I was a bit annoyed to be honest. I actually had to go to Germany to train people to do my job there just before I left and that was not the easiest thing to do.

So I was disappointed to leave and my last day was a downer – there were only two of us left. It had been a fun job and working nights was fine by me. By that time I had worked for five different contract banks.

But little did I know I would come back. It was just coincidence really that I did. I thought I d have get a stress-free job for a bit so I joined the local Tesco supermarket and did shelf-stacking and stock checking. But in 2007, I was looking around and answered the job call for a security guard. You can imagine my surprise at my first assignment – guarding the base at Daws Hill!

And so I came back in October 2007. My first shift was a night patrol with my colleague, Graham. And then my first main shift I went around once during the day and missed a couple of checkpoints. I hadn't been there for about 20 years. But the base hadn't changed at all really.

Having worked nights and seen or heard the animals during my time at the banking job really came in useful. Those of you who have ever walked in a forest at night alone will probably know what I mean. Every sound seems magnified.

On my solo security walks that made for a particularly eerie patrol – even in the daytime let alone at night. In the time I have been patrolling I have seen some Muntjac deer and a good few foxes. But the people – the people are long gone.

Mark Woodbridge stands in front of the former 'Gas Station' and Navy Exchange (NEX) Shoppette at RAF Daws Hill High Wycombe during the summer of 2009 while he was on patrol around the empty base. The top of the base's one remaining antenna, there since the 1970s, can just be seen in the background.

CHAPTER THIRTY-NINE

ON THE CASE

Robert Hanshew

Yeoman Second Class,
Naval Criminal Investigative Service (NCIS)
US Navy, London
1990 – 1994

Receiving orders to London, England in 1990 was, to me, the equivalent to receiving a Sailor of the Year award from the US Navy.

And it was all due to London's exclusive status as a duty assignment.

By that time I had been in the Navy since 1984 when I had enlisted at Pittsburgh, PA. Becoming a Yeoman through its technical school located at Meridian, Mississippi. I was fish out of water at that location. The South and I didn't mix – like fire and water. Luckily, the Navy first sent me to the Trident Training Facility at Bangor, Washington, close to Seattle, where I learned the basic functions on administrative organization and dealing with official Naval correspondence.

In August 1987, following that tour of duty, I was ordered to *USS DEWEY* (DDG-45). While on board, I was eventually promoted to Yeoman Second Class and became the Ship's Secretary, during which the ship departed on three Mediterranean Cruises and one Persian Gulf cruise. Upon decommissioning in August 1990 I had choice of orders, which I worked through the detailer (civilians should think of a detailer as a sort of headhunter/assigner) to be with the Naval Criminal Investigative Service (NCIS) Resident Agency in London, UK.

In London, NCIS was split into three locations. The main 'operational side' was stationed at a base called Eastcote, located in Middlesex. That base also was home to a wide variety of supply and fiscal support functions to all commands based in the European Theatre. To discuss funding and drop off fiscal and supply paperwork, I would sometimes travel there by the London Underground and take a quick, brisk 15-minute walk to the base and the NCIS offices, which were located in a one of interwoven building wings that connected through a central sloped hallway which also provided access building that housed eating and dining facilities.

At that time he Resident Agency of Special Agents was housed in government building not too far away from the Olympia Kensington Underground Station. The offices housed about six Special Agents and the administrative staff, which consulted of one civilian and one military employee. That latter one was my position when I first arrived. At the time, NCIS didn't have an office in the US Navy's Headquarters Building at 7th North Audley Street.

My own office and job was not to be in any of those locations. In a period of two months after I arrived a newly-created NCIS office of three personnel was organized. Through a lot of my hard work, skill and "painting," our new offices were took shape at the corner of Duke and Oxford Street, located above a bank and on the same floors as a British company called "The Retail Group" where I made a friend who remains close to this day, Mr. Mark Newbold. They took the three of us under their wing and we were invited to all their office parties, which was nice – and vice-versa. We couldn't have been more ideally located. We had wonderful views looking directly into Selfridges. It was a luxury.

The U.S. Navy Headquarters at 7th North Audley Street was a couple of minute's walk from our office and remained the central point for anyone stationed in the London area. Besides being the place where one would pick-up administrative papers and instructions, one could also receive haircuts by a cute Irish lass, open-up a bank account and shop for limited clothes and food at the small Navy Exchange in the basement. I don't know why but when I was stationed there and shopped in the Exchange milk was rationed!

Entering into the building was a sterile event but, as one turned to the right, which held an "events" room, it was clear to see the majesty of the building with the grand windows and lovely dining tables and chairs, which could be used for special events such as receptions and retirements. Most of the offices were used for basic office functions that monitored Naval facilities, bases and ships in the European Theatre. One of the interesting aspects is that the Navy didn't modify the building door and window frames that much and the dark wood still leaves an impression on me, even after many years of being away. The Library there was also quite interesting as one could occasionally find a good book to read, if one looked for an hour or two!

I wasn't an NCIS Investigator. Investigations were mainly handled by the team of agents based out of our office in Kensington near Shepherds Bush. Their job was to monitor if any U.S. Sailor or Marine got into either trouble and to prevent danger to them while serving abroad. In the case of criminal activity the agents' role was usually to interface with the local governments and eventually have the service member returned to the U.S. Navy for trial in a military court. They would also monitor if any crime (i.e. white-collar, physical, etc.) had been done by the military in the area and investigate if needed.

My assistance came in typing and proofreading the reports and then sending out information via their communication system, which was an e-mail system before it was widely used by the public. We used dial-ups and the system would break down with every other e-mail sent!

As stated, I initially served with the agents at Kensington. However, when the newly assigned Captain Dennis Neutze, Navy lawyer, Judge Advocate General Corps, arrived, I was reassigned to assist him and Commander John Crawford to review cases and maintain files on cases to see if any international law had been broken by all parties involved from the criminals to country officers to our Special Agents or even if other countries were stepping on international law in regard to the investigation of cases. Our Captain would look at the cases and address any international law issue, if there was one that came up during the tour of duty.

Due to these offices being new, we were slowly creating an office of U.S. Navy laws and regulations. Over a period of time we built something that was truly professional. Some of the building's décor dated from the mid-1950s and the railings were very interesting. Living close by I used to use the offices on weekends when I had to finish a report for studies or when I needed a quick trip to, as the British say, the loo!

Working in such a key location in London also had one additional (and often unexpected) perk. I got to see numerous celebrities such as Lauren Bacall (to whom I blew a kiss), Nile Rodgers (producer, musician), Robert Smith (from group The Cure), Rick Moranis (actor) and Robert Plant (from group Led Zeppelin) wandering through the area.

Too shy at the time I never approached them. But one day I saw Morrisey, the lead singer of a British band called The Smiths, while walking through the area. I didn't like the band – too much self-pittying and loathing to my mind – and so I gave him a quite dirty look, Perhaps to his credit he responded back with an equally dismayed glare.

Not wearing a uniform during any part of the duty was really a plus. It was interesting not to wear a uniform one single day of the rest of my tour. Unfortunately, over a period of time, I couldn't believe that I was still loosing pieces of my uniform that had been safety – or so I thought – s tucked away in a closet.

Overseas allowances were also healthy and for my last two years I was paid enough to live on Grosvenor Street – which was also near the American-Embassy and make my journey to work just a five-to-ten minute walk.

And for me another significant opportunity came through educational programs. For sailors and marines serving abroad there was The University of Maryland education, program, which then was based in the Headquarters Building, We used the American Embassy across the street as well. Taking advantage of this program I finished a degree in History while traveling to the European Continent for week-long classes.

Additionally, while serving in London, it offered me the chance to apply for university at Trinity College in Dublin, Ireland, to pursue a second degree. I attended there for one – and-a-half years studying Medieval History. Before meeting my then girlfriend, I used to make trips to Dublin and wanted to live there for awhile. With the money I saved and the G.I. Bill I was able to study there without working and enjoyed my time.

But by far the most important aspect of the London assignment was completely unexpected.

It was where I met Anna.

I had just finished my University of Maryland with a B.A. Degree in History in the late spring of 1994 and I took a walk into Grosvenor Square where I intended to just relax and read some music magazines.

It was then that I saw her – – my Polish beauty – coming through the park.. Other seats around the park were crowded and it was a case of will against hope that she and her friend would sit next to me.

They did.

Taking the opportunity I employed my sense of somewhat eccentric humor and managed to make them laugh, uncontrollably. Especially when I kept on dropping the magazines for a 'bad' attempt at humor, They then had to leave, but I kept watching for their return on other days. Would I see her again? The answer was yes! She was in the park most days for fresh air and to keep migraines at bay, so we'd meet and talk.

It took me two weeks but eventually I asked her out. We went out to my favorite restaurant in Soho where I knew the staff and used to have friendly banter back and forth with them. This seemingly eccentric behavior impressed her. It was 'Seinfeld' before 'Seinfeld'!

And then she left and returned to her home in Warsaw, Poland. I didn't follow her. I was committed to my History degree studies – this time at Trinity College, Dublin. But love is a pretty powerful thing and I followed her to Warsaw as soon as my degree was done.

In short we made some history of our own and more than 16 years and two boys later we're still making history!

While in Dublin I made some interesting friends such as Mr. Brendan Keenan who I consider a brother. We used to chat about love, life, poetry and history while sipping and eating in the Bewley's Tea Rooms.

Back to my time in London; it was clear that this tour would be my last duty station. How could any tour following that one top it? I've returned to the Navy, but as a civilian in D.C., now working for the U.S. Naval History and Heritage Command as a Photograph Curator.

When first checking in to my new assignment in London there was a long-serving civilian employee who stated: "Orders here will change your life." No words have ever been more true. Living in London taught me class, gave me an education and allowed me to meet the woman of my dreams.

From time to time I meet other people who served during the same time in London. All of us agree – it was the best tour that any sailor could have.

CHAPTER FORTY

SIGNING OFF

Capt Dave Dittmer, US Navy

Commander, US Naval Activities United Kingdom
RAF Daws Hill
Eastcote
RAF West Ruislip
Blenheim Crescent
US Naval Headquarters, London
August 2006 – 1 October 2007

It will probably go down as my most unusual and memorable tour of duty.

After all you don't expect to arrive at a new duty station and then have the unenviable task of taking all your children's friends and sending them away, firing everyone who works for you, closing your children's school and even 'evicting' your own family from the house they lived in.

But that was my job.

I had come to England to turn off the lights, lower the flag and lock the gates on what had been a lifetime – if not more – of US Naval involvement in and around London.

Since I had started in the submarine service in 1983 I had been looking for an overseas posting but had not been able to arrange one until after I finished my command of the *USS Nebraska* in 2002. And so, when an offer came through to go to Yokosuka in Japan, I was thrilled. I joined the US 7th Fleet staff along with my wife, Lauri, daughter Kathryn and son, Michael, on a two-year tour of duty and ended up getting extended for two extra years. During these four years we toured Asia thoroughly and were next looking to serve in Europe.

It was then that I heard from Capt. Steve Matts whom I had relieved as the Executive Officer of *USS Oklahoma City* in 1995. Capt. Matts was the Commander, US Naval Activities in the United Kingdom and was planning to finish up his tour in London in the summer of 2006.

One thing led to another and after a lot of phone calls the upshot was that while I was on the *USS Blue Ridge*, command ship of the 7th Fleet, I was directed to fly from Japan to relieve Capt. Matts as Commander, US Naval Activities in the UK (CNAUK). Even before I obtained my hard copy orders, Steve let me know that he had already commenced the base closure process – a part of a broader NAVEUR drawdown process.

But for a while it was touch and go for my new duty assignment. There was one school of thought in NAVEUR that Steve should be extended to complete the closure. Others contemplated 'gapping' the CNAUK Commander post and letting the new Executive Officer (XO), Commander Chuck Cordon, complete the wind up. In the end we did get our assignment to Europe and we had about a week for me to turnover with Steve and move into our new home before the official handover ceremony.

I recall that it was exceedingly hot that week and I began to think that I had been horribly misinformed about the British weather. Indeed it was so hot that there were long cracks in the parched earth of the back garden of our newly occupied quarters at RAF Daws Hill in High Wycombe, one of 70 homes that – in a touch of irony given the closure programme – we had just accepted back from the contractors after an extensive refurbishment programme.

It was so hot that they had instituted a smoking ban at the British Open Golf Championships in case someone's cigarette set fire to the grass!

My change of command ceremony was conducted under a huge marquee on the playing fields of Daws Hill and at the end of it I took over as Base Commander. It started to rain right after the change of command ceremony and didn't stop for what seemed like months. Now I felt like I was in the right place!

And now I was left to close down the remainder of US Naval Activities United Kingdom. Admiral Hank Ulrich, Commander of US Navy Forces in Europe, was driving hard to reduce as many of his bases as possible. He was established in the new Navy HQ in Naples, Italy, and was very aggressively reorienting the emphasis of the region.

He saw the focus of his command being "south and east" towards Africa and the Middle East. England was north and west. Our base at Keflavik in Iceland closed in 2006. The Navy's submarine repair base at LaMadalena in Sardinia, Italy, closed in late 2007 and, with the impending closure of CNAUK, it made three of seven US Navy bases in Europe closed within two years.

Admiral Ulrich's direction was clear and forceful: "Go to zero now!" meaning abandon our English 'ship' immediately. That precipitous course of action would have been very inefficient and harmful to our sailors, civilians and communities. I understood he was 'firing for effect,' so I steered a little less draconian course while meeting his intent to reduce the US Navy's costs and obligations in the region.

My mission was simple in its stated aim but complex in its execution. My command was split between offices in Blenheim Crescent in West Ruislip and RAF Daws Hill in High Wycombe.

It was my first time as a base commander and nothing I had done previously had prepared me for this challenge. I hadn't done anything with property management. I had to learn quickly the British Ministry of Defence regulations and procedures. I had to get my head around Navy Facilities Management regulations as well as learn the basics of civilian personnel management. And I even had to try to learn to speak the Queen's English!

At that point I had seven different properties under my command including:
* US Navy Headquarters, 7 North Audley Street
* RAF West Ruislip
* Eastcote
* Blenheim Crescent
* Romany House
* 2 Providence Court
* Edison House

Most of them were either already emptied or in a winding-up status.

When I took over there were about 150 personnel including 50 military and 100 civilians left in my command. A year before the command had had between 300 and 350 staff on station. Among those left on the military side were primarily Navy Masters at

Arms to oversee security.

As I learned the process wasn't just about handing each base back to the Ministry of Defence. First we had to 'advertise' the availability of the installation to see if other US government components might be interested in taking it on. When there were no expressions of interest then our primary option was to hand it over to the MoD. They were experiencing a cash crisis and were anxious to get all the property back they could so they could sell it off for development and use the funds to bolster their major area bases of RAF High Wycombe (Strike Command) and RAF Northholt.

Not all the installations I had would eventually be turned back to the MOD. The US military eventually would retain a small footprint just outside London.

The process of closing a base is very complicated. However in simple terms everyone accepted it worked like this: Imagine that you could pick the entire base up and turn it upside down and give it a little shake. Everything that drops out would belong to the US Military and we would have to package it up and ship it off to other Naval bases that wanted it. Everything else belonged to the MOD. That meant that a truck, cabinet or desk would be ours, but that the buildings, fixtures and fittings would be theirs…in the main.

Eastcote was empty and in the process of closing down. RAF West Ruislip was almost empty. The elementary school there had closed but we also had some 70 houses there to dispose of, too. They went to the RAF for housing although I was able to get an agreement that several US families who would remain on duty in the UK could stay in residence. Yes, although Admiral Ulrich wanted us to go to zero, there would eventually be over three hundred US Sailors remaining in the UK after CNAUK's closure serving with NATO and joint US-UK commands from Menwith Hill in Yorkshire to NATO HQ at Northwood just a few miles up the road from West Ruislip.

There was 2 Providence Court, a small three-storey office building across an alley from the US Navy HQ building at 7 North Audley St. downtown. This was home to the Personnel Support Detachment as well as a few housing maintenance staff who looked after our remaining buildings.

RAF Daws Hill, the largest property by acreage, was located at the top of the hill above High Wycombe. It was still home to many "cats and dogs" tenant commands including London Central High School, one of the only Department of Defense Dependents Schools in the world that offered boarding facilities .

The US Navy Headquarters at 7 North Audley Street was almost empty except for two communications groups: Naval Communication and Telecommunication Area Master Station Atlantic Detachment London (NCTAMSLANT Det London) which had about 10 people ensuring broadband and other communications links were maintained and the US Navy's Telephone Exchange which was manned by two people.

Romany House was a flag quarters that had been home to the respective families of the Navy Europe Commanders since WWII. It was a lovely house located on the Wentworth Golf Course southwest of London. The trouble was that it took about an hour to get to and from it to London. Its last occupants were the Duhrkopfs, the family of the US Navy Captain Defense Attaché. Neither his relief, Rear Admiral Henderson, nor the Defense Intelligence Agency that would have had to assume responsibility for the property, wanted it as the permanent home for the Defense Attaché.

We 'shopped' it around to the USAF and all other US government bodies but no other use made sense. It was even offered to me as my residence as I had negotiated follow-on orders to stay in the UK after the bases were closed. However my new duty station, the NATO Command in Northwood, would mean an hour commute on the M25 motorway each day so I regretfully declined. The Navy eventually sold off its last flag quarters in England in 2009 for over £4 million pounds.

Blenheim Crescent was a very useful three-level office building close to the West Ruislip underground station and it offered space we were going to need for remaining personnel during the shutdown process.

It wasn't just the bases and installations themselves that had to be managed. There were also a cluster of tenant commands at some of these installations who needed relocating or dispersing. Among them was the Canadian Support Element which was a support command for all the Canadian military personnel in Europe, along with a Defense Fuels Station, the US Post Office regional distribution center and London Central High School – all at RAF Daws Hill. Then there was the Naval Facilities and Engineering Command, the Navy's European Human Resources Center and its Regional Legal Service Office detachment at Blenheim Crescent.

Although not directly under my command, but complicating matters, were the three research and development commands – one from each of the US Military Services – that were based in an office building near Marylebone station in London, known as Edison House. They were in the mix because although they planned to remain in London, the long-term lease on their building was running out and the force protection rules introduced after 9-11 meant they couldn't stay in their current home as it had no protection from vehicle bombs.

Then there was the US Defense Contract Management Agency which had a single US military officer and a number of civilians in an office in Loudwater near High Wycombe. I needed to find a way to be as efficient as possible, saving the Navy and the nation money, while doing the right thing for the commands and personnel remaining in the UK after CNAUK's closure. It was a difficult balance to achieve while Admiral Ulrich continued to push to "go to zero now."

On 1 October 2006, just five weeks after my arrival, I signed over to the British Ministry of Defence's Defence Estates representatives our bases at Eastcote and the bulk of RAF West Ruislip.

Handing over Eastcote was very simple. We didn't have any US 'visibility' there anymore. Tenant organisations like Naval Criminal Investigation Service (NCIS), the Navy's Human Resources Office for Europe and the US Marines Det 1, which had their barracks there, had been among the last residents of the base but had long gone. I walked through it with LCDR Andre Coleman, my Public Works Officer, and we found just a few things left like a football table that had been in the Marines' recreation room. It was sad to see it empty and barren.

Of particular personal interest to me was that one of the base buildings had been home to a Boy Scouts of America troop. We were able to relocate the scout equipment and reinvigorate the Scouting programs at Daws Hill for a year with a Cub Pack, a Scout troop and a Venture Crew all providing opportunities for fun and development for the remaining dependents. We established our new Scouting headquarters in what had formerly been the Community Banking offices at RAF Daws Hill, High Wycombe.

We did the formal handover ceremony for Eastcote and RAF West Ruislip at West Ruislip near where the Navy Exchange (NEX) and Commissary used to be. As part of the ceremony we hauled down the American flag and then hauled up the RAF's flag so the MOD took control of the base for about five minutes. Then we hauled down that flag and the Defense Estates department took control before handing it over to a contractor for development.

I recall signing off a one or two-page 'Custody Document' as part of the handover to Defence Estates with Squadron Leader George Hannaford, the RAF Station Commander who had host nation responsibility for all the US Navy property in the area.

However it wasn't the end of the US Navy involvement with the West Ruislip base by any means. I had formally handed it over but I had a gentleman's agreement to continue to use the Chapel, housing, the gym facilities and the baseball field (which were on adjacent, but separate, sites to the main base) until the contractor was ready to develop the site. We continued to use those facilities until the following July after our final Fourth of July community celebration.

This community-wide 4th of July celebration had been announced as ending in 2006, but we decided to hold an additional 'final' event in 2007 continuing the tradition which had been in place for many years. It was a unique event at RAF West Ruislip which saw Navy personnel provide fund-raising food stalls, while we brought in events such as dog handling and motorbike displays, musical performances and a small circus.

The event, which in past years had been a two-day affair with US military families only on the first day, was brought to a conclusion by a major fireworks display. It was very popular with the local community and over the years up to a thousand people had turned out.

We decided to do it one more time as a single day event as a morale booster for our personnel and we received good support and funding from US Navy Europe. Sadly though it poured rain on the 4th of July 2007. I can remember standing there in my US Navy summer whites at the baseball field soaked to the skin and watching as the rockets shot up into the sky and became just a nebulous and fleeting flash of colour rather than the vibrant bursting displays you would expect.

On 1 October 2006 – the same day we handed back Eastcote and RAF West Ruislip – we completed the closure of the Dental and Medical facilities at Blenheim Crescent. Having this open office space provided us with an opportunity to solve some of the challenges for the remaining tenant commands and I made a strong case to NAVEUR and the US Air Force's Third Air Force UK based at RAF Mildenhall to retain this facility under US control.

I negotiated with the US Air Force who would be the remaining US service in the UK to have them assume responsibility for the building and we then brought the various tenant organizations and the Edison House groups that would be staying on into there. It was a complex shell game; balancing the office space being used for the dwindling closure crew of CNAUK with the new enduring residents.

Things changed during the process. One of the key changes was with London Central High School, where my daughter was in her senior year. When I first arrived in London the closure of London Central High School was actually targeted for June 2008. However, within days of my arrival, the US Navy and the USAF Europe (USAFE) agreed to close the school in one year. So my daughter's class of 2007 became the last-ever graduating class.

The school closed in June 2007 and, during that summer, school equipment was rapidly transferred to other bases in Europe. With the closure of its boarding facilities, DoD dependents from locations around the world without their own local DoD schools now needed to transfer to private boarding or, in some cases, local national schools.

During the closure process I realized I was also dealing with a lot of history and a unique legacy.

A number of the US Navy civilians had worked for us for a lifetime and had relocated to CNAUK from closures in the 1990s of our US Navy bases in Scotland at Prestwick, our submarine base at Holy Loch that had operated from 1961-1992, at RAF Edzell (the High Frequency Direction Finding (HFDF) USN base from 1960-1997) and others. I had people under my command whom had been working for the US Navy in the UK for upwards of 50 years.

The bases themselves had 'history' mostly with the USAF before the Navy took them over late in the Cold War. RAF Daws Hill and Eastcote in particular had connections going back to WWII. But it was the US Navy Headquarters building downtown on Grosvenor Square that had the most personal history. It went back to the days in WWII when General Eisenhower started planning the D-Day invasion and had been part of US Navy operations since Admiral Harold Rainsford Stark had moved in to become Commander of US Naval Forces in Europe in early 1942. This building was the headquarters where he oversaw the build-up of US Naval operations in the Eastern Atlantic for D-Day and the eventual re-conquest of the European continent.

Now I would walk through corridors and my footsteps were the only sound I heard. The only personnel left were about 10 communicators and telephone operators. What once had been a thriving headquarters supervising the defense of a continent was now an empty shell.

It was a huge building with lots of history and plenty of 'ghosts'. I visited the building four times before it was handed over. Externally it looked great, but internally the layout was discombobulated. It was a real rabbit warren. Anyone who served in there – at least in later years – will know that there were stairs that ran up to brick walls and corridors that went different ways. There was the Crown & Anchor pub/club downstairs and there was a Navy Exchange in the basement. The admiral's flag quarters were nice but dated.

What could we do with the building? I recall that a position paper was submitted to the Pentagon that advocated tuning it into a Morale Welfare and Recreation (MWR) hotel facility similar to what had happened in Tokyo with the New Sanno Hotel. However, force protection considerations killed this proposal – it was too close to a main street.

There was a 999-year lease and we paid a 'peppercorn' rent to the landlord Grosvenor Estates. However, it turned out that it wasn't technically the Navy's building to hand over! During WWII the only US Government agency that could negotiate for overseas properties was the US State Department so when it came to selling it off and the Pentagon started processing the request the State Department quickly raised its hand and laid claim to the ownership. They claimed they were considering using it for additional space or storage for the US Embassy's impending move (it was announced in 2009 that the Embassy would move to a new site at Nine Elms near Battersea). However that storage plan was soon discarded. The remaining lease on the 100,000 sq ft building was sold for around £250 million in May 2007 to a development consortium with the intention of turning it into up-market apartments.

Daws Hill High Wycombe also had legacy because of its large WWII bunker which had been headquarters for the USAF's 8th Air Force Bomber Command and then was utilized by the USAF during the Cold War as a major command and control facility.

Amazingly, despite being the Base Commander there, it took me four months to get permission to get in to visit 'my' bunker. The British Ministry of Defence, which by 2007 had responsibility for it, attributed the delay to 'atmospheric contamination' and had to run some air tests. It turned out that they were concerned about the oxygen levels in the isolated, multi-storey underground facility and had to make sure it was safe to walk through without breathing gear.

Several things apart from the WWII legacy amazed me about the bunker. During my walk-through I learned that it had been a nuclear command and control facility during the height of the Cold War. However there were unusual markings on some of the walls – and it turned out that it had been used as a 'set' for a horror movie in the early 2000s. The US Marines had apparently also used it as a training/exercise area before they left. The US Navy personnel on base at one point had also used part of it as a haunted house for the base's children for several years at Halloween. Admittedly it was trashed and pretty scary.

I know today that the MOD has now handed back the bunker site to the Wycombe Abbey School that had owned it before the war. It's a complex process and this was just the next stage. Perhaps they will fill it in and seal it up and then one day it will be rediscovered and become the modern day equivalent of the pyramids – just in High Wycombe!

The handover of Daws Hill comprised three parts. One was the bunker. The other was the main technical site and base housing and the third included the sports fields out by the main M40 motorway that ran by the base. The latter fields had been owned by the estate of Lord Carrington, a former British foreign minister.

Even with the handover at RAF Daws Hill in late September 2007, American involvement with the base had not ended. The London Youth Baseball League had established an operation at High Wycombe in the 1990s and wanted to continue to play. I managed to get the agreement of the MoD and the Carrington Estate for London Youth Baseball to continue using the fields until such time as it was redeveloped. They were still using it in the spring of 2013.

Then, of course, there was the US Navy's own legacy: The US Naval Headquarters at 7th North Audley Street.

We made a decision in April 2007 to disestablish the US military security at bases and turned it over to MOD Police. I redeployed our Master of Arms personnel to ROTA in Spain; Naples in Italy and to the US. The military personnel count went down from 50 to about 25.

On 14 September 2007, when the CNAUK disestablishment ceremony took place, we had just 25 military and 30 US civilian personnel left.

The ceremony took place at RAF Daws Hill and coincided with the handover of the base. There was a blessing by Navy Reserve Chaplain Jerry Bruce, the husband of one of the DoDDS school teachers on the base, along with brief speeches by myself, Rear Admiral Groothousen (the new Commander of US Navy Region Europe) and the British Lord Lieutenant for Buckinghamshire. A cake cutting and reception followed in the Daws Hill Pinetree base club. And that was it. I had signed the Custody Document and the US flag was hauled down for the last time on Daws Hill.

My primary job was done.

Despite CNAUK being consigned to history, there is still one aspect of the base in which I remained involved for several more years. One responsibility that accompanied my role was to become a trustee for the US Navy pension scheme for US Navy civilian

UK employees and pretty much on from the closure I had responsibility for that until it was wound up in Spring 2012.

To this day there are more than 500 UK citizens who have pension rights based on their service at US Navy commands from Northern Ireland to Cornwall. Our total US liabilities are over $100 million and part of my remaining duty was to transition that over to an insurance company to preserve the employee benefits so well earned by our British co-workers.

Looking back I'm proud that although the command was directed to divest the US Navy of everything immediately and "go to zero now," we were able to involve the entire command to solve creatively the problems we had and to find new homes for the people and commands that would stay behind. We did the right thing in face of great pressure and not only left a legacy of teamwork with our employees, but a strong sense of pride in the relationship with our British neighbours and partner defense forces.

Capt. Dave Dittmer, USN, the last man at the helm of the COMNAVACTUK presence in London.

CHAPTER FORTY-ONE

END GAME – THE DEMOLITION OF EASTCOTE

Danny Shrubb

Crusher Operator
Former Eastcote Base
January – December 2007

(Editor's Note: Danny Shrubb was part of a team from leading demolition specialists Syd Bishop & Sons, headed by Tom Bishop. The company was given the contract to demolish the entire Eastcote base site. From 15 January to 12 December 2007 a demolition team, under Contracts Manager Steve Bishop and managed by Tony Moore, worked on the site; first stripping out doors and other recyclable items and then using a combination of diggers, EX 250's and, finally, a C12 EXTEC crushing machine, to render the entire site to about a one-meter high pile of rubble which would form the hardcore 'backall' for the site. Their distinctive machines sport the logo "Watch It Come Down".

During 2007 as they were closing on the final stages of demolishing Block 2 – one of the last demolition phases as they worked from the Lime Grove end to Field End Road, the author made a number of site visits to watch the demolition team at work and interviewed Mr. Shrubb about his job as well as take a guided tour around the crusher.)

I joined this firm in 2004. I was driving 360s – diggers – previously. But here they gave me 'the Crusher' – and it was brand new. It's a C12 EXTEC. They bought it when I joined the firm. And they painted the nickname of the machine on the side – "Buckaroo" it was named after the board game of the same name in which you place items on a toy donkey. I was forever putting different pieces of equipment on it. Hooks. Chains. Buckets. Everything. It has a Deutch V6 engine and it's brilliant. In the three years I have had it before coming on this job at Eastcote I haven't had to put a drop of water in to it.

With this job you see the entire remains and contents of a building go by. When I first started here a dead fox as well as dead birds came by – they'd obviously gotten in the buildings and died there. But I've also seen heavy duty combination locks, reinforced doors, a few mobile phones – not my own – though I have lost a couple down through the jaws. But, in the main, this is where old buildings come to die…and be reborn as hardcore or material for recycling.

The crusher does what it says on the tin. The biggest bit of building you'll get into the crusher is about a 3'x2' block of bricks. It then munches it all down to rubble of about 75 millimeters in size.

And the business end of Buckaroo essentially is what I call the "jaws"; two steel plates with metal teeth through which the material passes for crushing. But like any teeth they wear out – on this machine they usually last about 10 months. The biggest problem is reinforced concrete –the steel from it gets trapped and you have to go down with a strap – it's like a big sling – and pull it out. Wood can also be a problem. When it becomes a problem you stop and reverse the jaws and that usually sorts it.

Buckaroo is completely focused on recycling the rubble passes along the hopper – and goes down the jaw and up the conveyor. There's a magnet conveyor that takes the metal out and throws it to the right of where I am standing. This is not an area for mobile phones or watches. I've broken three phones to date by getting them too close to the magnet. It just wipes them out.

The big problems and delays are if, for instance, we get a piece of angle iron and it splits the feeder belt. They normally have to send a specialist out to replace that and it costs in the region of £4,000 (about $6,200) a time. Fortunately I've only ever had one split to date but not on this site. The magnetic belts cost about £500 ($820) a time if they split.

Beyond that Buckaroo is a big 'drinker' of diesel fuel. Each week I have to make sure that I also find the machine's 36 'grease nipples' and put about 30 milliliters of grease in each.

Demolishing this site at Eastcote was a bit more complicated than some projects. The buildings are older with a range of materials. There were some reinforced buildings – I think they were reinforced gun and ammunition storage bunkers. But when they meet the business end of our machines there is only going to be one winner!

We work as a team but it's a bit like a football team. Each of us has a different role and responsibility. I've worked with some of the guys for awhile. Tom Bishop is our site boss. Clive Collard drives an EX250 and does demolition/material sorting and loading of the bins. We're there as a team but while working we are pretty much on our own during the job, coming together only at breaks or lunch.

Often when the bulldozed building reaches Buckaroo it's like a big soup of materials – everything is mixed together and it wants sorting out. It makes it difficult. On this job it's been about 60% hardcore and 40% wood, metal and plastics. We try and separate them all out.

I guess my favourite music to crush things by would be *Chasing Cars* by Snow Patrol and songs by the Kaiser Chiefs and the Killers among others. Music is nice but you can't be complacent. You have to keep a careful watch.

And as for the most unusual thing I ever had come through the crusher – though not on this site – were cannon balls. They were in the soil on a site. Buckaroo spit those right back out.

"Watch It Come Down" is the motto on their machines but these are the men who made it come down — the crew from Syd Bishop & Sons and other contractors at former entrance gates located in the middle of Eastcote base (US side) on 15 May 2007. The gates flanked a public footpath that ran right through the middle of the base. By then Blocks 1 and 2 had been demolished but the demolition of the large red brick carport building seen behind them –attached to Block 4 (and which may have been built to handle mortuary vehicles), had to be delayed for a few months after environmentally protect Great Crested Newts had been discovered close to the building. Danny Shrubb is centre at the very back of the picture.

Standing tall no more: Contractors from Syd Bishop & Sons finish razing Eastcote to the ground.

The Lime Grove entrance to Eastcote on 15 May 2007. Block 3 which was used by the British for various functions was off to the right behind the remaining building.

Above: Job almost done. The wing nearest Field End Road, which in later years contained a MWR gym for the Navy and Marines as well as the armory is all that stands of Block 1 in May 2007. Note the pull-up bars and their height above the ground. *Refer also back to pictures in Chapter 18.*

OTHER MEMORIES

During the long process of compiling material and writing this book the author was fortunate enough to encounter or contact other people who didn't have the time or weren't able to contribute full chapter memories or who, in passing, reminded me of key items, funny or particular memories. He has tried to include some of them here in a round-up chapter. For instance:

Nick Wingert – a member of the 892nd Military Police Company, was one of the very first US servicemen to arrive at Camp Lynn (aks Daws Hill, High Wycombe) in 1942 as the base was being built. The then 34-year-old returned to High Wycombe, as the base barber from 1955-1972 and then again as the 84-year-old guest of honour in 1992 for a 50th anniversary commemoration event when the author was honored to interview him.

"There were so many generals," Wingert who was armed with a shotgun and a pistol for his WWII base guarding duties, said. "We had to do a lot of saluting. We would get jam and peanut butter – but it drew all the wasps from the woods around there. You always had to check what you were eating."

Wingert recalled seeing 8th Air Force commander General Doolittle a number of times and also some scary moments on guard duty.

"One night there was a noise in the darkness and I challenged the 'intruder'," he said. "Fortunately it identified itself as a cow! Another time I was bending down to pick something up when a bullet whistled through the branches overhead. It turned out some of our own boys were out hunting."

William J. Kilty, a Department of Defense Dependents Schools (DODDS) educator and administrator, was Principal of London Central High School at RAF Daws Hill, High Wycombe, in the late 1970s and early 1980s. Kilty was there at the time of change for the base and for the school and for two particularly historical occasions.

"I had what I consider two face-to-face encounters with history," he says. "The first, in 1993, was as a guest to the dedication of a memorial plaque to the men and women of the 8th Air Force in WWII. It took place near what was then the Base Commander's office near the front of the High Wycombe base. Attending were General James Doolittle, the USAF 8th Air Force bomber commander who was based at Daws Hill and Sir Arthur "Bomber" Harris who was, in essence, his counterpart at the RAF.

"A plaque was planted in front of a small pine tree. The inscription read: "This pinetree is a living memorial to the Americans and British Airmen who served High Wycombe Air Station from WWII to the Cold War. High Wycombe has played a pivotal part in protecting those ideals we as nations so deeply cherish. Men and women. 7520th Air Base Squadron 30 June 1993."

"I also had a second chance to be witness to a little bit of history in the making. I was invited to a dedication of some new school buildings at the Wycombe Abbey School, the private girls school located on the hillside below our base. The base land once belonged to the Carrington family and Lord Carrington was also directly involved with the Wycombe Abbey School. However, the ribbon cutting ceremony was delayed because Lord Carrington arrived late. He was very apologetic for being late explaining that he had been summoned by then British Prime Minister Margaret Thatcher and offered the position of Foreign Secretary!

"Over the years, the British media has taken an interest on the Yanks in their midst. When I was at West Ruislip, we had a BBC crew film our school and kids for a program called *Braden's Week*. I knew that presenter Bernard Braden was anti-Vietnam but I let them film because the film showed that the Americans as decent people with similar aspirations for their children as their own and were not crazed war-mongers!

"Another media event – but one of a more comical nature – was the LCHS cheerleaders appearing on the BBC's *The Generation Game* presented by popular presenter Larry Grayson. It was a bit of a silly game show program where our cheerleaders showed how to do various cheers and then Larry had members of the audience come on stage and try and do the same moves. However it attracted upwards of 18 million viewers.

"During my time at London Central I was able to go into the secret bunker just once in about 1976. My recollection is of a passage (tunnel) to the large room with the wall maps and vacuum message tubes.

"With regards to the base bunker – I remember the 'pile driving summer' – I believe it was 1983/84 – when I was about the only one working at the High Wycombe campus for London Central High School 1 and my constant companion was the din of the piles being driven into the ground as the underground bunker headquarters was being upgraded as part of the Ground Launched Cruise Missile (GLCM) support program.

"The Cruise Missile program had ramifications well beyond the High Wycombe base and the school. Especially in the dramatic changes that occurred to a post-WWII sleepy base at RAF Greenham Common in Berkshire which, along with the base at RAF Molesworth in Cambridgeshire, was significantly upgraded to house the nuclear cruise missiles and their launcher systems. The arrival of the weapons system also had a direct bearing on our increased enrolment for a few years until a high school could be built at Greenham Common.

"Of course there have been many changes to the mission and function of the military over the years that naturally affected our schools. I arrived in the area in 1968 and over the years witnessed the many closures and relocations, new construction, renovation, etc. I really have felt a witness to history in the making on many occasions. I was there!

"Change was a part of our daily life and I feel privileged to have been a part of one of the great post World War II experiments in efforts to have a more peaceful world by being strong militarily but also extending the hand of friendship to all those who would accept it."

Lawrence Lotito, who had also been stationed at USAF Daws Hill, High Wycombe. In his case he arrived in April 1951 and started work at the USAF 3rd Air Force Headquarters at South Ruislip as part of the 28th Weather Squadron.

"I was recalled to active duty on 25th April 1951 and released in August 1955. I was a veteran of WWII from mid 1942-1946 and then had gone to work with Tans-World Airlines in September 1946. They loaned me to the Irish Meteorological Service as with the postwar burgeoning airline activity across the Atlantic, the Irish were short of forecasters. You will realize that in these early days, the piston aircraft had limited range and as Shannon Airport was the western-most airport of Europe, all the airlines stopped at Shannon for refueling. Even then, they normally could not proceed non-stop to New York, but had to land for refueling at Gander, Newfoundland. I worked at Shannon from January 1947 until I was recalled to active duty at Ruislip.

"Because of the Irish experience, I was the most familiar with the North Atlantic and

European weather of all the officers at Ruislip, so I was designated the squadron and hence was sent on various assignments. I came with my family initially had a cottage in Ickenham but moved into a nice apartment in Harrow-on-the-Hill that was available when a Major moved out. Our daughter went to St. Dominic's school where the future King of Jordan was also a scholar.

"Winters were a problem for London forecasts. This was before the Clean Air Act and the 'pea soup' fogs were unbelievable. You could stand under a street lamp and you would be unable to determine if it was lit! Walking across the street you would become disoriented. The basic problem was that when Londoners got back from work they lighted their fireplaces in the home and the smoky coal they used acted as 10 million smudge pots. Our apartment had a fairly long hallway from the entrance – about 40 feet. In the worst foggy days we could not see the door, which meant it was mostly smoke! On foggy days we always forecast worsening conditions at 5 p.m.

"Our office was at South Ruislip but we later moved High Wycombe, where our office was underground and we couldn't even see the weather we forecast. Our weather office in Ruislip and at High Wycombe was the centre for all the USAF Stations in England and the maps were drawn and faxed to all these bases – a very primitive fax machine made up of an 18 or 20-inch cylinder, 6 or 8-inch diameters, with a pen that went across it as it rotated. At the receiver end burn the image onto some sort of sensitive paper.

"Colonel Dole was the 28th Weather Squadron Commander then and Lt. Colonel Cartwright was the commander at High Wycombe and also Dole's Deputy. The Squadron had an official squadron newsletter known as 'The Thunderhead'. At one point the newsletter ran the following about me: "Captain Lotito has just returned from a 30-day leave to the Continent. He has been holding forth with some very good advice as to where the best places for enlargement of the waistline are to be found."

SSgt Alex Salinas who joined the USAF in 1953 and who vividly remembers 'GI parties' cleaning out barracks top-to-bottom with toothbrushes for barracks inspections. By September he was on his way to England via New York on board the *USS Taylor*. He would eventually work at three bases – South Ruislip, West Ruislip and High Wycombe – during assignments in the 1950s, 60s and 1970s.

"I found it hard to believe an airman in the Air Force was going on assignment via a ship," he says. "It took us nine days to arrive on the shores of England. I got on a bus in Shaftsbury (the USAF in-processing base) that took me to South Ruislip Air Base, 3rd Air Force Headquarters. Our living quarters was a one-level open bay brick building that didn't have a barracks look. I worked in three different buildings and my first job was to operate a small switchboard for an air rescue unit.

"I was assigned then to Bovingdon Airdrome in Hertfordshire in 1954. I met a fine lady, Barbara, at one of the popular dances there. I was posted back to South Ruislip where I worked at the base switchboard with a bunch of good-looking civilian young ladies but I was dating Barbara so it was 'no touch'!

"The switchboard was located on the first or second building on the right from the main gate. I was also assigned duties at the base communications centre such as sending and receiving numerous teletype messages. Off duty we would go to the Eastcote Arms which held dances. It was popular. Barbara and I finally tied the knot on New Year's Eve 1955 at a Catholic church in Harrow Wealdstone near the base. It was mid 1957 when we arrived back in the US with the addition of a cute baby girl (Bettina) who was born in the South Ruislip base hospital. Both she and our son, Michael, came back with us on another tour of duty in the 1960s and again in the 1970s."

Kathy McDurmon was a student at London Central High School.

"I have memories of going to the base snack-bar and gathering with friends to eat our favorite egg salad sandwiches and french fries," she says. "Going to the theatre there on Saturdays to meet friends and watch good 'ole movies like the 'original' *Romeo and Juliet*. I remember going to a sweetheart dance on base with Guy Robarge and wearing a metallic silver mini dress and red/white two-tone shoes.

"I actually got to meet and get autographs from the members of the group Iron Butterfly who came on a visit. I think it was at the base PX. At the time I had not even heard of them, I laugh now every time I hear a song of theirs on the radio."

Bill Douglas, the son of Assistant Deputy Chief of Staff for Personnel and later the DCS/Personnel for 3rd AF, Colonel Richard M. Douglas, USAF, was a student at LCHS and went on to become a Lt. Colonel in the USAF himself.

"Not on a regular basis, but fairly frequently I rode the school bus to South Ruislip from CHS rather than home to Beaconsfield (strangely the bus stop was by our house) usually to go to the BX or hang out with friends at the snack bar," he says. "First thing that I had to do was to check in with Dad's office (building on left with offices to the right of the center door in attached picture) to let them know I was riding home with him. He had two or three staff members as I recall, a secretary and two to three airmen in an outer office area, one of whom (at least part of the time) was Larry Hagman."

Li'anne Drysdale recalled her Junior High time at Bushey Hall in 1963 and 10th and 11th Grades at Bushy Park.

"I graduated from high school in Germany in 1963 but before that I went to 7th grade at Bushey Hall and 10th and 11th grades at London Central High School. They were very looooong bus rides from Cookham Dean, Ealing and then Northwood, were we lived over the years. I remember the Columbia Club well. Peter (my brother) and I used to play the slot machines. We were just Junior High and High School ages!"

Danielle (Saint Germain) Austin a Class of '71 LCHS student recalled her time as a student.

"We lived in the first house on Kings End — there is a block of flats on the street first, then our house," she recalled. "It was a red brick (what else) with big bay window and hydrangea bushes under the window if it and they are still there. It had a nice turn around area in front for our large Plymouth Fury 3 car that seated seven or eight; a large, green tank-looking vehicle that scared everyone but was great on trips to Scotland and Wales.

"The Priory was past the Swan Pub just before Eastcote Road. It looks like it is no longer a large restaurant, but two smaller buildings. There was a cinema near there. It used to be the only place open after 1pm on Wednesdays and Sundays other than the tobacconist who would cover up everything that couldn't be sold with paper bags. I remember trying to buy a pair of tights (pantyhose) on the wrong day at the wrong time.

"I was 16 years old in 1968 when a group of about 10 of us would walk to the base for "Teen Club" on Friday nights. Sometimes we would go to Uxbridge Station and ride home together as far as we could. I lived in Ruislip near the tube so I didn't get much of a ride.

"I remember the Battle of the Bands when an unknown group of four young men called The Corporation took the stage and won the $50 prize for a recording contract to record one song. That song was *Horse With No Name* and the group later became

America. They also played at my senior prom in 1971 at a luxury estate that I can't remember the name of just now. We always had the proms at a nice hotel or estate in those days, never in the gym. I got my first real kiss outside of the Teen Club in fall of 1968. One never forgets one's first boyfriend or kiss.

"I mostly remember my own town of Ruislip where I lived for five years and my parents stayed for 10. I actually had a job at the Priory Restaurant on the High Street as a waitress! And I remember daring others and being dared to walk through the cemetery. Sorry I don't remember more about the base. I think I was too wrapped up in 'teenagerhood' to pay much attention to my surroundings."

Merilee Freistedt recalls coming to London with her husband Cal Freisted, a Capt. in the USAF Medical Service Corps, in the summer of 1962. Her husband passed away in March 2009.

"Cal (Calvin) was then Capt., USAF Medical Service Corps, the supply officer for the AF hospital at Ruislip from June 1962 to May 1965. The hospital had a primary warehouse at West Ruislip, and Cal was injured while moving some supplies shortly after we arrived to join him in August. A very heavy roll of aluminum fell on his foot, smashing the big toe's bone into more than five pieces. I remember that it was a tense time in world affairs due to the Cuban Missile Crisis and though our furniture had not yet arrived, Cal, myself and our children were on alert to be sent home on 24 hours' notice.

Bob Miner, who served at US Navy Headquarters London, recalls the details of the lease for the 7 North Audley Street better than most – he had to ensure they Navy was living up to its tenancy requirements as Planning Officers for COMNAVACTS UK during the 1980s.

"The lease was about 20 pages long and the two principal parties were THE MOST NOBLE HUGH RICHARD ARTHUR DUKE OF WESTMINSTER, D.S.O. and THE HONOURABLE JOHN WILLIAM BAILEY (Junior Counselor of the US Embassy of 1 Grosvenor Square)," he says. "It was signed on 28 November 1947 and was good for 999 years (until 2946). The lease had required a cash payment of £164,150 pounds and an annual rent of £100 pounds. One of my jobs was being sure we met the provisions of the lease.

"Along with the Office of General Counsel I dealt with the Grosvenor Estate to resolve any problems. We really didn't have too many problems as long as we performed the required painting of the building trim and 'pointing' of the bricks. During my time I had to remove a few window air conditioning units.

"Another challenge was to find proper space for the antennas required by the Naval Communications function which then occupied the top floor of 7 NA.

"The antennas seemed to be getting larger and larger. At one time, the Navy made an agreement with the Grosvenor Estate that any antenna could be put on the roof as long as it could not be seen from the steps of the Britannia Hotel, which was located along the south side of Grosvenor Square. An antenna with which I had to deal was about 12 feet in diameter and had to be placed directly facing south. It would have been very visible.

"We worked with a British architect who came up with a solution. We fabricated a wall made of fiberglass that duplicated the brick of the building exterior. The Grosvenor people accepted this concept and everyone was happy.

"I can think of two incidents which stirred up 7NA at the time they happened. One incident occurred when the Marines at the entrance of the building mishandled a canister of gas. It discharged and portions of the building had to be evacuated. Another time, a

Marine up on the Marine Barracks floor thought he saw an intruder and fired his pistol. Public Works had to patch up a few holes in the wall as a result."

"There were many other stories about "7NA" as I knew it. One of the interesting people that I dealt with was Peggy Morrisey. She was the 'Protocol Officer' for CINCUS-NAVEUR during my time there. She was nicknamed 'The Admiral' and at least part of her role was to keep a focus on preserving the integrity of the building. My colleagues and I in the Public Works section of COMNAVACTS (which saw us work side by side with British staff to undertake development/refurbishment projects for the bases) were continually modifying the building to accommodate various functions and organizations which were always in a state of flux. One memorable example – we wanted to increase the capacity of the elevators in the building. And we could – with the exception of what was known as the "Eisenhower Elevator" which was used by a few high ranking officers in CINCUSNAVEUR. That couldn't be touched.'

Andrew Kay worked on the security staff at the US Embassy but also found himself taking a great focal interest in the Chapel at RAF West Ruislip.

"My wife and I were married there on December 19, 1987," he said. "She was a Navy HM2 stationed at the Navy clinic in the US Embassy in Grosvenor Square London and I worked in the Embassy on the local security staff. The Chaplain at that time was Chaplain Aufderheide who performed at our wedding. We are still hoping to make a pilgrimage back to London sometime in the near future, but apparently the Navy Chapel is one place we won't be able to visit."

Skip McIndoe, arrived as a Captain in the USAF in 1964 and left in 1969 as a Lt. Colonel.

"I was Chief of Hospital Services, 7520 USAF Hospital, Headquarters 3rd Air Force, at South Ruislip and, additionally, Commander of a 60-bed surface transportable hospital (260 vehicles and 210 personnel)," he says.

"We bought a house in Uxbridge over looking the Colnbrook Valley. Our back garden was beautiful and our outdoor clothes-line drying tree was embedded in an old concrete platform that had an antiaircraft gun during the second world war. Every Sunday I would go to the local pub and play darts with my neighbor. I was pretty bad to start with but after five years we rarely had to pay for our beer. One other of my best memories was using the Zwang (a USAF-owned motorboat) on the Thames with the family for several weekends.

"I became a Fellow of the Royal Society of Medicine during my tour and spent that last year in UK doing a post-doctoral program in the Metabolic Unit of the Hammersmith Hospital in London. The USAF hospital actually had 80 beds. It was a full-service hospital. By that I mean we were staffed In the major specialties. The hospital was part of the base medical services. The Director of Base Medical Services was responsible for the hospital, Aerospace Medical Services (the Flight Surgeon), Dental Services and its Prosthetics Lab, Veterinary Services (food inspection), Biomedical and Environmental Services. The hospital staff was broken down into various departments. The commander was at the top and was a physician. His executive was a Medical Service Corps Officer. He was in charge of all administrative services in the hospital: Registrar, Finance, Maintenance etc.

"The professional divisions were the emergency room and ambulance services, the outpatient clinic staffed by one medical officer and three British doctors. The internal medicine service had two medial officers, but there were actually six internists at the hospital for a couple of years. Among them: The Surgeon, Hq 3d AF, the Hospital Commander, the Flight Surgeon, a British doctor who worked in the out patient department and me – the Chief of Hospital Services and Chief of Internal Medicine.

"Surgical Services had two AF surgeons and an Anesthesiologist and his staff of Nurse Anesthetists. There were two pediatricians, one officer and one British pediatrician. The Obstetrics and Gynecology Service had two AF obstetricians and two British doctors.

"We also had an AF Psychiatrist and Pathologist as well as a Civil Service Mortician and a Mortuary. Hospital Services was also responsible for the nursing services, physical therapy, corpsmen and food services.

"Our Transportable Hospital was kept at another base where most of the materials were kept from France after De Gaulle threw us out. I seem to recall it was Greenham Common where we had vehicles. We never deployed. A convoy that size would have totally tied up the British road system."

Joe Boltz, a USAF security policeman stationed at High Wycombe from Dec 1984 to May 1987, particularly remembers his tour of duty.

"It was great to not only have an opportunity to work closely with the British Ministry of Defence Police but, as part of that link, to ride along occasionally in their "Police Wagons" in London," he says. "My duty saw me not only patrol High Wycombe Air Station, but also Carpenders Park and the West Ruislip Elementary School. At one point I was dispatched to pick up some AWOL (Absent Without Leave) servicemen in Dover on the south coast of England (they were coming over by Hovercraft).

"I also remember my first two weeks – not spent on security duty but on "leaf raking" the base in order to keep it looking beautiful for our frequent VIP visitors," he adds. "My other great memories were watching the Chicago Bears win the 1985 Super Bowl while at "The Club" on base. And then there were my visits to The Angel pub in High Wycombe where I made some great English friends. It was a great base."

Staff Sgt. Rick Rickard was stationed at High Wycombe from 1966 to 1969 as part of the the the Field Representative of the Joint Chiefs of Staff to Europe team and can remember when the new barracks were being built.

"I seem to remember that the facility underground was on two or three levels. I think it was a sort of maze with rooms off stark corridors. There was a communications center, a computer room and other functions I don't remember. I do remember the computer as being a large IBM 360 I think. I was not into computers at the time so don't remember much about it. I do remember that it had large (15 inches or so) discs that stored data. I would not be surprised if the home computer I now use is more powerful than that great huge machine. During the winter months we never saw the sun during the week. It was dark when we went in and dark when we came out."

Janie Fagen nee Smith had arrived in July 1988 as the Navy Housing Director for London, RAF West Ruislip and RAF Hendon in England.

"My office was at 14 Upper Brook Street. There used to be a brass plaque outside the door that said 'Octopus Books Limited.'" she said. "The Navy PSD office had the first two floors and I had the third. Upon arrival I discovered to my dismay that a normal change of occupancy maintenance and paint at RAF Hendon took up to three months and the costs involved were incredible! The US military standard is five days! I immediately met with the locally-provided Property Services Agency (PSA) agents who were then located at the base at High Wycombe. I told them that it was totally unacceptable and would not be tolerated. I was not very popular with them after that. I later heard the rumor around

PSA was that I was the meanest thing in shoe leather! Ha! The best I could do was to decrease it to 30 days while I was there but I continued to try!"

Derek Addison, a British musician who played in a number of groups, recalls performing at and going to a number of the bases in Middlesex. Addison who has written the book *Bushey Hall and the Forties Experience* with Tony Rock, said that it was a terrific time.

"South Ruislip being a pretty large base boasted an Airman's Club, a NCO's club and an Officers Club," he recalled. "In 1961 I used to sit in with a band called "The Alabama Hay Riders". Saturday night was Country & Western night at the club (it was later dropped for rock & roll) The band originally consisted of all Americans who were very good. After a while English musicians replaced the Americans who rotated back to the United States. West Ruislip had, I think, just an Airman's Club back then. My wife and I spent many great nights there in the seventies.

They would get some great groups over from the States that only played the bases. I did a couple of gigs with my own band at Douglas House in 1965 and then at Uxbridge in 1967.

"In 1965 Carl Perkins did a couple of shows at South Ruislip backed by a British band called The "Nashville Teens". This was his first appearance in this country. I did get to meet him and had a long chat with him. It was great to be one of the few 'Brits' to meet the greatest rockabilly of all time!

Dick Wilson, one of the contributors to the book, had another particularly interesting recollection about the residence halls at one of the first locations for London Central High School.

"The girls' dorm was surrounded by an eight-foot cyclone fence, topped by three strands of barbed wire. The barbed wire slanted *in*!"

ACKNOWLEDGEMENTS, FURTHER READING AND VIEWING

There is a wealth of information out there about the military and various units, equipment and operations during the World Wars and the Cold War, but it tends to focus on airfields and not so much on the US command and support bases in and around London. However should you wish to continue your research and reading or continue to evolve memories or thinking about these locations then some of the following websites and periodicals might be ones to add to your reading list.

If you want to read further on the subject or get in touch with people or add to the knowledge base about these places – and I hope you do – then there's plenty out there. So of relevance and of definite interest:

The American – (www.theamerican.co.uk) used to be a twice-monthly newspaper but is now a monthly magazine dedicated to the American community in the UK.

The Eagle (a weekly Third Air Force Publication that was also known as The UK Eagle) was published by the Third Air Force at USAF South Ruislip.

The London Consolidator (published by Kaiser & Cate Military Publications).

Leisure London – In the latter years of the US Navy presence in the UK, Leisure London was the Morale Welfare & Recreation Department's publication.

Overseas Weekly – was published between 1950 and 1975 and was described by Time magazine as "the least popular publication at the Pentagon". One of the more detailed and useful websites about it is one kept by vets of the US Army 3rd Armoured Division at: http://www.3ad.com/history/at.ease/overseas.weekly.htm

The Philadelphia Inquirer has a special interview with Paul Knauf who served in the communications centre in Selfridges during WWII. Knauff's World War II photograph collection is located in the Special Collections Department, University of Delaware Library, Newark, Delaware.

Stars & Stripes – the US military forces newspaper. (www.stripes.com) remains one of the primary sources of information for all things US military with a heritage stretching back to when the paper was founded in 1861. The Stripes stopped printing a special UK section it had for a number of years in 2013. For editions from at least 1948-1999 visit Heritage Newspaper's site where you can register: www.starsandstripes.newspaperarchive.com .

Yank Magazine – published by the US Army starting in June 1942 through to December 1945. http://www.wartmepress.com/archives.asp?TID=Yank%20Magazine%20-%20All%20Editions&MID=Army&q=41&FID=36

On the website front do not miss:

http://www.awon.org/discus/messages/14/499.html?1326778222 — (a highly informative website for GI babies and their families looking for biological fathers) with people still looking for WWII GI parents/grandparents as recently as March 2013.

America footfall games held in the United Kingdom – http://www.britballnow. co.uk/History/Military%20football.htm is a brief but informative website page on US military American Footall games in the UK up until the 1990s.

Battle of Britain War Monument – http://www.bbm.org.uk/ – The actual monument, located by the Thames, includes the names of US pilots who participated in the Battle of Britain. Participants are listed by countries of origin.

Bushy Tales can be found through http://BushyPark.org – the website managed and edited by Gary Schroeder and dedicated to all who attended London Central High School in Bushy Park, London England from 1952 to 1962. It provides wonderful insight into and memories about the life of American military dependent students during those years.

Baseball in Wartime — Gary Bedingfield's detailed website http://www.baseballin-wartime.com/baseball_in_wwii/baseball_in_wwii.htm is a fascinating look at the game and its wartime playing.

Wartime Babies — At http://www.bbc.co.uk/history/ww2peopleswar/stories/06/a3797706.shtml – My Father's "Brown Babies by Rachel James is one of many wartime stories featured as part of the BBC's WW2 People's War archives contributed by the public.

British Pathe Films — http://www.britishpathe.com/record.php?id=43094 – has footage of High Wycombe Air Station in 1964. It starts out with Americans visiting a local senior persons home before showing 'Yank Tanks' and a visit to the base.

Enigma at Eastcote — http://www.codesandciphers.org.uk/virtualbp/tbombe/the-bmb.htm – this site set up by Tony Sale, the original curator of the Bletchley Park Museum, has one of the best descriptions of the American-run Enigma system known as 'Atalanta'.

Central High / London Central High – http://www.harrold.org/rfhextra/brats.html is run by Robert Harrold and offers information about the school, about RAF Upper Heyford and other 'ghost bases'.

Control towers — mhttp://www.controltowers.co.uk/USAFlist.htm for those seeking information about the wider US military presence in the UK from 1947 to the present day this is a useful website.

US Embassy in World War II — http://www.fdrlibrary.marist.edu/archives/pdfs/findingaids/findingaid_winant.pdf – contains the papers of John Gilbert Winant during his time as US Ambassador to the Court of St. James from 1941 – 1946.

London Borough of Hillingdon — http://www.hillingdon.gov.uk/index. jsp?articleid=24835 – is the place to read about the Mayor of the London Borough of Hillingdon's chain of office and its US Navy connection.

Overseas Brats — http://www.overseasbrats.com – a website dedicated to children of military personnel stationed overseas run by Joe Condrill.

RAF Hendon Museum – http://www.rafmuseum.org.uk/online-exhibitions/amer-icans-in-the-raf/eagle-squadrons.cfm – RAF Hendon Museum's on-line information about the Eagle Squadrons.

London Central High School and **Carpenders Park** —http://londoncentral.org/photos/carpenderspark1.htm – a sitewhich is largely dedicated to the former Department of Defense Dependents School but which also covers memories and pictures of the Carpenders Park housing area near Watford.

Bushy Hall — www.londoncentral.org./royreeves.htm in regards to Bushy Hall.

US Marines — www.msg-history.com - link to those invoved with Marine Security Guard, London

The Naval Historical Collection & Research Association — http://www.nhcra-online.org/

RAF Northolt - http://www.northolt.biz/ is an unofficial enthusiasts website for matters to do with the RAF Northolt base.

US Marines in London - http://www.nps.gov/history/history/online_books/npswapa/extContent/usmc/pcn-190-003125-00/sec7.htm for the history of the US Marines in London.

Pathe Films - www.pathe.co.uk - with its magnificent collection of newsreels particularly from the 1950s and 60s includes some of USAF bases.

Pathe Films - http://www.britishpathe.com/video/u-s-eagle-club-aka-american-eagle-club A look at the American Eagle Club in London.

Ruislip - www.ruislip.co.uk - a definitive website through which I met several contributors to this book and through which others can reconnect. It's the best focal on-line location for exchanging memories and information about USAF South Ruislip, DOE Eastcote, RAF West Ruislip and RAF Uxbridge. There are also onward Facebook connections for some of the bases, units and Carpenders Park.

Silver Dollar Pizza - http://www.silverdollar.force9.co.uk/about.html - for the history of the Silver Dollar Pizza restaurant in Rayner's Lane. The Silver Dollar still opens but is mostly open from 6:30 p.m. on weekdays.

32nd AAA - http://www.usarmygermany.com/Sont.htm?http&&&www.usarmygermany.com/Units/Air%20Defense/USAREUR_32nd%20AADCOM.htm - a useful history of - and information about - the US Army's 32nd Anti-Aircraft Artillery (AAA) Brigade which provided anti-aircraft air defense and chemical smoke defense for USAF bases in England for a period.

Aircraft - www.warbirdregistrry.org - is one of the great reference sites for aircraft of all ages.

GI Brides - http://uswarbrides.com/WW2warbrides/facts.html - (a highly informative website for GI Brides)

RAF Hendon - http://www.vr-24.org is the website of the association founded to preserve the memory of VR-24 including their operations at RAF Hendon.

As for books and other publications:

3AF Brief: Short History and Chronology of the USAF in the United Kingdom. — Historical Division, Third Air Force, 1967, AFHRC By Charles H. Hildreth.

8th AF News, Journal of the Eight Air Force Historical Society, May 1976.

8th Air Force Remembered. An Illustrated Guide to the Memorabilia and Main Airfields of the Eighth Air Force in England in WWII, by George H. Fox, ISO Publications (1991).

A Canterbury Tale - Memories of a Classic Wartime Movie by Paul Tritton (Tritton Productions October 2000).

A Different War: Marines in Europe and North Africa by Lieutenant Colonel Harry W. Edwards, USMC (Ret). Published as a pamphlet for US Marines by the History and Museums Division, Headquarters, U.S. Marine Corps, Washington, D.C., as a part of the U.S. Department of Defense. Circa 1994.

A Hell of a War by Douglas Fairbanks Junior, Thomas Dunne/St. Martins (1993)

Robson (1995).

A History of the Third Air Force, 1940-1988 – Office of History, Third Air Force. By Jerome E. Schroeder.

Air Force Spoken Here – *General Ira Eaker and the Command of the Air* by James Parton, Air University Maxwell AFB (1987).

Airfields of the Eighth Air Force Then and Now by Roger Freeman, Battle of Britain Prints International (1978).

American Naval Planning Section, London by Edward Denby (Office of Naval Records and Library/Office of Naval Intelligence, 1923)

American Pilots in the RAF (The WWII Eagle Squadrons) by Philip D.Caine (Brassey's 1991)

And They Thought We Wouldn't Fight by Floyd Gibbons 2010 [EBook #31086] www. gutenberg.org

Anglo-American Naval Relations, 1917 – 1919 by Michael Simpson (Scholar Press for Navy Records Society, Aldershot) 1991.

Anglo-American Strategic Air Power Co-operation in the Cold War and Beyond by RAF Group Capt Christopher Finn and USAF Lt. Col. Paul D. Berg Air & Space Power Journal – Winter 2004.

The Army Air Forces in World War II prepared by the USAF Historical Division.

Army Air Forces Stations by Captain Barry Anderson, USAF Historical Research Centre (January 1985).

Around Ruislip, Eastcote, Northwood, Ickenham & Harefield – by Eileen Bowlt, Stroud, (2007). Sutton Publishing.

Bomber Commander – *The Life of James H. Doolittle* by Lowell Thomas and Edward Jablonski. 1977 Published by Sidwick & Jackson.

British Military Airfield Architecture by Paul Francis (Patrick Stephens Limited, 1996). While not really focused on USAF operations this remains a seminal book on base architecture, design and building.

Bushey Hall And The Forties Experience by Derek Addison and Tony Rock (Anchorprint Group Limited, 2013.

Bushy Park at War (Planning D-Day). Available from The Royal Parks.

The Children They Left Behind by Janet Baker. Originally published in Volume 4, Number 2, April, 1996 edition of a quarterly publication *Lest We Forget*.

Citizens of London: The Americans Who Stood with Britain in Its Darkest, Finest Hour by Lynne Olson. Hardcover, (Random House, 2010). Also: http://lynneolson.com/citizens-of-london/

Cobra by David Whitehead CEng MIEE – is the story of a prototype electronic relay rack that helped break German North Atlantic naval codes in WWII. but also features some of the details about Eastcote and the British Government Communications agency GCHQ.

Cold War: Building for Nuclear Confrontation 1946-89 by P.S. Barnwell, Wayne D. Cocroft and Roger C. Thomas.

Don't Shoot, It's Only Me by Bob Hope with Melville Shavelson. (GB Putnams'sSons, New York, 1990).

Douglas House by Gerald A. Collins. 1st World Publishing.

Eighth Air Force – *The American Bomber Crews in Britain* by Donald Air Miller. (Arum, 2007).

Eisenhower was my Boss by Kay Summersby (Prentice Hill, 1948).

The Few (American "Knights of the Air" Who Risked Everything to Save Britain in the Summer of 1940) by Alex Kershaw (Da Capo Press, 2006). A book about the Americans who joined the RAF and fought in the Battle of Britain.

Filer's Files: Worldwide Reports of UFO Sightings by Major. George Filer, USAF (Ret.) and David Twichell. The book includes Maj. Fikers' account of his encounter with a UFO after flying out of the USAF Base at Sculthorpe in 1962. http://www.ufoimplications.com/filer.html

Force for Freedom – the USAF in the UK since 1948 by Michael J.F. Bowyer (Patrick Stephens Limited, 1994).

Four Miles High – the US 8th Air Force 1st, 2nd and 3rd Air Divisions in World War 2 (Patrick Stephens Limited, 1992).

From the Faculty Lounge – Memories of London Central High School by Sean C. Kelly (Long Dash Publishing 2007), 2nd Edition – Bayberry Books 2009).

From Somewhere in England – The Life and times of 8th AF Bomber fighter and ground crews in WWII by D.A. Lande (Motor Books International, 1990).

GI Brides by Dunan Barrett and Nuala Calvi (Harper Books, 2013)

Got Any Gum Chum? GIs in Wartime Britain 1942 – 1945 by Helen D. Millgate (Sutton, 2001).

History of the Headquarters and Headquarters Squadron (and) WAC Detachment, Eighth Air Force (Alden Press, Oxford, 1945).

Instructions for American Servicemen in Britain 1942, Issued by the US War Department and the Department of the Navy, Washington. Reprinted by the Bodleian Library (2004).

Locations of United States Military Units in the United Kingdom (16 July 1948 – 31 December 1967 — Historical Division, Third Air Force, AFHRC By Richard H. Willard (1967).

London, England – The U.S. Navy Dimension by Jack Carter – historical document from the US Navy.

Masters of the Air: America's Bomber Boys Who Fought the Air War Against Germany by Donald L Miller (Simon & Schuster, 2006).

My War by Andy Rooney (Essay Productions, 1995, 2000).

One Hundred Years of Service: A History of the Army and Air Force Exchange Service by Col. Carol A Habgood and Lt. Col. Marcia Skaer, HQ AAFES, Dallas October 1994).

Over Here – The GIs in Wartime Britain by Juliet Gardiner (Collins & Brown, 1992).

RAF Hendon – the Birthplace of Aerial Power by Andrew Renwick (Royal Air Force Museum).

Rich Relations (The American Occupation of Britain, 1942-1945) by David Reynolds (Random House, 1995).

Star Spangled Square: The Saga of 'Little America' in London by Geoffrey Williamson. (Bles, London) 1956).

Subterranean Britain: Cold War Bunkers by Nick Catford (Folly Books Ltd., 2010) . A fascinating book by the man who has headed up the Subterranean Britain organization.

The Eagle Squadrons (Yanks in the RAF 1940-1942) by Vern Haugland.

The G.I.s: The Americans in Britain 1942-1945 by Norman Longmate (Hutchinson of London, 1975).

The Life and Times of a Cold War Serviceman – August 1928 – 30 November 1969 by Maurice F. Mercure (Trafford Publishing, 2012).

The Mighty Eighth – A History of the U.S. 8th Army Air Force by Roger Freeman(Macdonald London, 1970).

The Mighty Eighth in the Second World War by Graham Smith (Countryside Books, 2001).

The Moon's A Ballon by David Niven (Hamish Hamilton, 1971).

The Spying Game (The Secret History of British Espionage) by Michael Smith (Politicos, 2003). http://www.michaelsmithwriter.com/books_spies.html

The Story of Rainbow Corner – The American Red Cross Club Near Piccadilly Circus, London by Verbon F. Gay for the American Red Cross.

The Unsinkable Aircraft Carrier: The Implications of American Military Power in Britain by Duncan Campbell. (Michael Joseph, 1984). A definitive book on US Military in the UK particulary during the Cold War.

The Victory at Sea by William Sowden with Burton J Hendrick (Doubleday, New York, 1920).

The Writing 69th – http://www.greenharbor.com/wr69/wr69.html – has details about book The Writing 69th by Jim Hamilton (Green Harbor).

Three Years with Eisenhower by Capt. Harry C. Butcher, USNR (Windmills Press 1948).

To the Limit of Their Endurance by Lt. Col. Lynn Farnol is the story of the VIII Fighter Command. Published by Sunflower University Press.

When Jim Crow Met John Bull by Graham A. Smith and published by I.B. Tauris. A book that explores the issues surrounding black GIs in the UK.

Warlords: An Extraordinary Re-creation of World War II Through the Eyes and Minds of Hitler, Churchill, Roosevelt and Stalin — by Simon Berthon and Joanna Potts. (Da Capo Press, 2007).

Women's Army Corps by Mattie E. Treadwell (published by U.S. Army Center of Military History,1954).

Yank – The Army Weekly by Steve Kluger, 1991.

In addition there are an increasing number of YouTube videos with segments featuring some of the bases or American activities. They include:

http://www.youtube.com/watch?v=WykiI7_Dj0U&feature=endscreen&NR=1   A video featuring the London Central High School buildings and students at Bushey (sic) Park. Undated.

http://www.youtube.com/watch?feature=endscreen&NR=1&v=pMnyhBxFEP0   a video featuring Bushey Hall in 1964.

http://www.youtube.com/watch?v=HDBCtD6n3yE – First American Mechanized Home Guard Squadron in Britain.

http://www.youtube.com/watch?v=qGdl4P5Dba0 features the closing ceremony for London Central High School at RAF Daws Hill in 2007.

http://www.youtube.com/watch?v=HDBCtD6n3yE The Universal News features the visit by Winston Churchill to the American Mechanized Home Guard and a summary of the squadron's role and its reasoning from General Wade H. Hayes.

And in television/film terms:

A Welcome to Britain – the 1943 film starring Bob Hope and Burgess Meredith mentioned in the book. The film was made by the Crown Film Unit based out of Pinewood Studios for the War Office and was given to the American Office for War Information.

American Invasion – a special 2012 documentary for the Yesterday Channel hosted by Michael Brandon.

City at War: London Calling was a special TV documentary hosted by Walter Cronkite which aired in May 2007 on PBS. It brought the legendary reporter back to London 60 years after he first came to cover WWII. It is a co-production of Colonial Pictures and Thirteen/WNET New York in association with ITV London.

Combat America – the 1943 film made by Clark Gable that follows the 351st Bombardment Group from training to combat over Europe.

Know Your Ally (1944) – Director Walter Houston, produced for the War Department. Part of an intended series: Know Your Allies & Know Your Enemies.

Target For Today – First Motion Picture Unit/18th Army Air Force Base Unit – USAAF (1944). Might also be able to be seen on You Tube: www.youtube.com/watch?v=RtB6i90X3w The documentary, made by the acclaimed FMPU, focused on the planning and missions in October 1943. (The First Motion Picture Unit also made *Memphis Belle: The Story of a Flying Fortress* among around 400 information and propaganda films it made.

WITH THANKS

I am grateful to a long, long list of people for their help provision of contacts, leads, pictures, information and encouragement during the time I was researching and writing *Home Bases*. They include all of those who contributed chapters, memories and suggestions and all those, including family members and friends, who enabled them to contribute.

I'm particularly appreciative of the efforts of Ian Dennison and Peter Allan of Ministry of Defense Lands Agency (now part of the Defense Infrastructure Organisation) and their colleagues who have assisted me with permitting base photography and preservation of historical items. Additionally to Ian Palmer, Nick Newman and also George Hannaford – all RAF station commanders who, at various points in their careers, helped facilitate visits; particularly to Daws Hill, High Wycombe.

Joining them on the list:

The demolition team at Syd Bishop & Sons who allowed me to don safety equipment and watch them in action at key stages in the process of the demolition of Eastcote. In particular the kindness of Tom & Steve Bishop and on-site to Tony, Danny, Clive and all the team for putting up with my photography and my numerous questions.

I'm also extremely grateful to Eileen Bowlt, the esteemed local historian in the Ruislip area, who shared with me detailed memories she has gathered about South Ruislip, West Ruislip and Eastcote among other locations in which she takes detailed interest. She and her fellow members of the Ruislip & District Natural History Society (http://www.ruislip-districtnhs.co.uk/)have and share unparalleled amount of information about the local area. Also to Susan Toms & Jean Dixon who were both incredibly helpful on matters relating to Eastcote and GCHQ.

RAF 11 Group Curators Chris Wren and Daniel Stirland and colleagues at the Battle of Britain Bunker at the former RAF Uxbridge. They all have the exceptional ability to bring history to life on his underground bunker tours as he talks about Britain's famous battle for survival in the early days of WWII.

Guy Gusterson, Land Director and colleagues at VSM Estates, who originally enabled access to West Ruislip and Eastcote and also took note of my various preservation requests and allowed me to save base signage and other memorabilia. David Peycke Senior Land Manager, Taylor Wimpey West London, who was instrumental in helping with the Daws Hill site.

Ed & Patricia Côté and the family of Paul Côté who not only helped arrange the interview with Ed's father, Paul, but also invited myself and family to their home in California on several occasions and proved most engaging hosts.

Press officer Jonathan Weisgard of Hammersmith & Fulham Council. Nicholas Baldwin, Archivist at Museum & Archives Service, Great Ormond Street Hospital for Children NHS Trust. James Wyatt and his team at Barton Wyatt for help on Romany House.

Bill Douglas, LCHS student whose father was both Assistant Deputy Chief of Staff for Personnel and later the DCS/Personnel for 3rd AF, Colonel Richard M. Douglas, USAF. The ambassadorial Terry Prather of Majlar Productions with regards to Larry Hagman.

Fr. Edward Dalton White Jr., the husband of Iris White.

Ray Brodie, Park Manager of Bushy Park on behalf of the Royal Parks, who connected me with Douglas Rowland & Paul Evans, volunteers working in the history Room at Bushy Park and to Cliff Williams who has built up amazing historical knowledge of Bushy Park. The knowledgeable Alan Hayes of the *Uxbridge Gazette*.

Bob Miner who managed to dig up some of the original leasehold information and about 7 North Audley Street along with historical data about Eastcote. Michael Dowend and Carlos Jimenez for help with barracks information.

Paul Francis with whom I got to spend a wonderful couple of hours at West Ruislip and Eastcote. Paul is a legend among the airfield and base fraternity for his unrivalled knowledge of the buildings, their architecture, their changing uses and particular unique features. If anybody knows the design and building details of a base then he does!

Jason Lawrence, legal adviser for the US Military, and Carol Poynton of the OLB 422nd CES Housing Office with assistance in particular regards to Romany House. Frank Arre of the Navy History organization.

Chris Jepsen, Assistant Archivist at Orange County Archives. Lewis Wyman, Reference Librarian, Manuscript Division, Library of Congress. James Tobias, Historical Resources Brand, U.S. Army Center of Military History and Oscar De Jesus, Multi-Media Specialist, USO.

A particular note of thanks to Heather Piper, editor of *Ickenham Church News*, who put in touch with Henry Farwell, and also to the very knowledgeable Susan Kazara who connected me with several of the book's veteran contacts. Joining her in that thanks should be Adriane de Savorgnani for her help in putting me in touch with several other interviewees.

A nod must go to my educational predecessors, so to speak, William Cooper and Ron Crowe and Alvin Collins and Robert Harrold, fellow 'Bobcats' from London Central High School from earlier years who either shared their recollections or stirred awareness amongst other Bobcat to mention their memories. It's worth reiterating Robert's website information at http://www.harrold.org/rfhextra/brats.html is a terrific chance to find bout the early days of LCHS and about other bases. Alice Orton of publishers IB Tauris. Edith Lobo of Warner Chappel. Ruth White of CALA Homes.

Extended thanks, too, must be directed towards various researchers at institutions including, the RAF, MOD including Stuart Hadaway (AHB2 (RAF) at the Air Historical Branch of the RAF. Curti Ph.D. Public Service Coordinator and Research Associate at the National Security Archive of the United States. To Dave Birrell, Director of Library, Museums and Displays at the Bomber Command Museum of Canada for his help with Joe McCarthy. To Nina Smetek, Press Officer at the Imperial War Museum, London. Allan Leventhall and the Norfolk Library Service. Mark Dowd of Topfoto/Granger Picture Trust.

At the RAF Museum in London, Andrew Renwick, Curator of Photographs and to Peter Elliott, the Senior Keeper at Department of Research & Information Services. Also Tom Miller for his help with the Columbia Club, author Derek Addison for discussing Bushey Hall, Ed McManus of the Battle of Britain Monument, Andrew Hodges. Also Oscar de Jesus in regards to Bob Hope. American military history blogger Chris Kelly.

CMSgt. John Tway, 3 AF Public Affairs at Ramstein Air Base, Germany. James Tobias of the Historical Resources Branch, US Army Centre of Military History at Fort McNair. Robert Sullivan, National Museum of the Marine Corps,

Mark Howell of RAF Mildenhall, USAF / USAFE 100 ARW. Jonathan Hood, 3rd Air Force. Also for the prompt and expedient help of LCDR Dave Benham, USN, APR Public Affairs Officer, Navy Region Europe Africa Southwest Asia and Scott Campbell of the same office for proficient and detailed help; Lt. Callie Ferrari and LCDR Sarah Higgins of Navy Office of Information East. Melissa Bohan, Public Affairs Office, Arlington National Cemetery. MSgt George Trammell.

And I extend particular gratitude to John Hadfield, a slow-motion cameraman extraordinaire and the Airfield Society representative who took responsibility for gathering information on Daws Hill, High Wycombe and has one of the most comprehensive collections of memorabilia about the place today. We spent a happy day together going through the WWII 8th Air Force Bomber Command bunker at High Wycombe in the mid – 2000s – and still enjoy the memories of a unique visit.

Likewise to long-time friend and airbase aficionado Aldon Ferguson who has written two inspirational books about Burtonwood – the giant US Military depot once located halfway between Liverpool and Manchester.

Due for grateful thanks, also, is Max Lederer, Publisher of Stars & Stripes, and his colleagues at the armed forces newspaper who, through the kind offices of Liliana Vivanco (Archivist for the Stars & Stripes Library) and Jennifer Stepp, provided direction, permission and advice during my research over the years.

Sgt. Stew Thorpe of the RAF deserves grateful thanks for contributing detailed information about the USAF bunker (the Forgotten Bunker) at RAF Uxbridge. Also with the RAF Sgt Mark Bristow, Station Historian RAF Northolt and, additionally, the RAF Air Historical Branch team members including Anna Gibbs. Sonia Southerton of the MOD/DIO. Sqdn Leader Richard Willis of RAF Northolt.

The kindly Bruno Barber and David Jarvis of Selfridges for sorting out a visit 60 feet down to their sub-sub-basement – home of the top secret SIGSALY communications unit during WWII.

Others who need deserve my grateful thanks include:

Martina Oliver and team of Getty Images including Kelly Schmidt for connecting me with the WWII front cover image from among many in their incredible collection.

Robert Neil of Sir Robert McAlpine Limited for his help in regards to construction history for the bunker at Daws Hill, High Wycombe. Neil Hamilton who has helped preserve the history of Carpenders Park through his gathered information and photographs which are on the website: (www.londoncentral.org.carpndrs/cphistory.htm). Neil was there on patrol as a British policeman and it has been his diligence that has helped others to rekindle memories of the base housing.

John Allum (London Borough of Hillingdon Planning Department) who pointed me in the right direction on historical property applications; Johanna Michalski, Councillors' Support and Development Manager at the London Borough of Hillingdon; Susan Robbins Watson, the Archivist Manager, Historical Programs and Archives at the American Red Cross and her colleague Lindsay Flanagan Huban, the Red Cross Historical Programs and Collections Associate.

Mike Mockford (Trustee, Curator and Archivist of the Medmenham Collection). Curtis Haack of Nevada Microwave & Tower / CW Communications Consulting. Richard White of Wycombe Borough Council and Unison Wycombe District Branch. Craig P. Rahanian the superintendent for the American Battle Monuments Commission at Brookwood American Cemetery and Memorial in Surrey who is entrusted with the care of one of the memorial's to America's fallen military personnel. I am especially grateful to Francis McLennan and the team at the Lincolnsfields Centre (at Bushey Hall, Watford) for inviting me to see the remains of the base and the preserved 1940s House which is open once a month for public tours. The Lincolnsfield team continues to fight for sponsorship to preserve the memories and the buildings and would appreciate hearing from anyone who wants to help.

Paul B. Brown of the National Archives & Records Administration. Diana Swick. Jan Gillies and Janice Zerbrowski, Tracy Blackwood and Ronald D. Blackwood who all helped or did their best to help connect me with interviewees. Rebecca Steed, Media Relations officer at Fort Leavenworth. Bishop Higgins of the US Military Archdiocese.

Jim Rush of Sun City Civic Association Board of Directors and Katy Bennett, City Clerk at the City of Menifee in California. Writer and artist Marcia Gawekci for her kindness in connecting me with the Folsom family. Richard Prather and other members of the VR-24 Association.

Captain Dave Dittmer who kindly took time to meet with me and follow-up and also invite my wife and I to the memorable and family-encompassing retirement ceremony held at The Hard in Portsmouth Harbour, England.

Leading British restaurateur and developer Richard Carling and his team as well as Steven Sharp and Terry Menzies of Price Anderson and Adam Lawry, the project manager of Hanover Estate. They collectively enabled a visit to the empty former US Navy Headquarters building on Audley Street in 2009. Frank Arr of the US Navy History Organisation www.navyhistory.org. Robert Hanshew of Naval History and Heritage Command for his contribution and suggestion.

Gary Schroeder and his helpful research regarding South Ruislip. Jay Mercer who runs the web http://londoncentral.org/ a web site that hosts both a detailed history and pictures of London Central High School and some of the bases many other facilities as well as providing a connecting point for many former alumni.

Alex Kilpatrick, who provided the link with his father and Gregg Gunsch who provide the link to him. Also to Rick Boomer for his recollections on the bunker at High

Wycombe.

Rachel Wells, Central Chancery, St James's Palace, who helped with information about the George Medal presentations to Airman Leeming and SSgt. Kilpatrick in 1953.

Roger Weitkamp and Jeffrey Simmons of Coca Cola. Susan Scott, Hotel *Archivist* for The Savoy Hotel on the Strand for her momentous efforts in tracking down information about the US media contingent. Fred Vermorel, who has researched the murder of Jean Mary Townsend. Edward McManus of the Battle of Britain London Monument.

DeCA historian Dr. Peter Skirbunt. DECA has produced the *Illustrated History of American Military Commissaries.*

Family friend Skip McIndoe who served at South Ruislip from 1964 to 1969. Bill Frey, son of Jaunita Folsom. Steve Crook who runs one of the finest sites to be found about the British film-makers Powell & Pressburger.

Also on the list: Otto Kosa and his wonderful collection of photos of main gates of military bases. Arlie Burns another photographer who showed me the 1950 pictures by Jim Conway of South Ruislip and of a visit by Eleanor Roosevelt. Don Campbell, who shot some of the pictures of South Ruislip you can find at www.ruislip.co.uk; Gordon Bell of Bell Fischer Landscape Architects and Howard Cloke for his time and effort and providing connection to the Cloke family memories.

My friend David Murdoch for his great recollections of various local activities with 'the Yanks'. Michael Salinas who connected me with the memories of his father, Alex. Armonde Casagrande of the 343rd Corps of Engineers, US Army for his memories of Nettlebed. Likewise Malcolm and Barbara Lewis who have helped preserve the 343rd legacy locally.

Various staff members at USAF Third Air Force, RAF Museum London, the USAF Air Force Historical Research Centre (Maxwell AFT Alabama). In particular Scott D. Campbell, CNREURAFSWA DPAO. LCDR, Dave Benham, USN, APR, Public Affairs Office, Navy Region Europe Africa Southwest Asia. Holly Reed, National Archives & Records Administration (Still Pictures Reference).

Reverend David Osborne and Izabela Przybylska St. Clement Danes Church. Armonde Casagrande and Malcolm and Babarba Lewis for their assistance regarding the 343 castle monument.

Sarah Padley, Sales & Marketing, Empire Test Pilots' School at QinetiQ Ltd. Nena Couch and Orville W. Martin, respective Curator and Assistant Curator at The Jerome Lawrence and Robert E. Lee Theatre Research Institute, The Ohio State University Libraries, on matters relating to Burgess Meredith.

Ruth Juarez and the staff at the US Military Vicar General's office. Marie Chevant of The Dorchester Collection. Michael Faley and his info on the 100th Bomb Group. Author Michael Smith — an expert on spying, cryptology and the hush-hush world. Photo and Library staff Staff at the *Lynn News* and the *EADT.*

Sonia Southern of the MOD, C2 Pragati Baddam DIO, Rosemary Dixon of Archant, the staff at Norfolk Library system including AlanLeventhall and Clare Everitt. Thanks to either time, information and pictures from George Trussell and Tom Culver. Also to William Brown, Press Officer DCMA and Capt. Carolyn Glover of the 48th TFW. Shona Lowe, PAO at NATO Maritime Northwood. Staff at English Heritage. Alex Ritchie, Business Advice Manager, Archives Sector Development at the National Archives. George Marcec and PAO colleagues at the US Army Garrison, Fort Leavenworth, Kansas. Kathy Strauss, Audiovisual Archivist at the Dwight D. Eisenhower Presidential Library & Museum. Rosie Bagley Marketing & PR Executive, at the Grosvenor House, A JW Marriott Hotel, for her help on the US Officers Mess.

And to those who have helped, directly or indirectly – whether they know it or not – including Andrew Climance, Tim Harper, William Kilty, Kay Galloway, Suzanne Burke, Jack & Peg Wernette, Penny Dove-Winstock Crook, Ed Brennan, Peter Drysdale, Stephen Kane, Paul McGrath, Enda Fahy, Nigel Lewis, Graham Parker, Donald Shaw, Ivor Peters, Jackie Whitaker, Linsey Wooldridge, Mandy Bhullar, Eric Musgrave, Melanie Hopper, Melissa Byrne, Dan Innes and Kelly Bradshaw and also Neil Chapman and Adrian Wright. And happy memories and thanks to Tosh Lee who, sadly, passed away late last year.

There will be many others whom have passed on – sadly way too early in their lives. Among those I'd particularly like to pay tribute to are Jane Elizabeth France (LCHS Class of 1979) a truly dear friend and someone I'm privileged to have known. Also, Clifford Nnyoud Steger (LCHS Class of 1990), whom I met during a school reunion dinner in San Diego a few years ago. It was only a single – and way too short – conversation between alumni separated by a dozen years. But Cliff made a lasting impression. Both passed away – far too early in their lives – during 2012.

Special thanks to the ever positive Twila Pitcher Brand – a fellow Bobcat (Class of '79) and my main on-line link to the world of Facebook and our alma matter. Twila is doing her best to get me to join the rest of the Facebook world but until then I'm incredibly indebted to her for bridging the social media canyon!

Two eateries in particular proved - and continue to prove - ideal in offering both respite and sustenance over the years - Burgers in Marlow and Maison du Soleil in Ickenham. It's amazing how a cup of tea, a toasted croissant or Welsh rarebit (and an occasional cake!) can alter your perspective Respective thanks to proprietors Bernard Burger and Carlos Navarro and their staffs.

And then there is 'G', who, to my mind, knows far, far more about the US Military around London than anybody and who really should write *the* book.

Also to David 'Griff' Griffith, who knows far, far more about the US Military around London than anybody and who really should write the book.

Apologies – and there will be some I'm sure – If I have missed anybody among the massed ranks of those who helped me – directly or indirectly – over the years. I'm sure you'll let me know – but please know that I thank you even if not by name.

And finally some of the frontline enablers. My good friend and expert on things American in the UK Bob Pickens. Another good friend, Ian Harvey of the Family Historical Society, for his nifty ISBN and other help. Brian Lubrani on the computer front (he is one of the few people to have been up close and personal with my former Apple laptop and its legendary 'worn-clear' keys). And linked directly to that – the man who taught Typing to me at London Central High School in High Wycombe all those years ago – Ralph Ensz.

A fusillade of praise and grateful thanks to designer Jeremy Hopes not only for his book cover, design suggestions and technical delivery of *Home Bases*, but for his wonderful cups of tea and patience. Finally to Gabriel Stuart, publisher at Bayberry Books. Gabriel not only dealt deftly with my with my many questions and delays but then did the publishing bit.

Needless to say Jeremy, Brian and Gabriel, like all the aforementioned, come recommended. I thank – and salute – you all!

SIR, I HAVE A QUESTION

And finally, there are still some questions in my mind – about some of these places and bases.

There always seemed to be – at least to me – an unwritten rule in the US Military that as soon as a base could be made about as good as it could be then it would be closed. Case in point – USAF South Ruislip. It reportedly had one of the finest medical facilities outside the United States. One of my own brothers survived birth because they had a specialist incubator there – there were less than a handful in Britain at the time. But apparently – the story goes – that the decision was made to close the base and base hospital just months after they had created a world class cardiovascular unit.

West Ruislip may be another case – just a few years before it closed the US built a $5 million childcare centre and also refurbished the school and much of the housing!

As with military bases and life – particular US military bases – there were always 'tales'. Today, in the internet age, they are 'urban legends'. Whatever the case they are often intriguing and the grist for the rumor mill. Here's a few:

That the three-storey airmen's barracks at High Wycombe (later Mansfield and Trinity Halls – the respective boys and girls dorms for London Central High School) were really massive cover hatches for secret missile silos that would, come WWIII, slide down the hill towards the town of High Wycombe to allow the missiles to take flight.

That there was a ghost in one of dormitory residence halls at Daws Hill.

That there was a secret storeroom under the warehouses at West Ruislip filled with *Playboy* magazine from the 1960s. (Or so at least one demolition worker on the site thought.)

That there is (even though this author couldn't find it) some sort of a secret underground passage connecting the US Navy Headquarters from North Audley Street to the US Embassy and across Grosvenor Square to other buildings?

That there is a secret and now buried WWII map room in the grounds of Bushy Park.

That the nurse-turned-pizza parlour operator Major Patterson really did keep a shotgun under the counter at the Silver Dollar Pizza.

That the British police really have occasional late night drag races down the Victoria Road in front of the USAF Headquarters Base at South Ruislip in the 1950s and 1960s.

The list of stories and rumors is much, much longer and will, I am sure, provide much grist for years to come. However as my own mother always jokingly told me — one of the 'unwritten rules' of military life that she had heard from her colleague Hazel Barton: "If you know a good rumor, spread it!"

So my response is – as best as I know – at least one of those aforementioned rumors is untrue!

INVOCATION

The following invocation was made at the Inactivation Ceremony for the 7520th Air Base Squadron at High Wycombe Air Station on Monday, 10 May 1993.

The hand-written invocation was given to the author at the time but sadly the name of the Chaplain who gave the address has not been recorded though he was kind enough to give the author his original handwritten notes for it.

In many ways this invocation could apply to any or all of the US military bases and operations.

Lord God of the Universe, you revealed that after your work of creation you viewed your handiwork, sat that it was good and then rested from your labors.

The United States Air Force after 51 years of work at High Wycombe Air Station, looks back at the labors of so many, affirms that it was good we were here and now that world needs change and nations change it is time (for us) to rest from our mission here. It is time for those of us remaining to close this part of our history with our British friends and allies. It is the price of peace. Continue to bless this land and her people Lord. Bless us as we depart for new challenges in service to protect and defend our own lands, our cherished people, the foundations of freedom and a democratic way of life.

With symbolism and ceremony we proudly bid farewell, grateful for all those living and dead who served before us, knowing that in some way we the United States Air Force have crated the possibility of a better tomorrow. Go with us Lord, show us the way, help us to ;renew your world.

Amen.

LOCATOR MAP

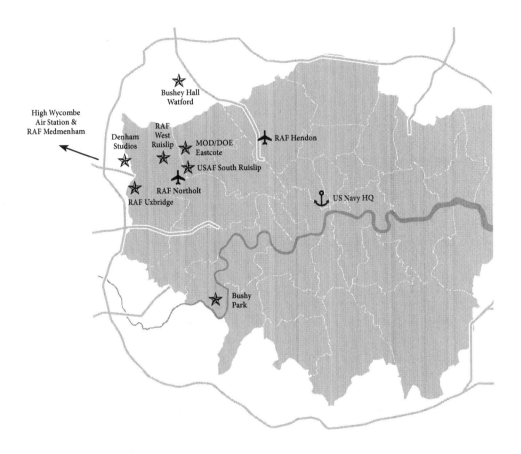

The above map indicates the locations of many of the US military bases in and around the London area and slightly further afield over the years. A ★ or a ⚓ represents a headquarters or base while the ✈ represents a flightline base.

ABOUT THE AUTHOR

Sean Kelly was born at USAF South Ruislip, England. He received his elementary education at the Department of Defense Dependents School at Eastcote, England, before attending both Junior High School and High School at London Central when the military base school was located at USAF's High Wycombe Air Station.

He graduated from LCHS in 1978 and studied at university in California where he obained a BSc before working in television production in Hollywood for 18 months.

His return to England in 1983 also saw him return to London Central where he worked for several years while also freelance writing for various publications in the US, UK, Europe and Japan on a range of subjects from military and retail to education and television.

Since then Sean has been involved in a range of journalism and public relations activities. He was deputy editor for a UK retail property industry trade magazine and has run successful PR and marketing campaigns for a number of UK shopping centers and property companies. He sits on the UK national awards committee judging excellence in customer service and is an enthusiastic proponent of urban regeneration...and remains fascinated by art, architecture, aviation and music.

This book follows his first book *From the Faculty Lounge*, which focused on the history of his former high school told primary from the point of view of teachers and staff.

Sean lives near London with his wife, Jacqueline, and their three children. He is currently working on a book about his father, the artist Francis Kelly, and is also exploring a fictional novel with a London setting.

www.ushomebases.com

Above: 1943 V-mail produced for US troops stationed in England in WWII. This one in particular was a wry take on the censorship requirements of the time. The artist, Lt. Dave Breger, was a popular cartoonist. 'V' was for Victory and this form of communication was originated by the US Postal Service in June 1942.

FOR YOUR OWN MEMORIES.

37772621R00250

Made in the USA
Lexington, KY
14 December 2014